Multilevel and Longitudinal Modeling Using Stata

Volume II: Categorical Responses, Counts, and Survival

Third Edition

Multilevel and Longitudinal Modeling Using Stata

Volume II: Categorical Responses, Counts, and Survival

Third Edition

SOPHIA RABE-HESKETH
University of California, Berkeley
Institute of Education, University of London

ANDERS SKRONDAL
Norwegian Institute of Public Health

A Stata Press Publication
StataCorp LP
College Station, Texas

Published by Stata Press, 4905 Lakeway Drive, College Station, Texas 77845
Typeset in LaTeX 2_ε
Printed in the United States of America

10 9 8 7 6 5 4 3 2 1

ISBN-10: 1-59718-108-0 (volumes I and II)
ISBN-10: 1-59718-103-X (volume I)
ISBN-10: 1-59718-104-8 (volume II)
ISBN-13: 978-1-59718-108-2 (volumes I and II)
ISBN-13: 978-1-59718-103-7 (volume I)
ISBN-13: 978-1-59718-104-4 (volume II)

To my children Astrid and Inge
Anders Skrondal

To Simon
Sophia Rabe-Hesketh

Contents

VIII   Models with nested and crossed random effects   871

16   Models with nested and crossed random effects   873

Tables

Figures

Part V

Models for categorical responses

10 Dichotomous or binary responses

10.1 Introduction

Dichotomous or binary responses are widespread. Examples include being dead or alive, agreeing or disagreeing with a statement, and succeeding or failing to accomplish something. The responses are usually coded as 1 or 0, where 1 can be interpreted as the answer "yes" and 0 as the answer "no" to some question. For instance, in section 10.2, we will consider the employment status of women where the question is whether the women are employed.

We start by briefly reviewing ordinary logistic and probit regression for dichotomous responses, formulating the models both as generalized linear models, as is common in statistics and biostatistics, and as latent-response models, which is common in econometrics and psychometrics. This prepares the foundation for a discussion of various approaches for clustered dichotomous data, with special emphasis on random-intercept models. In this setting, the crucial distinction between conditional or subject-specific effects and marginal or population-averaged effects is highlighted, and measures of dependence and heterogeneity are described.

We also discuss special features of statistical inference for random-intercept models with clustered dichotomous responses, including maximum likelihood estimation of model parameters, methods for assigning values to random effects, and how to obtain different kinds of predicted probabilities. This more technical material is provided here because the principles apply to all models discussed in this volume. However, you can skip it (sections 10.11 through 10.13) on first reading because it is not essential for understanding and interpreting the models.

Other approaches to clustered data with binary responses, such as fixed-intercept models (conditional maximum likelihood) and generalized estimating equations (GEE) are briefly discussed in section 10.14.

10.2 Single-level logit and probit regression models for dichotomous responses

In this section, we will introduce logit and probit models without random effects that are appropriate for datasets without any kind of clustering. For simplicity, we will start by considering just one covariate x_i for unit (for example, subject) i. The models can be specified either as generalized linear models or as latent-response models. These two approaches and their relationship are described in sections 10.2.1 and 10.2.2.

10.2.1 Generalized linear model formulation

As in models for continuous responses, we are interested in the expectation (mean) of the response as a function of the covariate. The expectation of a binary (0 or 1) response is just the probability that the response is 1:

$$E(y_i|x_i) \;=\; \Pr(y_i = 1|x_i)$$

In linear regression, the conditional expectation of the response is modeled as a linear function $E(y_i|x_i) \;=\; \beta_1 + \beta_2 x_i$ of the covariate (see section 1.5). For dichotomous responses, this approach may be problematic because the probability must lie between 0 and 1, whereas regression lines increase (or decrease) indefinitely as the covariate increases (or decreases). Instead, a nonlinear function is specified in one of two ways:

$$\Pr(y_i = 1|x_i) \;=\; h(\beta_1 + \beta_2 x_i)$$

or

$$g\{\Pr(y_i = 1|x_i)\} \;=\; \beta_1 + \beta_2 x_i = \nu_i$$

where ν_i (pronounced "nu") is referred to as the *linear predictor*. These two formulations are equivalent if the function $h(\cdot)$ is the inverse of the function $g(\cdot)$. Here $g(\cdot)$ is known as the *link function* and $h(\cdot)$ as the *inverse link function*, sometimes written as $g^{-1}(\cdot)$. An appealing feature of generalized linear models is that they all involve a linear predictor resembling linear regression (without a residual error term). Therefore, we can handle categorical explanatory variables, interactions, and flexible curved relationships by using dummy variables, products of variables, and polynomials or splines, just as in linear regression.

Typical choices of link function for binary responses are the logit or probit links. In this section, we focus on the logit link, which is used for logistic regression, whereas both links are discussed in section 10.2.2. For the logit link, the model can be written as

$$\text{logit}\,\{\Pr(y_i = 1|x_i)\} \equiv \ln\underbrace{\left\{\frac{\Pr(y_i = 1|x_i)}{1 - \Pr(y_i = 1|x_i)}\right\}}_{\text{Odds}(y_i=1|x_i)} \;=\; \beta_1 + \beta_2 x_i \qquad (10.1)$$

The fraction in parentheses in (10.1) represents the odds that $y_i=1$ given x_i, the expected number of 1 responses per 0 response. The odds—or in other words, the expected number of successes per failure—is the standard way of representing the chances against winning in gambling. It follows from (10.1) that the logit model can alternatively be expressed as an exponential function for the odds:

$$\text{Odds}(y_i = 1|x_i) \;=\; \exp(\beta_1 + \beta_2 x_i)$$

Because the relationship between odds and probabilities is

$$\text{Odds} = \frac{\text{Pr}}{1 - \text{Pr}} \quad \text{and} \quad \text{Pr} = \frac{\text{Odds}}{1 + \text{Odds}}$$

the probability that the response is 1 in the logit model is

$$\text{Pr}(y_i = 1 | x_i) \;=\; \text{logit}^{-1}(\beta_1 + \beta_2 x_i) \equiv \frac{\exp(\beta_1 + \beta_2 x_i)}{1 + \exp(\beta_1 + \beta_2 x_i)} \tag{10.2}$$

which is the inverse logit function (sometimes called logistic function) of the linear predictor.

We have introduced two components of a generalized linear model: the linear predictor and the link function. The third component is the distribution of the response given the covariates. Letting $\pi_i \equiv \text{Pr}(y_i = 1 | x_i)$, the distribution is specified as Bernoulli(π_i), or equivalently as binomial(1, π_i). There is no level-1 residual ϵ_i in (10.1), so the relationship between the probability and the covariate is deterministic. However, the responses are random because the covariate determines only the probability. Whether the response is 0 or 1 is the result of a Bernoulli trial. A Bernoulli trial can be thought of as tossing a biased coin with probability of heads equal to π_i. It follows from the Bernoulli distribution that the relationship between the conditional variance of the response and its conditional mean π_i, also known as the *variance function*, is $\text{Var}(y_i | x_i) = \pi_i(1 - \pi_i)$. (Including a residual ϵ_i in the linear predictor of binary regression models would lead to a model that is at best weakly identified[1] unless the residual is shared between units in a cluster as in the multilevel models considered later in the chapter.)

The logit link is appealing because it produces a linear model for the log of the odds, implying a multiplicative model for the odds themselves. If we add one unit to x_i, we must add β_2 to the log odds or multiply the odds by $\exp(\beta_2)$. This can be seen by considering a 1-unit change in x_i from some value a to $a+1$. The corresponding change in the log odds is

$$\ln\{\text{Odds}(y_i = 1 | x_i = a + 1)\} \;-\; \ln\{\text{Odds}(y_i = 1 | x_i = a)\}$$
$$= \{\beta_1 + \beta_2(a + 1)\} - (\beta_1 + \beta_2 a) = \beta_2$$

Exponentiating both sides, we obtain the *odds ratio* (OR):

$$\exp\left[\ln\{\text{Odds}(y_i = 1 | x_i = a + 1)\} \;-\; \ln\{\text{Odds}(y_i = 1 | x_i = a)\}\right]$$

$$= \frac{\text{Odds}(y_i = 1 | x_i = a + 1)}{\text{Odds}(y_i = 1 | x_i = a)} = \frac{\text{Pr}(y_i = 1 | x_i = a + 1)}{\text{Pr}(y_i = 0 | x_i = a + 1)} \bigg/ \frac{\text{Pr}(y_i = 1 | x_i = a)}{\text{Pr}(y_i = 0 | x_i = a)}$$

$$= \exp(\beta_2)$$

---

1. Formally, the model is identified by functional form. For instance, if x_i is continuous, the level-1 variance has a subtle effect on the shape of the relationship between $\text{Pr}(y_i = 1 | x_i)$ and x_i. With a probit link, single-level models with residuals are not identified.

Consider now the case where several covariates—for instance, x_{2i} and x_{3i}—are included in the model:

$$\text{logit}\left\{\Pr(y_i = 1 | x_{2i}, x_{3i})\right\} = \beta_1 + \beta_2 x_{2i} + \beta_3 x_{3i}$$

In this case, $\exp(\beta_2)$ is interpreted as the odds ratio comparing $x_{2i} = a + 1$ with $x_{2i} = a$ for given x_{3i} (controlling for x_{3i}), and $\exp(\beta_3)$ is the odds ratio comparing $x_{3i} = a + 1$ with $x_{3i} = a$ for given x_{2i}.

The predominant interpretation of the coefficients in logistic regression models is in terms of odds ratios, which is natural because the log odds is a *linear* function of the covariates. However, economists instead tend to interpret the coefficients in terms of marginal effects or partial effects on the response probability, which is a *nonlinear* function of the covariates. We relegate description of this approach to display 10.1, which may be skipped.

For a *continuous* covariate x_{2i}, economists often consider the partial derivative of the probability of success with respect to x_{2i}:

$$\Delta(x_{2i}|x_{3i}) \equiv \frac{\partial \Pr(y_i = 1 | x_{2i}, x_{3i})}{\partial x_{2i}} = \beta_2 \frac{\exp(\beta_1 + \beta_2 x_{2i} + \beta_3 x_{3i})}{\{\exp(\beta_1 + \beta_2 x_i + \beta_3 x_{3i})\}^2}$$

A small change in x_{2i} hence produces a change of $\beta_2 \frac{\exp(\beta_1 + \beta_2 x_{2i} + \beta_3 x_{3i})}{\{\exp(\beta_1 + \beta_2 x_{2i} + \beta_3 x_{3i})\}^2}$ in $\Pr(y_i = 1 | x_{2i}, x_{3i})$. Unlike in linear models, where the partial effect simply becomes β_2, the derivative of the nonlinear logistic function is not constant but depends on x_{2i} and x_{3i}.

For a *binary* covariate x_{3i}, economists consider the difference

$$\Delta(x_{3i}|x_{2i}) \equiv \Pr(y_i = 1 | x_{2i}, x_{3i} = 1) - \Pr(y_i = 1 | x_{2i}, x_{3i} = 0)$$
$$= \frac{\exp(\beta_1 + \beta_2 x_{2i} + \beta_3)}{1 + \exp(\beta_1 + \beta_2 x_{2i} + \beta_3)} - \frac{\exp(\beta_1 + \beta_2 x_{2i})}{1 + \exp(\beta_1 + \beta_2 x_{2i})}$$

which, unlike linear models, depends on x_{2i}.

The partial effect at the average (PEA) is obtained by substituting the sample means $\overline{x}_{2\cdot} = \frac{1}{N}\sum_{i=1}^{N} x_{i2}$ and $\overline{x}_{3\cdot} = \frac{1}{N}\sum_{i=1}^{N} x_{i3}$ for x_{i2} and x_{i3}, respectively, in the above expressions. Note that for binary covariates, the sample means are proportions and subjects cannot be at the average (because the proportions are between 0 and 1).

The average partial effect (APE) overcomes this problem by taking the sample means of the individual partial effects, $\text{APE}(x_{2i}|x_{3i}) = \frac{1}{N}\sum_{i=1}^{N}\Delta(x_{2i}|x_{3i})$ and $\text{APE}(x_{3i}|x_{2i}) = \frac{1}{N}\sum_{i=1}^{N}\Delta(x_{3i}|x_{2i})$. Fortunately, the APE and PEA tend to be similar.

Display 10.1: Partial effects at the average (PEA) and average partial effects (APE) for the logistic regression model, $\text{logit}\left\{\Pr(y_i = 1 | x_{2i}, x_{3i})\right\} = \beta_1 + \beta_2 x_{2i} + \beta_3 x_{3i}$, where x_{2i} is continuous and x_{3i} is binary.

To illustrate logistic regression, we will consider data on married women from the Canadian Women's Labor Force Participation Dataset used by Fox (1997). The dataset `womenlf.dta` contains women's employment status and two explanatory variables:

- `workstat`: employment status
  (0: not working; 1: employed part time; 2: employed full time)

- `husbinc`: husband's income in $1,000

- `chilpres`: child present in household (dummy variable)

The dataset can be retrieved by typing

> `. use http://www.stata-press.com/data/mlmus3/womenlf`

Fox (1997) considered a multiple logistic regression model for a woman being employed (full or part time) versus not working with covariates `husbinc` and `chilpres`

$$\text{logit}\{\Pr(y_i=1|\mathbf{x}_i)\} \;=\; \beta_1 + \beta_2 x_{2i} + \beta_3 x_{3i}$$

where $y_i = 1$ denotes employment, $y_i = 0$ denotes not working, x_{2i} is `husbinc`, x_{3i} is `chilpres`, and $\mathbf{x}_i = (x_{2i}, x_{3i})'$ is a vector containing both covariates.

We first merge categories 1 and 2 (employed part time and full time) of `workstat` into a new category 1 for being employed,

> `. recode workstat 2=1`

and then fit the model by maximum likelihood using Stata's `logit` command:

```
. logit workstat husbinc chilpres
Logistic regression                             Number of obs   =        263
                                                LR chi2(2)      =      36.42
                                                Prob > chi2     =     0.0000
Log likelihood = -159.86627                     Pseudo R2       =     0.1023
```

workstat	Coef.	Std. Err.	z	P>|z|	[95% Conf. Interval]	
husbinc	-.0423084	.0197801	-2.14	0.032	-.0810768	-.0035401
chilpres	-1.575648	.2922629	-5.39	0.000	-2.148473	-1.002824
_cons	1.33583	.3837632	3.48	0.000	.5836674	2.087992

The estimated coefficients are negative, so the estimated log odds of employment are lower if the husband earns more and if there is a child in the household. At the 5% significance level, we can reject the null hypotheses that the individual coefficients β_2 and β_3 are zero. The estimated coefficients and their estimated standard errors are also given in table 10.1.

Table 10.1: Maximum likelihood estimates for logistic regression model for women's labor force participation

	Est	(SE)	OR $=\exp(\beta)$	(95% CI)
β_1 [_cons]	1.34	(0.38)		
β_2 [husbinc]	−0.04	(0.02)	0.96	(0.92, 1.00)
β_3 [chilpres]	−1.58	(0.29)	0.21	(0.12, 0.37)

Instead of considering changes in log odds, it is more informative to obtain odds ratios, the exponentiated regression coefficients. This can be achieved by using the `logit` command with the `or` option:

```
. logit workstat husbinc chilpres, or
Logistic regression                             Number of obs   =        263
                                                LR chi2(2)      =      36.42
                                                Prob > chi2     =     0.0000
Log likelihood = -159.86627                     Pseudo R2       =     0.1023

    workstat | Odds Ratio   Std. Err.      z    P>|z|     [95% Conf. Interval]
-------------+----------------------------------------------------------------
     husbinc |  .9585741    .0189607    -2.14   0.032     .9221229    .9964662
    chilpres |  .2068734    .0604614    -5.39   0.000     .1166621    .3668421
       _cons |   3.80315    1.45951      3.48   0.000     1.792601    8.068699
```

Comparing women with and without a child at home, whose husbands have the same income, the odds of working are estimated to be about 5 ($\approx 1/0.2068734$) times as high for women who do not have a child at home as for women who do. Within these two groups of women, each $1,000 increase in husband's income reduces the odds of working by an estimated 4% $\{-4\% = 100\%(0.9585741 - 1)\}$. Although this odds ratio looks less important than the one for `chilpres`, remember that we cannot directly compare the magnitude of the two odds ratios. The odds ratio for `chilpres` represents a comparison of two distinct groups of women, whereas the odds ratio for `husbinc` merely expresses the effect of a $1,000 increase in the husband's income. A $10,000 increase would be associated with an odds ratio of 0.66 ($= 0.958741^{10}$).

The exponentiated intercept, estimated as 3.80, represents the odds of working for women who do not have a child at home and whose husbands' income is 0. This is not an odds ratio as the column heading implies, but the odds when all covariates are zero. For this reason, the exponentiated intercept was omitted from the output in earlier releases of Stata (until Stata 12.0) when the `or` option was used. As for the intercept itself, the exponentiated intercept is interpretable only if zero is a meaningful value for all covariates.

In an attempt to make effects directly comparable and assess the relative importance of covariates, some researchers standardize all covariates to have standard deviation 1, thereby comparing the effects of a standard deviation change in each covariate. As discussed in section 1.5, there are many problems with such an approach, one of them being the meaningless notion of a standard deviation change in a dummy variable, such as `chilpres`.

The standard errors of exponentiated estimated regression coefficients should generally not be used for confidence intervals or hypothesis tests. Instead, the 95% confidence intervals in the above output were computed by taking the exponentials of the confidence limits for the regression coefficients β:

$$\exp\{\widehat{\beta} \pm 1.96 \times \mathrm{SE}(\widehat{\beta})\}$$

In table 10.1, we therefore report estimated odds ratios with 95% confidence intervals instead of standard errors.

To visualize the model, we can produce a plot of the predicted probabilities versus `husbinc`, with separate curves for women with and without children at home. Plugging in maximum likelihood estimates for the parameters in (10.2), the predicted probability for woman i, often denoted $\widehat{\pi}_i$, is given by the inverse logit of the estimated linear predictor

$$\widehat{\pi}_i \equiv \widehat{\Pr}(y_i = 1|x_i) = \frac{\exp(\widehat{\beta}_1 + \widehat{\beta}_2 x_{2i} + \widehat{\beta}_3 x_{3i})}{1 + \exp(\widehat{\beta}_1 + \widehat{\beta}_2 x_{2i} + \widehat{\beta}_3 x_{3i})} = \mathrm{logit}^{-1}(\widehat{\beta}_1 + \widehat{\beta}_2 x_{2i} + \widehat{\beta}_3 x_{3i})$$

$$(10.3)$$

and can be obtained for the women in the dataset by using the `predict` command with the `pr` option:

```
. predict prob, pr
```

We can now produce the graph of predicted probabilities, shown in figure 10.1, by using

```
. twoway (line prob husbinc if chilpres==0, sort)
> (line prob husbinc if chilpres==1, sort lpatt(dash)),
> legend(order(1 "No child" 2 "Child"))
> xtitle("Husband's income/$1000") ytitle("Probability that wife works")
```

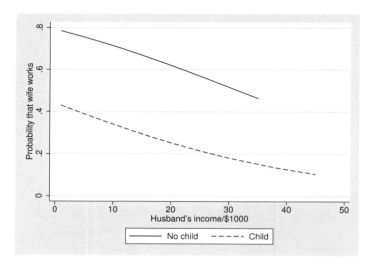

Figure 10.1: Predicted probability of working from logistic regression model (for range of `husbinc` in dataset)

The graph is similar to the graph of the predicted means from an analysis of covariance model (a linear regression model with a continuous and a dichotomous covariate; see section 1.7) except that the curves are not exactly straight. The curves have been plotted for the range of values of `husbinc` observed for the two groups of women, and for these ranges the predicted probabilities are nearly linear functions of `husbinc`.

To see what the inverse logit function looks like, we will now plot the predicted probabilities for a widely extended range of values of `husbinc` (including negative values, although this does not make sense). This could be accomplished by inventing additional observations with more extreme values of `husbinc` and then using the `predict` command again. More conveniently, we can also use Stata's useful `twoway` plot type, `function`:

```
. twoway (function y=invlogit(_b[husbinc]*x+_b[_cons]), range(-100 100))
> (function y=invlogit(_b[husbinc]*x+_b[chilpres]+_b[_cons]),
> range(-100 100) lpatt(dash)),
> xtitle("Husband's income/$1000") ytitle("Probability that wife works")
> legend(order(1 "No child" 2 "Child")) xline(1) xline(45)
```

The estimated regression coefficients are referred to as `_b[husbinc]`, `_b[chilpres]`, and `_b[_cons]`, and we have used Stata's `invlogit()` function to obtain the predicted probabilities given in (10.3). The resulting graph is shown in figure 10.2.

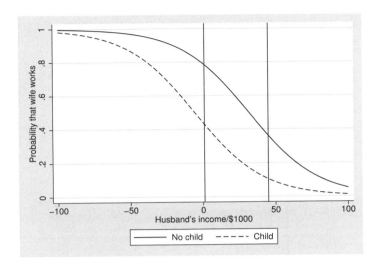

Figure 10.2: Predicted probability of working from logistic regression model (extrapolating beyond the range of `husbinc` in the data)

The range of `husbinc` actually observed in the data lies approximately between the two vertical lines. It would not be safe to rely on predicted probabilities extrapolated outside this range. The curves are approximately linear in the region where the linear predictor is close to zero (and the predicted probability is close to 0.5) and then flatten as the linear predictor becomes extreme. This flattening ensures that the predicted probabilities remain in the permitted interval from 0 to 1.

We can fit the same model by using the `glm` command for generalized linear models. The syntax is the same as that of the `logit` command except that we must specify the logit link function in the `link()` option and the binomial distribution in the `family()` option:

```
. glm workstat husbinc chilpres, link(logit) family(binomial)
Generalized linear models                    No. of obs      =        263
Optimization     : ML                        Residual df     =        260
                                             Scale parameter =          1
Deviance         =  319.7325378              (1/df) Deviance =   1.229741
Pearson          =  265.9615312              (1/df) Pearson  =   1.022929
Variance function: V(u) = u*(1-u)            [Bernoulli]
Link function    : g(u) = ln(u/(1-u))        [Logit]
                                             AIC             =   1.238527
Log likelihood   = -159.8662689              BIC             = -1129.028
```

		OIM				
workstat	Coef.	Std. Err.	z	P>|z|	[95% Conf.	Interval]
husbinc	-.0423084	.0197801	-2.14	0.032	-.0810768	-.0035401
chilpres	-1.575648	.2922629	-5.39	0.000	-2.148473	-1.002824
_cons	1.33583	.3837634	3.48	0.000	.5836674	2.087992

To obtain estimated odds ratios, we use the `eform` option (for "exponentiated form"), and to fit a probit model, we simply change the `link(logit)` option to `link(probit)`.

10.2.2   Latent-response formulation

The logistic regression model and other models for dichotomous responses can also be viewed as latent-response models. Underlying the observed dichotomous response y_i (whether the woman works or not), we imagine that there is an unobserved or latent continuous response y_i^* representing the propensity to work or the excess utility of working as compared with not working. If this latent response is greater than 0, then the observed response is 1; otherwise, the observed response is 0:

$$y_i = \begin{cases} 1 & \text{if } y_i^* > 0 \\ 0 & \text{otherwise} \end{cases}$$

For simplicity, we will assume that there is one covariate x_i. A linear regression model is then specified for the latent response y_i^*

$$y_i^* = \beta_1 + \beta_2 x_i + \epsilon_i$$

where ϵ_i is a residual error term with $E(\epsilon_i|x_i) = 0$ and the error terms of different women i are independent.

The latent-response formulation has been used in various disciplines and applications. In genetics, where y_i is often a phenotype or qualitative trait, y_i^* is called a *liability*. For attitudes measured by agreement or disagreement with statements, the latent response can be thought of as a "sentiment" in favor of the statement. In economics, the latent response is often called an *index function*. In discrete-choice settings (see chapter 12), y_i^* is the *difference in utilities* between alternatives.

Figure 10.3 illustrates the relationship between the latent-response formulation, shown in the lower graph, and the generalized linear model formulation, shown in the upper graph in terms of a curve for the conditional probability that $y_i = 1$. The regression line in the lower graph represents the conditional expectation of y_i^* given x_i as a function of x_i, and the density curves represent the conditional distributions of y_i^* given x_i. The dotted horizontal line at $y_i^* = 0$ represents the threshold, so $y_i = 1$ if y_i^* exceeds the threshold and $y_i = 0$ otherwise. Therefore, the areas under the parts of the density curves that lie above the dotted line, here shaded gray, represent the probabilities that $y_i = 1$ given x_i. For the value of x_i indicated by the vertical dotted line, the mean of y_i^* is 0; therefore, half the area under the density curve lies above the threshold, and the conditional probability that $y_i = 1$ equals 0.5 at that point.

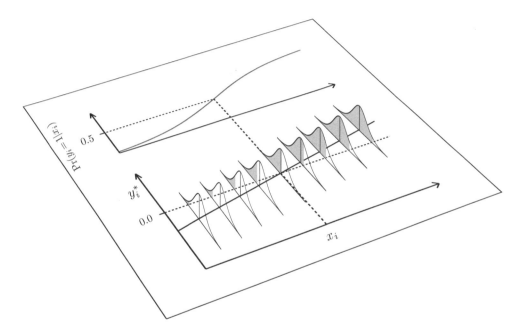

Figure 10.3: Illustration of equivalence of latent-response and generalized linear model formulations for logistic regression

We can derive the probability curve from the latent-response formulation as follows:

$$\begin{aligned}
\Pr(y_i = 1 | x_i) &= \Pr(y_i^* > 0 | x_i) = \Pr(\beta_1 + \beta_2 x_i + \epsilon_i > 0 | x_i) \\
&= \Pr\{\epsilon_i > -(\beta_1 + \beta_2 x_i) | x_i\} = \Pr(-\epsilon_i \leq \beta_1 + \beta_2 x_i | x_i) \\
&= F(\beta_1 + \beta_2 x_i)
\end{aligned}$$

where $F(\cdot)$ is the cumulative density function of $-\epsilon_i$, or the area under the density curve for $-\epsilon_i$ from minus infinity to $\beta_1 + \beta_2 x_i$. If the distribution of ϵ_i is symmetric, the cumulative density function of $-\epsilon_i$ is the same as that of ϵ_i.

Logistic regression

In logistic regression, ϵ_i is assumed to have a standard logistic cumulative density function given x_i,

$$\Pr(\epsilon_i < \tau | x_i) = \frac{\exp(\tau)}{1 + \exp(\tau)}$$

For this distribution, ϵ_i has mean zero and variance $\pi^2/3 \approx 3.29$ (note that π here represents the famous mathematical constant pronounced "pi", the circumference of a circle divided by its diameter).

Probit regression

When a latent-response formulation is used, it seems natural to assume that ϵ_i has a normal distribution given x_i, as is typically done in linear regression. If a standard (mean 0 and variance 1) normal distribution is assumed, the model becomes a probit model,

$$\Pr(y_i = 1 | x_i) = F(\beta_1 + \beta_2 x_i) = \Phi(\beta_1 + \beta_2 x_i) \tag{10.4}$$

Here $\Phi(\cdot)$ is the standard normal cumulative distribution function, the probability that a standard normally distributed random variable (here ϵ_i) is less than the argument. For example, when $\beta_1 + \beta_2 x_i$ equals 1.96, $\Phi(\beta_1 + \beta_2 x_i)$ equals 0.975. $\Phi(\cdot)$ is the inverse link function $h(\cdot)$, whereas the link function $g(\cdot)$ is $\Phi^{-1}(\cdot)$, the inverse standard normal cumulative distribution function, called the *probit link* function [the Stata function for $\Phi^{-1}(\cdot)$ is `invnormal()`].

To understand why a *standard* normal distribution is specified for ϵ_i, with the variance θ fixed at 1, consider the graph in figure 10.4. On the left, the standard deviation is 1, whereas the standard deviation on the right is 2. However, by doubling the slope of the regression line for y_i^* on the right (without changing the point where it intersects the threshold 0), we obtain the same curve for the probability that $y_i = 1$. Because we can obtain equivalent models by increasing both the standard deviation and the slope by the same multiplicative factor, the model with a freely estimated standard deviation is not identified.

This lack of identification is also evident from inspecting the expression for the probability if the variance θ were not fixed at 1 [from (10.4)],

$$\Pr(y_i = 1 | x_i) = \Pr(\epsilon_i \leq \beta_1 + \beta_2 x_i) = \Pr\left(\frac{\epsilon_i}{\sqrt{\theta}} \leq \frac{\beta_1 + \beta_2 x_i}{\sqrt{\theta}}\right) = \Phi\left(\frac{\beta_1}{\sqrt{\theta}} + \frac{\beta_2}{\sqrt{\theta}} x_i\right)$$

where we see that multiplication of the regression coefficients by a constant can be counteracted by multiplying $\sqrt{\theta}$ by the same constant. This is the reason for fixing the standard deviation in probit models to 1 (see also exercise 10.10). The variance of ϵ_i in logistic regression is also fixed but to a larger value, $\pi^2/3$.

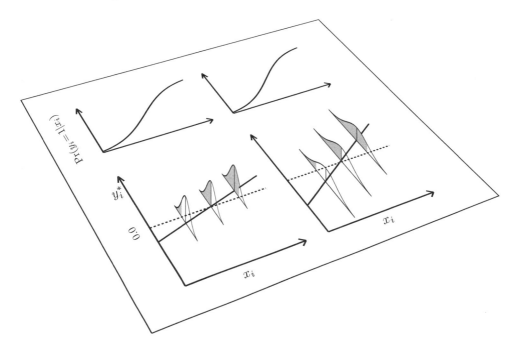

Figure 10.4: Illustration of equivalence between probit models with change in residual standard deviation counteracted by change in slope

A probit model can be fit to the women's employment data in Stata by using the `probit` command:

```
. probit workstat husbinc chilpres
Probit regression                               Number of obs   =        263
                                                LR chi2(2)      =      36.19
                                                Prob > chi2     =     0.0000
Log likelihood = -159.97986                     Pseudo R2       =     0.1016
```

workstat	Coef.	Std. Err.	z	P>|z|	[95% Conf. Interval]
husbinc	-.0242081	.0114252	-2.12	0.034	-.0466011 -.001815
chilpres	-.9706164	.1769051	-5.49	0.000	-1.317344 -.6238887
_cons	.7981507	.2240082	3.56	0.000	.3591028 1.237199

These estimates are closer to zero than those reported for the logit model in table 10.1 because the standard deviation of ϵ_i is 1 for the probit model and $\pi/\sqrt{3} \approx 1.81$ for the logit model. Therefore, as we have already seen in figure 10.4, the regression coefficients in logit models must be larger in absolute value to produce nearly the same curve for the conditional probability that $y_i = 1$. Here we say "nearly the same" because the shapes of the probit and logit curves are similar yet not identical. To visualize the subtle difference in shape, we can plot the predicted probabilities for women without children at home from both the logit and probit models:

```
. twoway (function y=invlogit(1.3358-0.0423*x), range(-100 100))
> (function y=normal(0.7982-0.0242*x), range(-100 100) lpatt(dash)),
> xtitle("Husband's income/$1000") ytitle("Probability that wife works")
> legend(order(1 "Logit link" 2 "Probit link")) xline(1) xline(45)
```

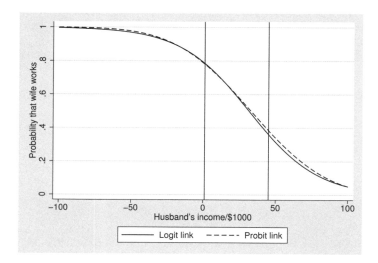

Figure 10.5: Predicted probabilities of working from logistic and probit regression models for women without children at home

Here the predictions from the models coincide nearly perfectly in the region where most of the data are concentrated and are very similar elsewhere. It is thus futile to attempt to empirically distinguish between the logit and probit links unless one has a huge sample.

Regression coefficients in probit models cannot be interpreted in terms of odds ratios as in logistic regression models. Instead, the coefficients can be interpreted as differences in the population means of the *latent response* y_i^*, controlling or adjusting for other covariates (the same kind of interpretation can also be made in logistic regression). Many people find interpretation based on latent responses less appealing than interpretation using odds ratios, because the latter refer to observed responses y_i. Alternatively, the coefficients can be interpreted in terms of average partial effects or partial effects at the average as shown for logit models[2] in display 10.1.

---

2. For probit models with continuous x_{2i} and binary x_{3i}, $\Delta(x_{2i}|x_{3i}) = \beta_2\,\phi(\beta_1 + \beta_2 x_{2i} + \beta_3 x_{3i})$, where $\phi(\cdot)$ is the density function of the standard normal distribution, and $\Delta(x_{3i}|x_{2i}) = \Phi(\beta_1 + \beta_2 x_{2i} + \beta_3) - \Phi(\beta_1 + \beta_2 x_{2i})$.

10.3 Which treatment is best for toenail infection?

Lesaffre and Spiessens (2001) analyzed data provided by De Backer et al. (1998) from a randomized, double-blind trial of treatments for toenail infection (dermatophyte ony-chomycosis). Toenail infection is common, with a prevalence of about 2% to 3% in the United States and a much higher prevalence among diabetics and the elderly. The infection is caused by a fungus, and not only disfigures the nails but also can cause physical pain and impair the ability to work.

In this clinical trial, 378 patients were randomly allocated into two oral antifungal treatments (250 mg/day terbinafine and 200 mg/day itraconazole) and evaluated at seven visits, at weeks 0, 4, 8, 12, 24, 36, and 48. One outcome is onycholysis, the degree of separation of the nail plate from the nail bed, which was dichotomized ("moderate or severe" versus "none or mild") and is available for 294 patients.

The dataset `toenail.dta` contains the following variables:

- `patient`: patient identifier
- `outcome`: onycholysis (separation of nail plate from nail bed)
  (0: none or mild; 1: moderate or severe)
- `treatment`: treatment group (0: itraconazole; 1: terbinafine)
- `visit`: visit number (1, 2, ..., 7)
- `month`: exact timing of visit in months

We read in the toenail data by typing

```
. use http://www.stata-press.com/data/mlmus3/toenail, clear
```

The main research question is whether the treatments differ in their efficacy. In other words, do patients receiving one treatment experience a greater decrease in their probability of having onycholysis than those receiving the other treatment?

10.4 Longitudinal data structure

Before investigating the research question, we should look at the longitudinal structure of the toenail data using, for instance, the `xtdescribe`, `xtsum`, and `xttab` commands, discussed in *Introduction to models for longitudinal and panel data (part III)*.

Here we illustrate the use of the `xtdescribe` command, which can be used for these data because the data were intended to be balanced with seven visits planned for the same set of weeks for each patient (although the exact timing of the visits varied between patients).

Before using `xtdescribe`, we `xtset` the data with `patient` as the cluster identifier and `visit` as the time variable:

```
. xtset patient visit
       panel variable:  patient (unbalanced)
        time variable:  visit, 1 to 7, but with gaps
                delta:  1 unit
```

The output states that the data are unbalanced and that there are gaps. [We would describe the time variable `visit` as balanced because the values are identical across patients apart from the gaps caused by missing data; see the introduction to models for longitudinal and panel data (part III in volume I).]

To explore the missing-data patterns, we use

```
. xtdescribe if outcome < .
 patient:  1, 2, ..., 383                                    n =        294
   visit:  1, 2, ..., 7                                      T =          7
           Delta(visit) = 1 unit
           Span(visit)  = 7 periods
           (patient*visit uniquely identifies each observation)
 Distribution of T_i:   min      5%     25%     50%     75%     95%     max
                          1       3       7       7       7       7       7

   Freq.   Percent   Cum. |  Pattern
 ---------------------------+---------
     224     76.19  76.19  |  1111111
      21      7.14  83.33  |  11111.1
      10      3.40  86.73  |  1111.11
       6      2.04  88.78  |  111....
       5      1.70  90.48  |  1......
       5      1.70  92.18  |  11111..
       4      1.36  93.54  |  1111...
       3      1.02  94.56  |  11.....
       3      1.02  95.58  |  111.111
      13      4.42 100.00  |  (other patterns)
 ---------------------------+---------
     294    100.00         |  XXXXXXX
```

We see that 224 patients have complete data (the pattern "1111111"), 21 patients missed the sixth visit ("11111.1"), 10 patients missed the fifth visit ("1111.11"), and most other patients dropped out at some point, never returning after missing a visit. The latter pattern is sometimes referred to as *monotone missingness*, in contrast with *intermittent missingness*, which follows no particular pattern.

As discussed in section 5.8, a nice feature of maximum likelihood estimation for incomplete data such as these is that all information is used. Thus not only patients who attended all visits but also patients with missing visits contribute information. If the model is correctly specified, maximum likelihood estimates are consistent when the responses are missing at random (MAR).

10.5 Proportions and fitted population-averaged or marginal probabilities

A useful graphical display of the data is a bar plot showing the proportion of patients with onycholysis at each visit by treatment group. The following Stata commands can be used to produce the graph shown in figure 10.6:

```
. label define tr 0 "Itraconazole" 1 "Terbinafine"
. label values treatment tr
. graph bar (mean) proportion = outcome, over(visit) by(treatment)
> ytitle(Proportion with onycholysis)
```

Here we defined value labels for **treatment** to make them appear on the graph.

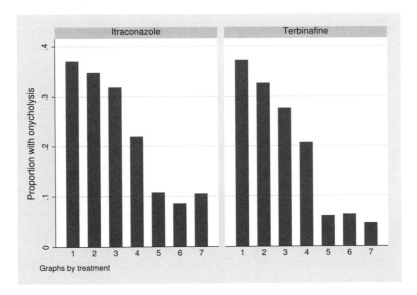

Figure 10.6: Bar plot of proportion of patients with toenail infection by visit and treatment group

We used the visit number **visit** to define the bars instead of the exact timing of the visit **month** because there would generally not be enough patients with the same timing to estimate the proportions reliably. An alternative display is a line graph, plotting the observed proportions at each visit against time. For this graph, it is better to use the average time associated with each visit for the x axis than to use visit number, because the visits were not equally spaced. Both the proportions and the average times for each visit in each treatment group can be obtained using the **egen** command with the **mean()** function:

```
. egen prop = mean(outcome), by(treatment visit)
. egen mn_month = mean(month), by(treatment visit)
. twoway line prop mn_month, by(treatment) sort
> xtitle(Time in months) ytitle(Proportion with onycholysis)
```

The resulting graph is shown in figure 10.7.

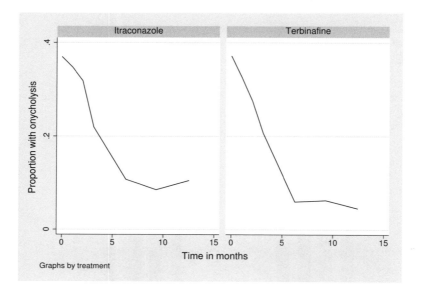

Figure 10.7: Line plot of proportion of patients with toenail infection by average time at visit and treatment group

The proportions shown in figure 10.7 represent the estimated average (or marginal) probabilities of onycholysis given the two covariates, time since randomization and treatment group. We are not attempting to estimate individual patients' personal probabilities, which may vary substantially, but are considering the population averages given the covariates.

Instead of estimating the probabilities for each combination of visit and treatment, we can attempt to obtain smooth curves of the estimated probability as a function of time. We then no longer have to group observations for the same visit number together— we can use the exact timing of the visits directly. One way to accomplish this is by using a logistic regression model with month, treatment, and their interaction as covariates. This model for the dichotomous outcome y_{ij} at visit i for patient j can be written as

$$\text{logit}\{\Pr(y_{ij}=1|\mathbf{x}_{ij})\} \;=\; \beta_1 + \beta_2 x_{2j} + \beta_3 x_{3ij} + \beta_4 x_{2j} x_{3ij} \tag{10.5}$$

where x_{2j} represents treatment, x_{3ij} represents month, and $\mathbf{x}_{ij} = (x_{2j}, x_{3ij})'$ is a vector containing both covariates. This model allows for a difference between groups at baseline β_2, and linear changes in the log odds of onycholysis over time with slope β_3 in the itraconazole group and slope $\beta_3 + \beta_4$ in the terbinafine group. Therefore, β_4, the difference in the rate of improvement (on the log odds scale) between treatment groups, can be viewed as the treatment effect (terbinafine versus itraconazole).

This model makes the unrealistic assumption that the responses for a given patient are conditionally independent after controlling for the included covariates. We will relax this assumption in the next section. Here we can get satisfactory inferences for marginal effects by using robust standard errors for clustered data instead of using model-based standard errors. This approach is analogous to pooled OLS in linear models and corresponds to the generalized estimating equations approach discussed in section 6.6 with an independence working correlation structure (see 10.14.2 for an example with a different working correlation matrix).

We start by constructing an interaction term, `trt_month`, for `treatment` and `month`,

```
. generate trt_month = treatment*month
```

before fitting the model by maximum likelihood with robust standard errors:

```
. logit outcome treatment month trt_month, or vce(cluster patient)
Logistic regression                             Number of obs   =       1908
                                                Wald chi2(3)    =      64.30
                                                Prob > chi2     =     0.0000
Log pseudolikelihood = -908.00747               Pseudo R2       =     0.0830
```

outcome	Odds Ratio	Robust Std. Err.	z	P>|z|	[95% Conf. Interval]	
treatment	.9994184	.2511294	-0.00	0.998	.6107468	1.635436
month	.8434052	.0246377	-5.83	0.000	.7964725	.8931034
trt_month	.934988	.0488105	-1.29	0.198	.8440528	1.03572
_cons	.5731389	.0982719	-3.25	0.001	.4095534	.8020642

Instead of creating a new variable for the interaction, we could have used factor-variables syntax as follows:

```
logit outcome i.treatment##c.month, or vce(cluster patient)
```

We will leave interpretation of estimates for later and first check how well predicted probabilities from the logistic regression model correspond to the observed proportions in figure 10.7. The predicted probabilities are obtained and plotted together with the observed proportions by using the following commands, which result in figure 10.8.

```
. predict prob, pr
. twoway (line prop mn_month, sort) (line prob month, sort lpatt(dash)),
> by(treatment) legend(order(1 "Observed proportions" 2 "Fitted probabilities"))
> xtitle(Time in months) ytitle(Probability of onycholysis)
```

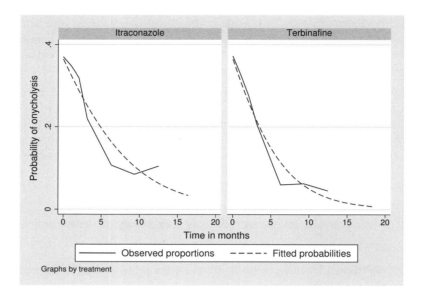

Figure 10.8: Proportions and fitted probabilities using ordinary logistic regression

The marginal probabilities predicted by the model fit the observed proportions reasonably well. However, we have treated the dependence among responses for the same patient as a nuisance by fitting an ordinary logistic regression model with robust standard errors for clustered data. We now add random effects to model the dependence and estimate the degree of dependence instead of treating it as a nuisance.

10.6 Random-intercept logistic regression

10.6.1 Model specification

Reduced-form specification

To relax the assumption of conditional independence among the responses for the same patient given the covariates, we can include a patient-specific random intercept ζ_j in the linear predictor to obtain a random-intercept logistic regression model

$$\text{logit}\{\Pr(y_{ij}=1|\mathbf{x}_{ij},\zeta_j)\} = \beta_1 + \beta_2 x_{2j} + \beta_3 x_{3ij} + \beta_4 x_{2j} x_{3ij} + \zeta_j \qquad (10.6)$$

The random intercepts $\zeta_j \sim N(0,\psi)$ are assumed to be independent and identically distributed across patients j and independent of the covariates $\mathbf{x}_{ij}$. Given ζ_j and $\mathbf{x}_{ij}$, the responses y_{ij} for patient j at different occasions i are independently Bernoulli distributed. To write this down more formally, it is useful to define $\pi_{ij} \equiv \Pr(y_{ij}|\mathbf{x}_{ij},\zeta_j)$, giving

$$\text{logit}(\pi_{ij}) = \beta_1 + \beta_2 x_{2j} + \beta_3 x_{3ij} + \beta_4 x_{2j} x_{3ij} + \zeta_j$$
$$y_{ij}|\pi_{ij} \sim \text{Binomial}(1, \pi_{ij})$$

This is a simple example of a *generalized linear mixed model* (GLMM) because it is a generalized linear model with both fixed effects β_1 to β_4 and a random effect ζ_j. The model is also sometimes referred to as a hierarchical generalized linear model (HGLM) in contrast to a hierarchical linear model (HLM). The random intercept can be thought of as the combined effect of omitted patient-specific (time-constant) covariates that cause some patients to be more prone to onycholysis than others (more precisely, the component of this combined effect that is independent of the covariates in the model— not an issue if the covariates are exogenous). It is appealing to model this unobserved heterogeneity in the same way as observed heterogeneity by simply adding the random intercept to the linear predictor. As we will explain later, be aware that odds ratios obtained by exponentiating regression coefficients in this model must be interpreted conditionally on the random intercept and are therefore often referred to as conditional or subject-specific odds ratios.

Using the latent-response formulation, the model can equivalently be written as

$$y_{ij}^* = \beta_1 + \beta_2 x_{2j} + \beta_3 x_{3ij} + \beta_4 x_{2j} x_{3ij} + \zeta_j + \epsilon_{ij} \tag{10.7}$$

where $\zeta_j \sim N(0, \psi)$ and the ϵ_{ij} have standard logistic distributions. The binary responses y_{ij} are determined by the latent continuous responses via the threshold model

$$y_{ij} = \begin{cases} 1 & \text{if } y_{ij}^* > 0 \\ 0 & \text{otherwise} \end{cases}$$

Confusingly, logistic random-effects models are sometimes written as $y_{ij} = \pi_{ij} + e_{ij}$, where e_{ij} is a normally distributed level-1 residual with variance $\pi_{ij}(1 - \pi_{ij})$. This formulation is clearly incorrect because such a model does not produce binary responses (see Skrondal and Rabe-Hesketh [2007]).

In both formulations of the model (via a logit link or in terms of a latent response), it is assumed that the ζ_j are independent across patients and independent of the covariates $\mathbf{x}_{ij}$ at occasion i. It is also assumed that the covariates at other occasions do not affect the response probabilities given the random intercept (called strict exogeneity conditional on the random intercept). For the latent response formulation, the ϵ_{ij} are assumed to be independent across both occasions and patients, and independent of both ζ_j and $\mathbf{x}_{ij}$. In the generalized linear model formulation, the analogous assumptions are implicit in assuming that the responses are independently Bernoulli distributed (with probabilities determined by ζ_j and $\mathbf{x}_{ij}$).

In contrast to linear random effects models, consistent estimation in random-effects logistic regression requires that the random part of the model is correctly specified in addition to the fixed part. Specifically, consistency formally requires (1) a correct linear predictor (such as including relevant interactions), (2) a correct link function, (3) correct specification of covariates having random coefficients, (4) conditional independence of responses given the random effects and covariates, (5) independence of the random effects and covariates (for causal inference), and (6) normally distributed random effects. Hence, the assumptions are stronger than those discussed for linear models in section 3.3.2. However, the normality assumption for the random intercepts seems to be rather innocuous in contrast to the assumption of independence between the random intercepts and covariates (Heagerty and Kurland 2001). As in standard logistic regression, the ML estimator is not necessarily unbiased in finite samples even if all the assumptions are true.

Two-stage formulation

Raudenbush and Bryk (2002) and others write two-level models in terms of a level-1 model and one or more level-2 models (see section 4.9). In generalized linear mixed models, the need to specify a link function and distribution leads to two further stages of model specification.

Using the notation and terminology of Raudenbush and Bryk (2002), the level-1 sampling model, link function, and structural model are written as

$$
\begin{aligned}
y_{ij} &\sim \text{Bernoulli}(\varphi_{ij}) \\
\text{logit}(\varphi_{ij}) &= \eta_{ij} \\
\eta_{ij} &= \beta_{0j} + \beta_{1j}x_{2j} + \beta_{2j}x_{3ij} + \beta_{3j}x_{2j}x_{3ij}
\end{aligned}
$$

respectively.

The level-2 model for the intercept β_{0j} is written as

$$
\beta_{0j} = \gamma_{00} + u_{0j}
$$

where γ_{00} is a fixed intercept and u_{0j} is a residual or random intercept. The level-2 models for the coefficients β_{1j}, β_{2j}, and β_{3j} have no residuals for a random-intercept model,

$$
\beta_{pj} = \gamma_{p0}, \quad p = 1, 2, 3
$$

Plugging the level-2 models into the level-1 structural model, we obtain

$$
\begin{aligned}
\eta_{ij} &= \gamma_{00} + u_{0j} + \gamma_{01}x_{2j} + \gamma_{02}x_{3ij} + \gamma_{03}x_{2j}x_{3ij} \\
&\equiv \beta_1 + \zeta_{0j} + \beta_2 x_{2j} + \beta_3 x_{3ij} + \beta_4 x_{2j}x_{3ij}
\end{aligned}
$$

Equivalent models can be specified using either the reduced-form formulation (used for instance by Stata) or the two-stage formulation (used in the HLM software of Raudenbush et al. 2004). However, in practice, model specification is to some extent influenced by the approach adopted as discussed in section 4.9.

10.7 Estimation of random-intercept logistic models

As of Stata 10, there are three commands for fitting random-intercept logistic models in Stata: `xtlogit`, `xtmelogit`, and `gllamm`. All three commands provide maximum likelihood estimation and use adaptive quadrature to approximate the integrals involved (see section 10.11.1 for more information). The commands have essentially the same syntax as their counterparts for linear models discussed in volume I. Specifically, `xtlogit` corresponds to `xtreg`, `xtmelogit` corresponds to `xtmixed`, and `gllamm` uses essentially the same syntax for linear, logistic, and other types of models.

All three commands are relatively slow because they use numerical integration, but for random-intercept models, `xtlogit` is much faster than `xtmelogit`, which is usually faster than `gllamm`. However, the rank ordering is reversed when it comes to the usefulness of the commands for predicting random effects and various types of probabilities as we will see in sections 10.12 and 10.13. Each command uses a default for the number of terms (called "integration points") used to approximate the integral, and there is no guarantee that a sufficient number of terms has been used to achieve reliable estimates. It is therefore the user's responsibility to make sure that the approximation is adequate by increasing the number of integration points until the results stabilize. The more terms are used, the more accurate the approximation at the cost of increased computation time.

We do not discuss random-coefficient logistic regression in this chapter, but such models can be fit with `xtmelogit` and `gllamm` (but not using `xtlogit`), using essentially the same syntax as for linear random-coefficient models discussed in section 4.5. Random-coefficient logistic regression using `gllamm` is demonstrated in chapters 11 (for ordinal responses) and 16 (for models with nested and crossed random effects) and using `xtmelogit` in chapter 16. The probit version of the random-intercept model is available in `gllamm` (see sections 11.10 through 11.12) and `xtprobit`, but random-coefficient probit models are available in `gllamm` only.

10.7.1 Using xtlogit

The `xtlogit` command for fitting the random-intercept model is analogous to the `xtreg` command for fitting the corresponding linear model. We first use the `xtset` command to specify the clustering variable. In the `xtlogit` command, we use the `intpoints(30)` option (`intpoints()` stands for "integration points") to ensure accurate estimates (see section 10.11.1):

```
. quietly xtset patient
. xtlogit outcome treatment month trt_month, intpoints(30)
Random-effects logistic regression              Number of obs     =       1908
Group variable: patient                         Number of groups  =        294

Random effects u_i ~ Gaussian                   Obs per group: min =          1
                                                               avg =        6.5
                                                               max =          7
                                                Wald chi2(3)      =     150.65
Log likelihood  = -625.38558                    Prob > chi2       =     0.0000
```

outcome	Coef.	Std. Err.	z	P>|z|	[95% Conf. Interval]	
treatment	-.160608	.5796716	-0.28	0.782	-1.296744	.9755275
month	-.390956	.0443707	-8.81	0.000	-.4779209	-.3039911
trt_month	-.1367758	.0679947	-2.01	0.044	-.270043	-.0035085
_cons	-1.618795	.4303891	-3.76	0.000	-2.462342	-.7752477
/lnsig2u	2.775749	.1890237			2.405269	3.146228
sigma_u	4.006325	.3786451			3.328876	4.821641
rho	.8298976	.026684			.7710804	.8760322

```
Likelihood-ratio test of rho=0: chibar2(01) =    565.24 Prob >= chibar2 = 0.000
```

The estimated regression coefficients are given in the usual format. The value next to **sigma_u** represents the estimated residual standard deviation $\sqrt{\widehat{\psi}}$ of the random intercept and the value next to **rho** represents the estimated residual intraclass correlation of the latent responses (see section 10.9.1).

We can use the **or** option to obtain exponentiated regression coefficients, which are interpreted as conditional odds ratios here. Instead of refitting the model, we can simply change the way the results are displayed using the following short **xtlogit** command (known as "replaying the estimation results" in Stata parlance):

```
. xtlogit, or
```

Random-effects logistic regression
Group variable: patient

Random effects u_i ~ Gaussian

	Number of obs	=	1908
	Number of groups	=	294

Obs per group:	min =	1
	avg =	6.5
	max =	7

Log likelihood = -625.38558

Wald chi2(3)	=	150.65
Prob > chi2	=	0.0000

outcome	OR	Std. Err.	z	P>|z|	[95% Conf. Interval]	
treatment	.8516258	.4936633	-0.28	0.782	.2734207	2.652566
month	.6764099	.0300128	-8.81	0.000	.6200712	.7378675
trt_month	.8721658	.0593027	-2.01	0.044	.7633467	.9964976
_cons	.1981373	.0852762	-3.76	0.000	.0852351	.4605897
/lnsig2u	2.775749	.1890237			2.405269	3.146228
sigma_u	4.006325	.3786451			3.328876	4.821641
rho	.8298976	.026684			.7710804	.8760322

Likelihood-ratio test of rho=0: chibar2(01) =   565.24 Prob >= chibar2 = 0.000

The estimated odds ratios and their 95% confidence intervals are also given in table 10.2. We see that the estimated conditional odds (given ζ_j) for a subject in the itraconazole group are multiplied by 0.68 every month and the conditional odds for a subject in the terbinafine group are multiplied by 0.59 ($= 0.6764099 \times 0.8721658$) every month. In terms of percentage change in estimated odds, $100\%(\widehat{OR} - 1)$, the conditional odds decrease 32% $[-32\% = 100\%(0.6764099 - 1)]$ per month in the itraconazole group and 41% $[-41\% = 100\%(0.6764099 \times 0.8721658 - 1)]$ per month in the terbinafine group. (the difference between the kind of effects estimated in random-intercept logistic regression and ordinary logistic regression is discussed in section 10.8).

Table 10.2: Estimates for toenail data

	Marginal effects		Conditional effects	
	Ordinary logistic	GEE† logistic	Random int. logistic	Conditional logistic
Parameter	OR (95% CI)	OR (95% CI)*	OR (95% CI)	OR (95% CI)
Fixed part				
$\exp(\beta_2)$ [treatment]	1.00 (0.74, 1.36)	1.01 (0.61, 1.68)	0.85 (0.27, 2.65)	
$\exp(\beta_3)$ [month]	0.84 (0.81, 0.88)	0.84 (0.79, 0.89)	0.68 (0.62, 0.74)	0.68 (0.62, 0.75)
$\exp(\beta_4)$ [trt_month]	0.93 (0.87, 1.01)	0.93 (0.83, 1.03)	0.87 (0.76, 1.00)	0.91 (0.78, 1.05)
Random part				
ψ			16.08	
ρ			0.83	
Log likelihood	−908.01		−625.39	−188.94•

† Using exchangeable working correlation
★ Based on the sandwich estimator
• Log conditional likelihood

10.7.2   Using xtmelogit

The syntax for xtmelogit is analogous to that for xtmixed except that we also specify
the number of quadrature points, or integration points, using the intpoints() option

```
. xtmelogit outcome treatment month trt_month || patient:, intpoints(30)
Mixed-effects logistic regression              Number of obs      =       1908
Group variable: patient                        Number of groups   =        294

                                               Obs per group: min =          1
                                                              avg =        6.5
                                                              max =          7

Integration points =   30                      Wald chi2(3)       =     150.52
Log likelihood = -625.39709                    Prob > chi2        =     0.0000
```

outcome	Coef.	Std. Err.	z	P>|z|	[95% Conf.	Interval]
treatment	-.1609377	.584208	-0.28	0.783	-1.305964	.984089
month	-.3910603	.0443957	-8.81	0.000	-.4780744	-.3040463
trt_month	-.1368073	.0680236	-2.01	0.044	-.270131	-.0034836
_cons	-1.618961	.4347772	-3.72	0.000	-2.471108	-.7668132

Random-effects Parameters	Estimate	Std. Err.	[95% Conf.	Interval]
patient: Identity				
sd(_cons)	4.008164	.3813917	3.326216	4.829926

```
LR test vs. logistic regression: chibar2(01) =    565.22 Prob>=chibar2 = 0.0000
```

The results are similar but not identical to those from xtlogit because the commands
use slightly different versions of adaptive quadrature (see section 10.11.1). Because the
estimates took some time to obtain, we store them for later use within the same Stata
session:

```
. estimates store xtmelogit
```

(The command estimates save can be used to save the estimates in a file for use in a
future Stata session.)

Estimated odds ratios can be obtained using the or option. xtmelogit can also be
used with one integration point, which is equivalent to using the Laplace approximation.
See section 10.11.2 for the results obtained by using this less accurate but faster method
for the toenail data.

10.7.3   Using gllamm

We now introduce the user-contributed command for multilevel and latent variable
modeling, called gllamm (stands for generalized linear latent and mixed models) by
Rabe-Hesketh, Skrondal, and Pickles (2002, 2005). See also http://www.gllamm.org
where you can download the gllamm manual, the gllamm companion for this book, and
find many other resources.

To check whether `gllamm` is installed on your computer, use the command

```
. which gllamm
```

If the message

```
command gllamm not found as either built-in or ado-file
```

appears, install `gllamm` (assuming that you have a net-aware Stata) by using the `ssc` command:

```
. ssc install gllamm
```

Occasionally, you should update `gllamm` by using `ssc` with the `replace` option:

```
. ssc install gllamm, replace
```

Using `gllamm` for the random-intercept logistic regression model requires that we specify a logit link and binomial distribution with the `link()` and `family()` options (exactly as for the `glm` command). We also use the `nip()` option (for the number of integration points) to request that 30 integration points be used. The cluster identifier is specified in the `i()` option:

```
. gllamm outcome treatment month trt_month, i(patient) link(logit) family(binomial)
> nip(30) adapt
number of level 1 units = 1908
number of level 2 units = 294

Condition Number = 23.0763

gllamm model

log likelihood = -625.38558
```

outcome	Coef.	Std. Err.	z	P>|z|	[95% Conf. Interval]	
treatment	-.1608751	.5802054	-0.28	0.782	-1.298057	.9763066
month	-.3911055	.0443906	-8.81	0.000	-.4781096	-.3041015
trt_month	-.136829	.0680213	-2.01	0.044	-.2701484	-.0035097
_cons	-1.620364	.4322409	-3.75	0.000	-2.46754	-.7731873

```
Variances and covariances of random effects
------------------------------------------------------------------------------

***level 2 (patient)

    var(1): 16.084107 (3.0626224)
------------------------------------------------------------------------------
```

The estimates are again similar to those from `xtlogit` and `xtmelogit`. The estimated random-intercept variance is given next to `var(1)` instead of the random-intercept standard deviation reported by `xtlogit` and `xtmelogit`, unless the `variance` option is used for the latter. We store the `gllamm` estimates for later use:

```
. estimates store gllamm
```

We can use the `eform` option to obtain estimated odds ratios, or we can alternatively use the command

```
gllamm, eform
```

to replay the estimation results after having already fit the model. We can also use the `robust` option to obtain robust standard errors based on the sandwich estimator. At the time of writing this book, `gllamm` does not accept factor variables (`i.`, `c.`, and `#`) but does accept `i.` if the `gllamm` command is preceded by the prefix command `xi:`.

10.8 Subject-specific or conditional vs. population-averaged or marginal relationships

The estimated regression coefficients for the random-intercept logistic regression model are more extreme (more different from 0) than those for the ordinary logistic regression model (see table 10.2). Correspondingly, the estimated odds ratios are more extreme (more different from 1) than those for the ordinary logistic regression model. The reason for this discrepancy is that ordinary logistic regression fits overall *population-averaged* or *marginal* probabilities, whereas random-effects logistic regression fits *subject-specific* or *conditional* probabilities for the individual patients.

This important distinction can be seen in the way the two models are written in (10.5) and (10.6). Whereas the former is for the overall or population-averaged probability, conditioning only on covariates, the latter is for the subject-specific probability, given the subject-specific random intercept ζ_j and the covariates. Odds ratios derived from these models can be referred to as population-averaged (although the averaging is applied to the probabilities) or subject-specific odds ratios, respectively.

For instance, in the random-intercept logistic regression model, we can interpret the estimated subject-specific or conditional odds ratio of 0.68 for `month` (a covariate varying *within* patient) as the odds ratio for each patient in the itraconazole group: the odds for *a given patient* hence decreases by 32% per month. In contrast, the estimated population-averaged odds ratio of 0.84 for `month` means that the odds of having onycholysis *among the patients* in the itraconazole group decreases by 16% per month.

Considering instead the odds for `treatment` (a covariate only varying *between* patients) when `month` equals 1, the estimated subject-specific or conditional odds ratio is estimated as 0.74 (=0.85×0.87) and the odds are hence 26% lower for terbinafine than for itraconazole for each subject. However, because no patients are given both terbinafine and itraconazole, it might be best to interpret the odds ratio in terms of a comparison between two patients j and j' with the same value of the random intercept $\zeta_j = \zeta_{j'}$, one of whom is given terbinafine and the other itraconazole. The estimated population-averaged or marginal odds ratio of about 0.93 (=1.00×0.93) means that the odds are 7% lower for the group of patients given terbinafine compared with the group of patients given itraconazole.

When interpreting subject-specific or conditional odds ratios, keep in mind that these are not purely based on within-subject information and are hence not free from subject-level confounding. In fact, for between-subject covariates like treatment group above, there is no within-subject information in the data. Although the odds ratios are interpreted as effects keeping the subject-specific random intercepts ζ_j constant, these random intercepts are assumed to be independent of the covariates included in the model and hence do not represent effects of unobserved *confounders*, which are by definition correlated with the covariates. Unlike fixed-effects approaches, we are therefore not controlling for unobserved confounders. Both conditional and marginal effect estimates suffer from omitted-variable bias if subject-level or other confounders are not included in the model. See section 3.7.4 for a discussion of this issue in linear random-intercept models. Section 10.14.1 is on conditional logistic regression, the fixed-effects approach in logistic regression that controls for subject-level confounders.

The population-averaged probabilities implied by the random-intercept model can be obtained by averaging the subject-specific probabilities over the random-intercept distribution. Because the random intercepts are continuous, this averaging is accomplished by integration

$$\Pr(y_{ij} = 1 | x_{2j}, x_{3ij})$$

$$= \int \Pr(y_{ij} = 1 | x_{2j}, x_{3ij}, \zeta_j) \phi(\zeta_j; 0, \psi) \, d\zeta_j$$

$$= \int \frac{\exp(\beta_1 + \beta_2 x_{2j} + \beta_3 x_{3ij} + \beta_4 x_{2j} x_{3ij} + \zeta_j)}{1 + \exp(\beta_1 + \beta_2 x_{2j} + \beta_3 x_{3ij} + \beta_4 x_{2j} x_{3ij} + \zeta_j)} \phi(\zeta_j; 0, \psi) \, d\zeta_j$$

$$\neq \frac{\exp(\beta_1 + \beta_2 x_{2j} + \beta_3 x_{3ij} + \beta_4 x_{2j} x_{3ij})}{1 + \exp(\beta_1 + \beta_2 x_{2j} + \beta_3 x_{3ij} + \beta_4 x_{2j} x_{3ij})} \tag{10.8}$$

where $\phi(\zeta_j; 0, \psi)$ is the normal density function with mean zero and variance ψ.

The difference between population-averaged and subject-specific effects is due to the average of a nonlinear function not being the same as the nonlinear function of the average. In the present context, the average of the inverse logit of the linear predictor, $\beta_1 + \beta_2 x_{2j} + \beta_3 x_{3ij} + \beta_4 x_{2j} x_{3ij} + \zeta_j$, is not the same as the inverse logit of the average of the linear predictor, which is $\beta_1 + \beta_2 x_{2j} + \beta_3 x_{3ij} + \beta_4 x_{2j} x_{3ij}$. We can see this by comparing the simple average of the logits of 1 and 2 with the logit of the average of 1 and 2:

```
. display (invlogit(1) + invlogit(2))/2
.80592783
. display invlogit((1+2)/1)
.95257413
```

We can also see this in figure 10.9. Here the individual, thin, dashed curves represent subject-specific logistic curves, each with a subject-specific (randomly drawn) intercept. These are inverse logit functions of the subject-specific linear predictors (here the linear predictors are simply $\beta_1 + \beta_2 x_{ij} + \zeta_j$). The thick, dashed curve is the inverse logit function of the average of the linear predictor (with $\zeta_j = 0$) and this is not the same as the average of the logistic functions shown as a thick, solid curve.

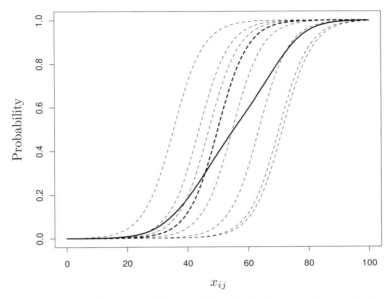

Figure 10.9: Subject-specific probabilities (thin, dashed curves), population-averaged probabilities (thick, solid curve), and population median probabilities (thick, dashed curve) for random-intercept logistic regression

The average curve has a different shape than the subject-specific curves. Specifically, the effect of x_{ij} on the average curve is smaller than the effect of x_{ij} on the subject-specific curves. However, the population median probability is the same as the subject-specific probability evaluated at the median of ζ_j ($\zeta_j = 0$), shown as the thick, dashed curve, because the inverse logit function is a strictly increasing function.

Another way of understanding why the subject-specific effects are more extreme than the population-averaged effects is by writing the random-intercept logistic regression model as a latent-response model:

$$y_{ij}^* = \beta_1 + \beta_2 x_{2j} + \beta_3 x_{3ij} + \beta_4 x_{2j} x_{3ij} + \underbrace{\zeta_j + \epsilon_{ij}}_{\xi_{ij}}$$

The total residual variance is

$$\mathrm{Var}(\xi_{ij}) = \psi + \pi^2/3$$

estimated as $\widehat{\psi} + \pi^2/3 = 16.08 + 3.29 = 19.37$, which is much greater than the residual variance of about 3.29 for an ordinary logistic regression model. As we have already seen in figure 10.4 for probit models, the slope in the model for y_i^* has to increase when the residual standard deviation increases to produce an equivalent curve for the marginal probability that the observed response is 1. Therefore, the regression coefficients of the random-intercept model (representing subject-specific effects) must be larger in absolute value than those of the ordinary logistic regression model (representing population-averaged effects) to obtain a good fit of the model-implied marginal probabilities to the corresponding sample proportions (see exercise 10.10). In section 10.13, we will obtain predicted subject-specific and population-averaged probabilities for the toenail data.

Having described subject-specific and population-averaged probabilities or expectations of y_{ij}, for given covariate values, we now consider the corresponding variances. The subject-specific or conditional variance is

$$\mathrm{Var}(y_{ij}|\mathbf{x}_{ij}, \zeta_j) = \Pr(y_{ij} = 1|\mathbf{x}_{ij}, \zeta_j)\{1 - \Pr(y_{ij} = 1|\mathbf{x}_{ij}, \zeta_j)\}$$

and the population-averaged or marginal variance (obtained by integrating over ζ_j) is

$$\mathrm{Var}(y_{ij}|\mathbf{x}_{ij}) = \Pr(y_{ij} = 1|\mathbf{x}_{ij})\{1 - \Pr(y_{ij} = 1|\mathbf{x}_{ij})\}$$

We see that the random-intercept variance ψ does not affect the relationship between the marginal variance and the marginal mean. This is in contrast to models for counts described in chapter 13, where a random intercept (with $\psi > 0$) produces so-called overdispersion, with a larger marginal variance for a given marginal mean than the model without a random intercept ($\psi = 0$). Contrary to widespread belief, overdispersion is impossible for dichotomous responses (Skrondal and Rabe-Hesketh 2007).

10.9   Measures of dependence and heterogeneity

10.9.1   Conditional or residual intraclass correlation of the latent responses

Returning to the latent-response formulation, the dependence among the dichotomous responses for the same subject (or the between-subject heterogeneity) can be quantified by the *conditional intraclass correlation* or *residual intraclass correlation* ρ of the latent responses y_{ij}^* given the covariates:

$$\rho \equiv \mathrm{Cor}(y_{ij}^*, y_{i'j}^*|\mathbf{x}_{ij}, \mathbf{x}_{i'j}) = \mathrm{Cor}(\xi_{ij}, \xi_{i'j}) = \frac{\psi}{\psi + \pi^2/3}$$

Substituting the estimated variance $\widehat{\psi} = 16.08$, we obtain an estimated conditional intraclass correlation of 0.83, which is large even for longitudinal data. The estimated intraclass correlation is also reported next to `rho` by `xtlogit`.

For probit models, the expression for the residual intraclass correlation of the latent responses is as above with $\pi^2/3$ replaced by 1.

10.9.2   Median odds ratio

Larsen et al. (2000) and Larsen and Merlo (2005) suggest a measure of heterogeneity for random-intercept models with normally distributed random intercepts. They consider repeatedly sampling two subjects with the same covariate values and forming the odds ratio comparing the subject who has the larger random intercept with the other subject. For a given pair of subjects j and j', this odds ratio is given by $\exp(|\zeta_j - \zeta_{j'}|)$ and heterogeneity is expressed as the median of these odds ratios across repeated samples.

The median and other percentiles $a > 1$ can be obtained from the cumulative distribution function

$$\Pr\{\exp(|\zeta_j - \zeta_{j'}|) \le a\} \;=\; \Pr\left\{\frac{|\zeta_j - \zeta_{j'}|}{\sqrt{2\psi}} \le \frac{\ln(a)}{\sqrt{2\psi}}\right\} \;=\; 2\,\Phi\left\{\frac{\ln(a)}{\sqrt{2\psi}}\right\} - 1$$

If the cumulative probability is set to $1/2$, a is the median odds ratio, $\text{OR}_{\text{median}}$:

$$2\,\Phi\left\{\frac{\ln(\text{OR}_{\text{median}})}{\sqrt{2\psi}}\right\} - 1 \;=\; 1/2$$

Solving this equation gives

$$\text{OR}_{\text{median}} \;=\; \exp\{\sqrt{2\psi}\,\Phi^{-1}(3/4)\}$$

Plugging in the parameter estimates, we obtain $\widehat{\text{OR}}_{\text{median}}$:

```
. display exp(sqrt(2*16.084107)*invnormal(3/4))
45.855974
```

When two subjects are chosen at random at a given time point from the same treatment group, the odds ratio comparing the subject who has the larger odds with the subject who has the smaller odds will exceed 45.83 half the time, which is a very large odds ratio. For comparison, the estimated odds ratio comparing two subjects at 20 months who had the same value of the random intercept, but one of whom received itraconazole (`treatment=0`) and the other of whom received terbinafine (`treatment=1`), is about $18 \;\{= 1/\exp(-0.1608751 + 20 \times -0.136829)\}$.

10.9.3   ❖ Measures of association for observed responses at median fixed part of the model

The reason why the degree of dependence is often expressed in terms of the residual intraclass correlation for the *latent* responses y_{ij}^* is that the intraclass correlation for the observed responses y_{ij} varies according to the values of the covariates.

One may nevertheless proceed by obtaining measures of association for specific values of the covariates. In particular, Rodríguez and Elo (2003) suggest obtaining the marginal association between the binary observed responses at the sample median value of the estimated fixed part of the model, $\widehat{\beta}_1 + \widehat{\beta}_2 x_{2j} + \widehat{\beta}_3 x_{3ij} + \widehat{\beta}_4 x_{2j} x_{3ij}$. Marginal association here refers to the fact that the associations are based on marginal probabilities (averaged over the random-intercept distribution with the maximum likelihood estimate $\widehat{\psi}$ plugged in).

Rodríguez and Elo (2003) have written a program called `xtrho` that can be used after `xtlogit`, `xtprobit`, and `xtclog` to produce such marginal association measures and their confidence intervals. The program can be downloaded by issuing the command

```
. findit xtrho
```

clicking on `st0031`, and then clicking on `click here to install`. Having downloaded `xtrho`, we run it after refitting the random-intercept logistic model with `xtlogit`:

```
. quietly xtset patient
. quietly xtlogit outcome treatment month trt_month, re intpoints(30)
. xtrho
```

Measures of intra-class manifest association in random-effects logit
Evaluated at median linear predictor

Measure	Estimate	[95% Conf.Interval]	
Marginal prob.	.250812	.217334	.283389
Joint prob.	.178265	.139538	.217568
Odds ratio	22.9189	16.2512	32.6823
Pearson's r	.61392	.542645	.675887
Yule's Q	.916384	.884066	.940622

We see that for a patient whose fixed part of the linear predictor is equal to the sample median, the marginal probability of having onycholysis (a measure of toenail infection) at an occasion is estimated as 0.25 and the joint probability of having onycholysis at two occasions is estimated as 0.18. From the estimated joint probabilities for the responses 00, 10, 01, and 11 in the 2×2 table for two occasions (with linear predictor equal to the sample median), `xtrho` estimates various measures of association for onycholysis for two occasions, given that the fixed part of the linear predictor equals the sample median.

The estimated odds ratio of 22.92 means that the odds of onycholysis at one of the two occasions is almost 23 times as high for a patient who had onycholysis at the other occasion as for a patient with the same characteristics who did not have onycholysis at the other occasion. The estimated Pearson correlation of 0.61 for the observed responses is lower than the estimated residual correlation for the latent responses of 0.83, as would be expected from statistical theory. Squaring the Pearson correlation, we see that onycholysis at one occasion explains about 36% of the variation in onycholysis at the other occasion.

We can use the `detail` option to obtain the above measures of associations evaluated at sample percentiles other than the median. We can also use Rodríguez and Elo's (2003) `xtrhoi` command to obtain measures of associations for other values of the fixed part of the linear predictor and/or other values of the variance of the random-intercept distribution.

Note that `xtrho` and `xtrhoi` assume that the fixed part of the linear predictor is the same across occasions. However, in the toenail example, `month` must change between any two occasions within a patient, and the linear predictor is a function of `month`. Considering two occasions with `month` equal to 3 and 6, the odds ratio is estimated as 25.6 for patients in the control group and 29.4 for patients in the treatment group. A do-file that produces the 2×2 tables by using `gllamm` and `gllapred` with the `ll` option can be copied into the working directory with the command

```
copy http://www.stata-press.com/data/mlmus3/ch10table.do ch10table.do
```

10.10   Inference for random-intercept logistic models

10.10.1   Tests and confidence intervals for odds ratios

As discussed earlier, we can interpret the regression coefficient β as the difference in log odds associated with a unit change in the corresponding covariate, and we can interpret the exponentiated regression coefficient as an odds ratio, $\text{OR} = \exp(\beta)$. The relevant null hypothesis for odds ratios usually is H_0: $\text{OR} = 1$, and this corresponds directly to the null hypothesis that the corresponding regression coefficient is zero, H_0: $\beta = 0$.

Wald tests and z tests can be used for regression coefficients just as described in section 3.6.1 for linear models. Ninety-five percent Wald confidence intervals for individual regression coefficients are obtained using

$$\widehat{\beta} \pm z_{0.975} \, \widehat{\text{SE}}(\widehat{\beta})$$

where $z_{0.975} = 1.96$ is the 97.5th percentile of the standard normal distribution. The corresponding confidence interval for the odds ratio is obtained by exponentiating both limits of the confidence interval:

$$\exp\{\widehat{\beta} - z_{0.975} \, \widehat{\text{SE}}(\widehat{\beta})\} \quad \text{to} \quad \exp\{\widehat{\beta} + z_{0.975} \, \widehat{\text{SE}}(\widehat{\beta})\}$$

Wald tests for linear combinations of regression coefficients can be used to test the corresponding multiplicative relationships among odds for different covariate values. For instance, for the toenail data, we may want to obtain the odds ratio comparing the treatment groups after 20 months. The corresponding difference in log odds after 20 months is a linear combination of regression coefficients, namely, $\beta_2 + \beta_4 \times 20$ (see section 1.8 if this is not clear). We can test the null hypothesis that the difference in log odds is 0 and hence that the odds ratio is 1 by using the `lincom` command:

```
. lincom treatment + trt_month*20
 ( 1)  [outcome]treatment + 20*[outcome]trt_month = 0
```

outcome	Coef.	Std. Err.	z	P>|z|	[95% Conf. Interval]
(1)	-2.896123	1.309682	-2.21	0.027	-5.463053 -.3291935

If we require the corresponding odds ratio with a 95% confidence interval, we can use the `lincom` command with the `or` option:

```
. lincom treatment + trt_month*20, or
 ( 1)   [outcome]treatment + 20*[outcome]trt_month = 0
```

outcome	Odds Ratio	Std. Err.	z	P>|z|	[95% Conf. Interval]	
(1)	.0552369	.0723428	-2.21	0.027	.0042406	.7195038

After 20 months of treatment, the odds ratio comparing terbinafine (`treatment`=1) with itraconazole is estimated as 0.055. Such small numbers are difficult to interpret, so we can switch the groups around by taking the reciprocal of the odds ratio, 18 (= 1/0.055), which represents the odds ratio comparing itraconazole with terbinafine. Alternatively, we can always switch the comparison around by simply changing the sign of the corresponding difference in log odds in the `lincom` command:

```
lincom -(treatment + trt_month*20), or
```

If we had used factor-variable notation in the estimation command, using the syntax `i.treatment##c.month`, then the `lincom` command above would have to be replaced with

```
lincom -(1.treatment + 1.treatment#c.month*20), or
```

Multivariate Wald tests can be performed by using `testparm`. Wald tests and confidence intervals can be based on robust standard errors from the sandwich estimator. At the time of printing, robust standard errors can only be obtained using `gllamm` with the `robust` option.

Null hypotheses about individual regression coefficients or several regression coefficients can also be tested using likelihood-ratio tests. Although likelihood-ratio and Wald tests are asymptotically equivalent, the test statistics are not identical in finite samples. (See display 2.1 for the relationships between likelihood-ratio, Wald, and score tests.) If the statistics are very different, there may be a sparseness problem, for instance with mostly "1" responses or mostly "0" responses in one of the groups.

10.10.2   Tests of variance components

Both `xtlogit` and `xtmelogit` provide likelihood-ratio tests for the null hypothesis that the residual between-cluster variance ψ is zero in the last line of the output. The p-values are based on the correct asymptotic sampling distribution (not the naïve χ_1^2), as described for linear models in section 2.6.2. For the toenail data, the likelihood-ratio statistic is 565.2 giving $p < 0.001$, which suggests that a multilevel model is required.

10.11   Maximum likelihood estimation

10.11.1 ❖ Adaptive quadrature

The marginal likelihood is the joint probability of all observed responses given the observed covariates. For linear mixed models, this marginal likelihood can be evaluated and maximized relatively easily (see section 2.10). However, in generalized linear mixed models, the marginal likelihood does not have a closed form and must be evaluated by approximate methods.

To see this, we will now construct this marginal likelihood step by step for a random-intercept logistic regression model with one covariate x_j. The responses are conditionally independent given the random intercept ζ_j and the covariate x_j. Therefore, the joint probability of all the responses y_{ij} ($i = 1, \ldots, n_j$) for cluster j given the random intercept and covariate is simply the product of the conditional probabilities of the individual responses:

$$\Pr(y_{1j}, \ldots, y_{n_jj} | x_j, \zeta_j) = \prod_{i=1}^{n_j} \Pr(y_{ij} | x_j, \zeta_j) = \prod_{i=1}^{n_j} \frac{\exp(\beta_1 + \beta_2 x_j + \zeta_j)^{y_{ij}}}{1 + \exp(\beta_1 + \beta_2 x_j + \zeta_j)}$$

In the last term

$$\frac{\exp(\beta_1 + \beta_2 x_j + \zeta_j)^{y_{ij}}}{1 + \exp(\beta_1 + \beta_2 x_j + \zeta_j)} = \begin{cases} \frac{\exp(\beta_1 + \beta_2 x_j + \zeta_j)}{1 + \exp(\beta_1 + \beta_2 x_j + \zeta_j)} & \text{if } y_{ij} = 1 \\ \frac{1}{1 + \exp(\beta_1 + \beta_2 x_j + \zeta_j)} & \text{if } y_{ij} = 0 \end{cases}$$

as specified by the logistic regression model.

To obtain the marginal joint probability of the responses, not conditioning on the random intercept ζ_j (but still on the covariate x_j), we integrate out the random intercept

$$\Pr(y_{1j}, \ldots, y_{n_jj} | x_j) = \int \Pr(y_{1j}, \ldots, y_{n_jj} | x_j, \zeta_j) \, \phi(\zeta_j; 0, \psi) \, d\zeta_j \qquad (10.9)$$

where $\phi(\zeta_j, 0, \psi)$ is the normal density of ζ_j with mean 0 and variance ψ. Unfortunately, this integral does not have a closed-form expression.

The marginal likelihood is just the joint probability of all responses for all clusters. Because the clusters are mutually independent, this is given by the product of the marginal joint probabilities of the responses for the individual clusters

$$L(\beta_1, \beta_2, \psi) = \prod_{j=1}^{N} \Pr(y_{1j}, \ldots, y_{n_jj} | x_j)$$

This marginal likelihood is viewed as a function of the parameters β_1, β_2, and ψ (with the observed responses treated as given). The parameters are estimated by finding the values of β_1, β_2, and ψ that yield the largest likelihood. The search for the maximum is iterative, beginning with some initial guesses or starting values for the parameters and updating these step by step until the maximum is reached, typically using a Newton–Raphson or expectation-maximization (EM) algorithm.

The integral over ζ_j in (10.9) can be approximated by a sum of R terms with e_r substituted for ζ_j and the normal density replaced by a weight w_r for the rth term, $r = 1, \ldots, R$,

$$\Pr(y_{1j}, \ldots, y_{n_j j} | x_j) \approx \sum_{r=1}^{R} \Pr(y_{1j}, \ldots, y_{n_j j} | x_j, \zeta_j = e_r) \, w_r$$

where e_r and w_r are called Gauss–Hermite quadrature locations and weights, respectively. This approximation can be viewed as replacing the continuous density of ζ_j with a discrete distribution with R possible values of ζ_j having probabilities $\Pr(\zeta_j = e_r)$. The Gauss–Hermite approximation is illustrated for $R = 5$ in figure 10.10. Obviously, the approximation improves when the number of points R increases.

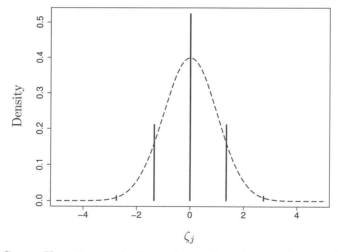

Figure 10.10: Gauss–Hermite quadrature: Approximating continuous density (dashed curve) by discrete distribution (bars)

The ordinary quadrature approximation described above can perform poorly if the function being integrated, called the *integrand*, has a sharp peak, as discussed in Rabe-Hesketh, Skrondal, and Pickles (2002, 2005). Sharp peaks can occur when the clusters are very large so that many functions (the individual response probabilities as functions of ζ_j) are multiplied to yield $\Pr(y_{1j}, \ldots, y_{n_j j} | x_j, \zeta_j)$. Similarly, if the responses are counts or continuous responses, even a few terms can result in a highly peaked function. Another potential problem is a high intraclass correlation. Here the functions being multiplied coincide with each other more closely because of the greater similarity of responses within clusters, yielding a sharper peak. In fact, the toenail data we have been analyzing, which has an estimated conditional intraclass correlation for the latent responses of 0.83, poses real problems for estimation using ordinary quadrature, as pointed out by Lesaffre and Spiessens (2001).

The top panel in figure 10.11 shows the same five-point quadrature approximation and density of ζ_j as in figure 10.10. The solid curve is proportional to the integrand for a hypothetical cluster. Here the quadrature approximation works poorly because the peak falls between adjacent quadrature points.

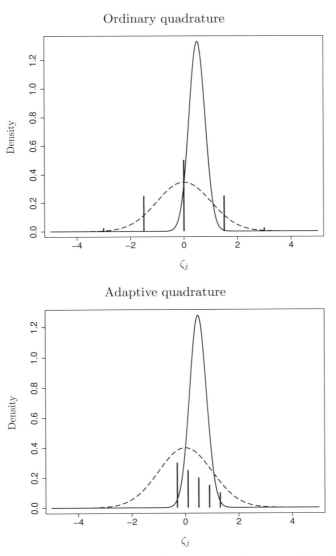

Figure 10.11: Density of ζ_j (dashed curve), normalized integrand (solid curve), and quadrature weights (bars) for ordinary quadrature and adaptive quadrature (*Source:* Rabe-Hesketh, Skrondal, and Pickles 2002)

The bottom panel of figure 10.11 shows an improved approximation, known as *adaptive quadrature*, where the locations are rescaled and translated,

$$e_{rj} = a_j + b_j e_r \tag{10.10}$$

to fall under the peak of the integrand, where a_j and b_j are cluster-specific constants. This transformation of the locations is accompanied by a transformation of the weights w_r that also depends on a_j and b_j. The method is called *adaptive* because the quadrature locations and weights are adapted to the data for the individual clusters.

To maximize the likelihood, we start with a set of initial or starting values of the parameters and then keep updating the parameters until the likelihood is maximized. The quantities a_j and b_j needed to evaluate the likelihood are functions of the parameters (as well as the data) and must therefore be updated or "readapted" when the parameters are updated.

There are two different implementations of adaptive quadrature in Stata that differ in the values used for a_j and b_j in (10.10). The method implemented in `gllamm`, which is the default method in `xtlogit` (as of Stata 10), uses the posterior mean of ζ_j for a_j and the posterior standard deviation for b_j. However, obtaining the posterior mean and standard deviation requires numerical integration so adaptive quadrature sometimes does not work when there are too few quadrature points (for example, fewer than five). Details of the algorithm are given in Rabe-Hesketh, Skrondal, and Pickles (2002, 2005) and Skrondal and Rabe-Hesketh (2004).

The method implemented in `xtmelogit`, and available in `xtlogit` with the option `intmethod(aghermite)`, uses the posterior mode of ζ_j for a_j and for b_j uses the standard deviation of the normal density that approximates the log posterior of ζ_j at the mode. An advantage of this approach is that it does not rely on numerical integration and can therefore be implemented even with one quadrature point. With one quadrature point, this version of adaptive quadrature becomes a Laplace approximation.

10.11.2 Some speed and accuracy considerations

As discussed in section 10.11.1, the likelihood involves integrals that are evaluated by numerical integration. The likelihood itself, as well as the maximum likelihood estimates, are therefore only approximate. The accuracy increases as the number of quadrature points increases, at the cost of increased computation time. We can assess whether the approximation is adequate in a given situation by repeating the analysis with a larger number of quadrature points. If we get essentially the same result, the lower number of quadrature points is likely to be adequate. Such checking should always be done before estimates are taken at face value. See section 16.4.1 for an example in `gllamm` and section 16.4.2 for an example in `xtmelogit`. For a given number of quadrature points, adaptive quadrature is more accurate than ordinary quadrature. Stata's commands therefore use adaptive quadrature by default, and we recommend using the `adapt` option in `gllamm`.

Because of numerical integration, estimation can be slow, especially if there are many random effects. The time it takes to fit a model is approximately proportional to the product of the number of quadrature points for all random effects (although this seems to be more true for `gllamm` than for `xtmelogit`). For example, if there are two random effects at level 2 (a random intercept and slope) and eight quadrature points are used for each random effect, the time will be approximately proportional to 64. Therefore, using four quadrature points for each random effect will take only about one-fourth (16/64) as long as using eight. The time is also approximately proportional to the number of observations and, for programs using numerical differentiation (`gllamm` and `xtmelogit`), to the square of the number of parameters. (For `xtlogit`, computation time increases less dramatically when the number of parameters increases because it uses analytical derivatives.)

For large problems, it may be advisable to estimate how long estimation will take before starting work on a project. In this case, we recommend fitting a similar model with fewer random effects, fewer parameters (for example, fewer covariates), or fewer observations, and then using the above approximate proportionality factors to estimate the time that will be required for the larger problem.

For random-intercept models, by far the fastest command is `xtlogit` (because it uses analytical derivatives). However, `xtlogit` cannot fit random-coefficient models or higher-level models introduced in chapter 16. For such models, `xtmelogit` or `gllamm` must be used. The quickest way of obtaining results here is using `xtmelogit` with one integration point, corresponding to the Laplace approximation. Although this method sometimes works well, it can produce severely biased estimates, especially if the clusters are small and the (true) random-intercept variance is large, as for the toenail data. For these data, we obtain the following:

```
. xtmelogit outcome treatment month trt_month || patient:, intpoints(1)
```

Mixed-effects logistic regression   Number of obs   =   1908
Group variable: patient   Number of groups   =   294

                                                 Obs per group: min =   1
                                                                avg =   6.5
                                                                max =   7

Integration points =   1   Wald chi2(3)   =   131.96
Log likelihood = -627.80894   Prob > chi2   =   0.0000

outcome	Coef.	Std. Err.	z	P>|z|	[95% Conf. Interval]	
treatment	-.3070156	.6899551	-0.44	0.656	-1.659303	1.045272
month	-.4000908	.0470586	-8.50	0.000	-.492324	-.3078576
trt_month	-.1372594	.0695863	-1.97	0.049	-.2736459	-.0008728
_cons	-2.5233	.7882542	-3.20	0.001	-4.06825	-.9783501

Random-effects Parameters	Estimate	Std. Err.	[95% Conf. Interval]	
patient: Identity				
sd(_cons)	4.570866	.7198949	3.356892	6.223858

LR test vs. logistic regression: chibar2(01) =   560.40 Prob>=chibar2 = 0.0000
Note: log-likelihood calculations are based on the Laplacian approximation.

We see that the estimated intercept and coefficient of treatment are very different from the estimates in section 10.7.1 using adaptive quadrature with 30 quadrature points. As mentioned in the previous section, gllamm cannot be used with only one quadrature point, and adaptive quadrature in gllamm typically requires at least five quadrature points.

Advice for speeding up estimation in gllamm

To speed up estimation in gllamm, we recommend using good starting values whenever they are available. For instance, when increasing the number of quadrature points or adding or dropping covariates, use the previous estimates as starting values. This can be done by using the from() option to specify a row matrix of starting values. This option should be combined with skip if the new model contains fewer parameters than supplied. You can also use the copy option if your parameters are supplied in the correct order yet are not necessarily labeled correctly. Use of these options is demonstrated in sections 11.7.2 and 16.4.1 and throughout this volume (see subject index).

The from() option can also be used with the xtmelogit command, together with the refineopts(iterate(0)) option, to prevent xtmelogit from finding its own starting values (see section 16.4.2). However, the time saving is not as pronounced as in gllamm.

In gllamm, there are two other methods for speeding up estimation: collapsing the data and using spherical quadrature. These methods, which cannot be used for xtlogit or xtmelogit, are described in the following two paragraphs.

For some datasets and models, you can represent the data using fewer rows than there are observations, thus speeding up estimation. For example, if the response is dichotomous and we are using one dichotomous covariate in a two-level dataset, we can use one row of data for each combination of covariate and response (00, 01, 10, 11) for each cluster, leading to at most four rows per cluster. We can then specify a variable containing level-1 frequency weights equal to the number of observations, or level-1 units, in each cluster having each combination of the covariate and response values. Level-2 weights can be used if several clusters have the same level-2 covariates and the same number of level-1 units with the same response and level-1 covariate pattern. The `weight()` option in `gllamm` is designed for specifying frequency weights at the different levels. See exercise 10.7 for an example with level-1 weights, and see exercises 10.3 and 2.3 for examples with level-2 weights. In exercise 16.11, collapsing the data reduces computation time by about 99%. If the dataset is large, starting values could be obtained by fitting the model to a random sample of the data.

For models involving several random effects at the same level, such as two-level random-coefficient models with a random intercept and slope, the multivariate integral can be evaluated more efficiently using *spherical quadrature* instead of the default Cartesian-product quadrature. For the random intercept and slope example, Cartesian-product quadrature consists of evaluating the function being integrated on the rectangular grid of quadrature points consisting of all combinations of $\zeta_{1j} = e_1, \ldots, e_R$ and $\zeta_{2j} = e_1, \ldots, e_R$, giving R^2 terms. In contrast, spherical quadrature consists of evaluating ζ_{1j} and ζ_{2j} at values falling on concentric circles (spheres in more dimensions). The important point is that the same accuracy can now be achieved with fewer than R^2 points. For example, when $R = 8$, Cartesian-product quadrature requires 64 evaluations, while spherical quadrature requires only 44 evaluations, taking nearly 30% less time to achieve the same accuracy. Here accuracy is expressed in terms of the degree of the approximation given by $d = 2R - 1$. For $R = 8$, $d = 15$. To use spherical quadrature, specify the `ip(m)` option in `gllamm` and give the degree d of the approximation by using the `nip(#)` option. Unfortunately, spherical integration is available only for certain combinations of numbers of dimensions (or numbers of random effects) and degrees of accuracy, d: For two dimensions, d can be 5, 7, 9, 11, or 15, and for more than two dimensions, d can be 5 or 7. See Rabe-Hesketh, Skrondal, and Pickles (2005) for more information.

10.12 Assigning values to random effects

Having estimated the model parameters (the β's and ψ), we may want to assign values to the random intercepts ζ_j for individual clusters j. The ζ_j are not model parameters, but as for linear models, we can treat the estimated parameters as known and then either estimate or predict ζ_j.

Such predictions are useful for making inferences for the clusters in the data, important examples being assessment of institutional performance (see section 4.8.5) or of abilities in item response theory (see exercise 10.4). The estimated or predicted values of ζ_j should generally not be used for model diagnostics in random-intercept logistic regression because their distribution if the model is true is not known. In general, the values should also not be used to obtain cluster-specific predicted probabilities (see section 10.13.2).

10.12.1 Maximum "likelihood" estimation

As discussed for linear models in section 2.11.1, we can estimate the intercepts ζ_j by treating them as the only unknown parameters, after estimates have been plugged in for the model parameters:

$$\text{logit}\{\Pr(y_{ij} = 1|\mathbf{x}_{ij}, \zeta_j)\} = \underbrace{\text{offset}_{ij}}_{\widehat{\beta}_1 + \widehat{\beta}_2 x_{2ij} + \cdots} + \zeta_j$$

This is a logistic regression model for cluster j with offset (a term with regression coefficient set to 1) given by the estimated fixed part of the linear predictor and with a cluster-specific intercept ζ_j.

We then maximize the corresponding likelihood for cluster j

$$\text{Likelihood}(y_{1j}, y_{2j}, \ldots, y_{n_j,j}|\mathbf{X}_j, \zeta_j)$$

with respect to ζ_j, where $\mathbf{X}_j$ is a matrix containing all covariates for cluster j. As explained in section 2.11.1, we put "likelihood" in quotes in the section heading because it differs from the marginal likelihood that is used to estimate the model parameters. Maximization can be accomplished by fitting logistic regression models to the individual clusters. First, obtain the offset from the `xtmelogit` estimates:

```
. estimates restore xtmelogit
(results xtmelogit are active now)
. predict offset, xb
```

Then use the `statsby` command to fit individual logistic regression models for each patient, specifying an offset:

```
. statsby mlest=_b[_cons], by(patient) saving(ml, replace): logit outcome,
> offset(offset)
(running logit on estimation sample)
        command:  logit outcome, offset(offset)
          mlest:  _b[_cons]
             by:  patient

Statsby groups
  ———+—— 1 ——+—— 2 ——+—— 3 ——+—— 4 ——+—— 5
......xx.......xx..xxx...x.x...xxxxx.xx...xxx.xxxx     50
xx.xxxxxxxxx.xxxx..xxxxxxxxx.x..xx..x.xxx.xxx.x...    100
xx.xxxxxxxxxx.xxx.x.x...x.xx.xxxxx.xx....xxx.x.xx    150
.x..x.xxxx..xxxxx.xx..xxxx..xxx.x.xxxxx.x.x.xxx...   200
.xxxxx.xx.xx..x.xxx...xx.x..xxxxx.x..x.x..x..xxxxx   250
x.xx.x..xxxxxx..x..x..xxx.x..xxxxxxxx.x.x...
```

Here we have saved the estimates under the variable name `mlest` in a file called `ml.dta` in the local directory. The `x`'s in the output indicate that the `logit` command did not converge for many clusters. For these clusters, the variable `mlest` is missing. This happens for clusters where all responses are 0 or all responses are 1 because the maximum likelihood estimate then is $-\infty$ and $+\infty$, respectively.

We now merge the estimates with the data for later use:

```
. sort patient
. merge m:1 patient using ml
. drop _merge
```

10.12.2 Empirical Bayes prediction

The ideas behind empirical Bayes prediction discussed in section 2.11.2 for linear variance-components models also apply to other generalized linear mixed models. Instead of basing inference completely on the likelihood of the responses for a cluster given the random intercept, we combine this information with the prior of the random intercept, which is just the density of the random intercept (a normal density with mean 0 and estimated variance $\widehat{\psi}$), to obtain the posterior density:

$$\text{Posterior}(\zeta_j | y_{1j}, \ldots, y_{n_j j}, \mathbf{X}_j) \; \propto \; \text{Prior}(\zeta_j) \times \text{Likelihood}(y_{1j}, \ldots, y_{n_j j} | \mathbf{X}_j, \zeta_j)$$

The product on the right is proportional to, but not equal to, the posterior density. Obtaining the posterior density requires dividing this product by a normalizing constant that can only be obtained by numerical integration. Note that the model parameters are treated as known, and estimates are plugged into the expression for the posterior, giving what is sometimes called an estimated posterior distribution.

The estimated posterior density is no longer normal as for linear models, and hence its mode does not equal its mean. There are therefore two different types of predictions we could consider: the mean of the posterior and its mode. The first is undoubtedly the most common and is referred to as empirical Bayes prediction [sometimes called expected a posterior (EAP) prediction], whereas the second is referred to as empirical Bayes modal prediction [sometimes called modal a posterior (MAP) prediction].

The empirical Bayes prediction of the random intercept for a cluster j is the mean of the estimated posterior distribution of the random intercept. This can be obtained as

$$\widetilde{\zeta}_j \;=\; \int \zeta_j \, \mathrm{Posterior}(\zeta_j | y_{1j}, \ldots, y_{n_j j}, \mathbf{X}_j) \, d\zeta_j$$

using numerical integration.

At the time of writing this book, the only Stata command that provides empirical Bayes predictions for generalized linear mixed models is the postestimation command `gllapred` for `gllamm` with the u option:

```
. estimates restore gllamm
. gllapred eb, u
```

The variable `ebm1` contains the empirical Bayes predictions. In the next section, we will produce a graph of these predictions, together with maximum likelihood estimates and empirical Bayes modal predictions.

The posterior standard deviations produced by `gllapred` in the variable `ebs1` represent the conditional standard deviations of the prediction errors, given the observed responses and treating the parameter estimates as known. The square of `ebs1` is also the conditional mean squared error of the prediction, conditional on the observed responses. As in section 2.11.3, we refer to this standard error as the *comparative standard error* because it can be used to make inferences regarding the random effects of individual clusters and to compare clusters.

We mentioned in section 2.11.3 that, for linear models, the posterior variance was the same as the unconditional mean squared error of prediction (MSEP). However, this is not true for generalized linear mixed models not having an identity link, such as the random-intercept logistic model discussed here.

There is also no longer an easy way to obtain the sampling standard deviation of the empirical Bayes predictions or diagnostic standard error (see section 2.11.3). The `ustd` option for standardized level-2 residuals therefore divides the empirical Bayes predictions by an approximation for this standard deviation, $\sqrt{\widehat{\psi} - \mathtt{ebs1}^2}$ (see Skrondal and Rabe-Hesketh [2004, 231–232] or Skrondal and Rabe-Hesketh [2009] for details).

10.12.3 Empirical Bayes modal prediction

Instead of basing prediction of random effects on the mean of the posterior distribution, we can use the mode. Such empirical Bayes modal predictions are easy to obtain using the `predict` command with the `reffects` option after estimation using `xtmelogit`:

```
. estimates restore xtmelogit
. predict ebmodal, reffects
```

To see how the various methods compare, we now produce a graph of the empirical Bayes modal predictions (circles) and maximum likelihood estimates (triangles) versus the empirical Bayes predictions, connecting empirical Bayes modal predictions and maximum likelihood estimates with vertical lines.

```
. twoway (rspike mlest ebmodal ebm1 if visit==1)
> (scatter mlest  ebm1 if visit==1, msize(small) msym(th) mcol(black))
> (scatter ebmodal ebm1 if visit==1, msize(small) msym(oh) mcol(black))
> (function y=x, range(ebm1) lpatt(solid)),
> xtitle(Empirical Bayes prediction)
> legend(order(2 "Maximum likelihood" 3 "Empirical Bayes modal"))
```

The graph is given in figure 10.12.

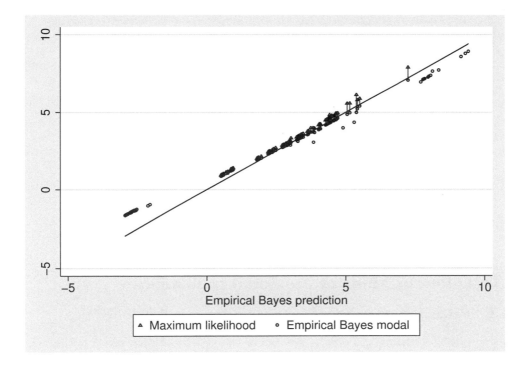

Figure 10.12: Empirical Bayes modal predictions (circles) and maximum likelihood estimates (triangles) versus empirical Bayes predictions

We see that the maximum likelihood predictions are missing when the empirical Bayes predictions are extreme (where the responses are all 0 or all 1) and that the empirical Bayes modal predictions tend to be quite close to the empirical Bayes predictions (close to the line).

We can also obtain standard errors for the random-effect predictions after estimation with **xtmelogit** by using the **predict** command with the **reses** (for "random-effects standard errors") option.

```
. predict se2, reses
```

These standard errors are the standard deviations of normal densities that approximate
the posterior at the mode. They can be viewed as approximations of the posterior
standard deviations provided by gllapred. Below we list the predictions and standard
errors produced by gllapred (ebm1 and ebs1) with those produced by predict after
estimation with xtmelogit (ebmodal and se2), together with the number of 0 responses,
num0, and the number of 1 responses, num1, for the first 16 patients:

```
. egen num0 = total(outcome==0), by(patient)
. egen num1 = total(outcome==1), by(patient)
. list patient num0 num1 ebm1 ebmodal ebs1 se2 if visit==1&patient<=12, noobs
```

patient	num0	num1	ebm1	ebmodal	ebs1	se2
1	4	3	3.7419957	3.736461	1.0534592	1.025574
2	4	2	1.8344596	1.934467	1.0192062	.9423445
3	6	1	.58899428	.9477552	1.3098199	1.131451
4	6	1	.60171957	.9552238	1.3148935	1.136338
6	4	3	3.2835777	3.253659	1.0118905	.9709948
7	4	3	3.4032244	3.367345	1.0307951	.9956154
9	7	0	-2.6807107	-1.399524	2.7073681	2.608825
10	7	0	-2.888319	-1.604741	2.6450981	2.503938
11	3	4	4.4649443	4.361801	1.0885138	1.072554
12	4	3	2.7279723	2.728881	.94173461	.8989795

We see that the predictions and standard errors agree reasonably well (except the extreme negative predictions). The standard errors are large when all responses are 0.

10.13   Different kinds of predicted probabilities

10.13.1   Predicted population-averaged or marginal probabilities

At the time of writing this book, population-averaged or marginal probabilities $\overline{\pi}(\mathbf{x}_{ij})$
can be predicted for random-intercept logistic regression models only by using gllapred
after estimation using gllamm. This is done by evaluating the integral in (10.8) nu-
merically for the estimated parameters and values of covariates in the data, that is,
evaluating

$$\overline{\pi}(\mathbf{x}_{ij}) \;\equiv\; \int \widehat{\Pr}(y_{ij} = 1 | x_{2j}, x_{3ij}, \zeta_j) \phi(\zeta_j; 0, \widehat{\psi}) \, d\zeta_j$$

To obtain these predicted marginal probabilities using gllapred, specify the options
mu (for the mean response, here a probability) and marginal (for integrating over the
random-intercept distribution):

```
. estimates restore gllamm
. gllapred margprob, mu marginal
(mu will be stored in margprob)
```

We now compare predictions of population-averaged or marginal probabilities from the ordinary logit model (previously obtained under the variable name `prob`) and the random-intercept logit model, giving figure 10.13.

```
. twoway (line prob month, sort) (line margprob month, sort lpatt(dash)),
> by(treatment) legend(order(1 "Ordinary logit" 2 "Random-intercept logit"))
> xtitle(Time in months) ytitle(Fitted marginal probabilities of onycholysis)
```

The predictions are nearly identical. This is not surprising because marginal effects derived from generalized linear mixed models are close to true marginal effects even if the random-intercept distribution is misspecified (Heagerty and Kurland 2001).

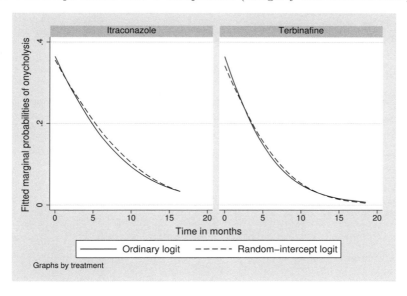

Figure 10.13: Fitted marginal probabilities using ordinary and random-intercept logistic regression

10.13.2 Predicted subject-specific probabilities

Predictions for hypothetical subjects: Conditional probabilities

Subject-specific or conditional predictions of $\widehat{\Pr}(y_{ij} = 1|x_{2j}, x_{3ij}, \zeta_j)$ for different values of ζ_j can be produced using `gllapred` with the `mu` and `us(varname)` options, where *varname*1 is the name of the variable containing the value of the first (here the only) random effect. We now produce predicted probabilities for ζ_j equal to 0, -4, 4, -2, and 2:

```
. generate zeta1 = 0
. gllapred condprob0, mu us(zeta)
(mu will be stored in condprob0)
. generate lower1 = -4
. gllapred condprobm4, mu us(lower)
(mu will be stored in condprobm4)
. generate upper1 = 4
. gllapred condprob4, mu us(upper)
(mu will be stored in condprob4)
. replace lower1 = -2
(1908 real changes made)
. gllapred condprobm2, mu us(lower)
(mu will be stored in condprobm2)
. replace upper1 = 2
(1908 real changes made)
. gllapred condprob2, mu us(upper)
(mu will be stored in condprob2)
```

Plotting all of these conditional probabilities together with the observed proportions and marginal probabilities produces figure 10.14.

```
. twoway (line prop mn_month, sort)
> (line margprob month, sort lpatt(dash))
> (line condprob0 month, sort lpatt(shortdash_dot))
> (line condprob4 month, sort lpatt(shortdash))
> (line condprobm4 month, sort lpatt(shortdash))
> (line condprob2 month, sort lpatt(shortdash))
> (line condprobm2 month, sort lpatt(shortdash)),
> by(treatment)
> legend(order(1 "Observed proportion" 2 "Marginal probability"
>               3 "Median probability" 4 "Conditional probabilities"))
> xtitle(Time in months) ytitle(Probabilities of onycholysis)
```

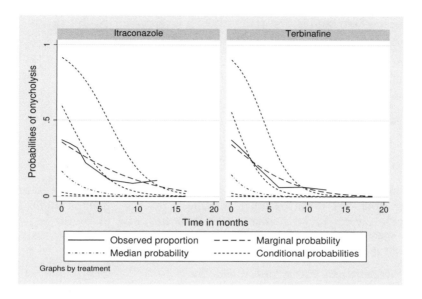

Figure 10.14: Conditional and marginal predicted probabilities for random-intercept logistic regression model

Clearly, the conditional curves have steeper downward slopes than does the marginal curve. The conditional curve represented by a dash-dot line is for $\zeta_j = 0$ and hence represents the population *median* curve.

Predictions for the subjects in the sample: Posterior mean probabilities

We may also want to predict the probability that $y_{ij} = 1$ for a given subject j. The predicted conditional probability, given the unknown random intercept ζ_j, is

$$\widehat{\Pr}(y_{ij} = 1 | \mathbf{x}_{ij}, \zeta_j) = \frac{\exp(\widehat{\beta}_1 + \widehat{\beta}_2 x_{2j} + \widehat{\beta}_3 x_{3ij} + \widehat{\beta}_4 x_{2j} x_{3ij} + \zeta_j)}{1 + \exp(\widehat{\beta}_1 + \widehat{\beta}_2 x_{2j} + \widehat{\beta}_3 x_{3ij} + \widehat{\beta}_4 x_{2j} x_{3ij} + \zeta_j)}$$

Because our knowledge about ζ_j for subject j is represented by the posterior distribution, a good prediction $\widetilde{\pi}_j(\mathbf{x}_{ij})$ of the unconditional probability is obtained by integrating over the posterior distribution:

$$\widetilde{\pi}_j(\mathbf{x}_{ij}) \equiv \int \widehat{\Pr}(y_{ij} = 1 | \mathbf{x}_{ij}, \zeta_j) \times \text{Posterior}(\zeta_j | y_{1j}, \ldots, y_{n_j j}, \mathbf{X}_j) \, d\zeta_j \quad (10.11)$$

$$\neq \widehat{\Pr}(y_{ij} = 1 | \mathbf{x}_{ij}, \widetilde{\zeta}_j)$$

This minimizes the mean squared error of prediction for known parameters. We cannot simply plug in the posterior mean of the random intercept $\widetilde{\zeta}_j$ for ζ_j in generalized linear mixed models. The reason is that the mean of a given nonlinear function of ζ_j does not in general equal the same function evaluated at the mean of ζ_j.

The posterior means of the predicted probabilities as defined in (10.12) can be obtained using `gllapred` with the `mu` option (and not the `marginal` option) after estimation using `gllamm`:

```
. gllapred cmu, mu
(mu will be stored in cmu)
Non-adaptive log-likelihood: -625.52573
  -625.3853   -625.3856   -625.3856
log-likelihood:-625.38558
```

As of September 2008, `gllapred` can produce predicted posterior mean probabilities also for occasions where the response variable is missing. This is useful for making forecasts for a patient or for making predictions for visits where the patient did not attend the assessment. As we saw in section 10.4, such missing data occur frequently in the toenail data.

Listing `patient` and `visit` for patients 2 and 15,

```
. sort patient visit
. list patient visit if patient==2|patient==15, sepby(patient) noobs
```

patient	visit
2	1
2	2
2	3
2	4
2	5
2	6
15	1
15	2
15	3
15	4
15	5
15	7

we see that these patients each have one missing visit: visit 7 is missing for patient 2 and visit 6 is missing for patient 15. To make predictions for these visits, we must first create rows of data (or records) for these visits. A very convenient command to accomplish this is `fillin`:

```
. fillin patient visit
. list patient visit _fillin if patient==2|patient==15, sepby(patient) noobs
```

patient	visit	_fillin
2	1	0
2	2	0
2	3	0
2	4	0
2	5	0
2	6	0
2	7	1
15	1	0
15	2	0
15	3	0
15	4	0
15	5	0
15	6	1
15	7	0

fillin finds all values of **patient** that occur in the data and all values of **visit** and fills in all combinations of these values that do not already occur in the data, for example, patient 2 and visit 7. The command creates a new variable, **_fillin**, taking the value 1 for filled-in records and 0 for records that existed before. All variables have missing values for these new records except **patient**, **visit**, and **_fillin**.

Before we can make predictions, we must fill in values for the covariates: **treatment**, **month**, and the interaction **trt_month**. Note that, by filling in values for covariates, we are not imputing missing data but just specifying for which covariate values we would like to make predictions.

We start by filling in the appropriate values for **treatment**, taking into account that **treatment** is a time-constant variable.

```
. egen trt = mean(treatment), by(patient)
. replace treatment = trt if _fillin==1
```

We proceed by filling in the average time (month) associated with the visit number for the time-varying variable **month** by using

```
. drop mn_month
. egen mn_month = mean(month), by(treatment visit)
. replace month = mn_month if _fillin==1
```

Finally, we obtain the filled-in version of the interaction variable, **trt_month**, by multiplying the variables **treatment** and **month** that we have constructed:

```
. replace trt_month = treatment*month
```

It is important that the response variable, `outcome`, remains missing; the posterior distribution should only be based on the responses that were observed. We also cannot change the covariate values corresponding to these responses because that would change the posterior distribution.

We can now make predictions for the entire dataset by repeating the `gllapred` command (after deleting `cmu`) with the `fsample` (for "full sample") option:

```
. drop cmu

. gllapred cmu, mu fsample
(mu will be stored in cmu)
Non-adaptive log-likelihood: -625.52573
 -625.3853  -625.3856  -625.3856
log-likelihood:-625.38558
. list patient visit _fillin cmu if patient==2|patient==15, sepby(patient) noobs
```

patient	visit	_fillin	cmu
2	1	0	.54654227
2	2	0	.46888925
2	3	0	.3867953
2	4	0	.30986966
2	5	0	.12102271
2	6	0	.05282663
2	7	1	.01463992
15	1	0	.59144346
15	2	0	.47716226
15	3	0	.39755635
15	4	0	.30542907
15	5	0	.08992082
15	6	1	.01855957
15	7	0	.00015355

The predicted forecast probability for visit 7 for patient 2 hence is 0.015.

To look at some patient-specific posterior mean probability curves, we will produce trellis graphs of 16 randomly chosen patients from each treatment group. We will first randomly assign consecutive integer identifiers (1, 2, 3, etc.) to the patients in each group, in a new variable, `randomid`. We will then plot the data for patients with `randomid` 1 through 16 in each group.

To create the random identifier, we first generate a random number from the uniform distribution whenever `visit` is 1 (which happens once for each patient):

```
. set seed 1234421

. sort patient

. generate rand = runiform() if visit==1
```

Here use of the `set seed` and `sort` commands ensures that you get the same values of
`randomid` as we do, because the same "seed" is used for the random-number genera-
tor. We now define a variable, `randid`, that represents the rank order of `rand` within
treatment groups and is missing when `rand` is missing:

```
. by treatment (rand), sort: generate randid = _n if rand<.
```

`randid` is the required random identifier, but it is only available when `visit` is 1 and
missing otherwise. We can fill in the missing values using

```
. egen randomid = mean(randid), by(patient)
```

We are now ready to produce the trellis graphs:

```
. twoway (line cmu month, sort) (scatter cmu month if _fillin==1, mcol(black))
> if randomid<=16&treatment==0,  by(patient, compact legend(off)
> l1title("Posterior mean probabilities"))
```

and

```
. twoway (line cmu month, sort) (scatter cmu month if _fillin==1, mcol(black))
> if randomid<=16&treatment==1,  by(patient, compact legend(off)
> l1title("Posterior mean probabilities"))
```

The graphs are shown in figure 10.15. We see that there is considerable variability in
the probability trajectories of different patients within the same treatment group.

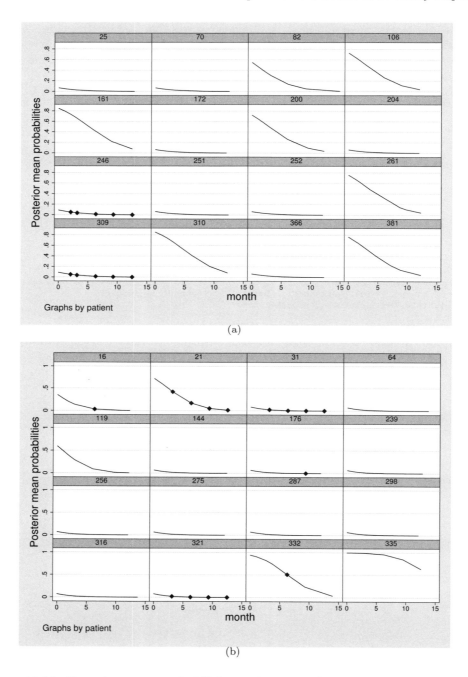

Figure 10.15: Posterior mean probabilities against time for 16 patients in the control group (a) and treatment group (b) with predictions for missing responses shown as diamonds

After estimation with `xtmelogit`, the `predict` command with the `mu` option gives the posterior mode of the predicted conditional probability $\widehat{\Pr}(y_{ij}|\mathbf{x}_{ij}, \zeta_j)$ instead of the posterior mean. This is achieved by substituting the posterior mode of ζ_j into the expression for the conditional probability. [The mode of a strictly increasing function of ζ_j (here an inverse logit), is the same function evaluated at the mode of ζ_j.]

10.14   Other approaches to clustered dichotomous data

10.14.1   Conditional logistic regression

Instead of using random intercepts for clusters (patients in the toenail application), it would be tempting to use fixed intercepts by including a dummy variable for each patient (and omitting the overall intercept). This would be analogous to the fixed-effects estimator of within-patient effects discussed for linear models in section 3.7.2. However, in logistic regression, this approach would lead to inconsistent estimates of the within-patient effects unless n is large, due to what is known as the *incidental parameter problem*. Roughly speaking, this problem occurs because the number of cluster-specific intercepts (the incidental parameters) increases in tandem with the sample size (number of clusters), so that the usual asymptotic, or large-sample results, break down. Obviously, we also cannot eliminate the random intercepts in nonlinear models by simply cluster-mean-centering the responses and covariates, as in (3.12).

Instead, we can eliminate the patient-specific intercepts by constructing a likelihood that is conditional on the number of responses that take the value 1 (a sufficient statistic for the patient-specific intercept). This approach is demonstrated in display 12.2 in the chapter on nominal responses. In the linear case, assuming normality, ordinary least-squares estimation of the cluster-mean-centered model is equivalent to conditional maximum likelihood estimation. In logistic regression, conditional maximum likelihood estimation is more involved and is known as *conditional logistic regression*. Importantly, this method estimates conditional or subject-specific effects. When using conditional logistic regression, we can only estimate the effects of within-patient or time-varying covariates. Patient-specific covariates, such as `treatment`, cannot be included. However, interactions between patient-specific and time-varying variables, such as `treatment` by `month`, can be estimated.

Conditional logistic regression can be performed using Stata's `xtlogit` command with the `fe` option or using the `clogit` command (with the `or` option to obtain odds ratios):

```
. clogit outcome month trt_month, group(patient) or
note: multiple positive outcomes within groups encountered.
note: 179 groups (1141 obs) dropped because of all positive or
      all negative outcomes.
```

```
Conditional (fixed-effects) logistic regression    Number of obs   =        767
                                                   LR chi2(2)      =     290.97
                                                   Prob > chi2     =     0.0000
Log likelihood = -188.94377                        Pseudo R2       =     0.4350
```

outcome	Odds Ratio	Std. Err.	z	P>|z|	[95% Conf. Interval]	
month	.6827717	.0321547	-8.10	0.000	.6225707	.748794
trt_month	.9065404	.0667426	-1.33	0.183	.7847274	1.047262

The subject-specific or conditional odds ratio for the treatment effect (treatment by time interaction) is now estimated as 0.91 and is no longer significant at the 5% level. However, both this estimate and the estimate for `month`, also given in the last column of table 10.2 on page 526, are quite similar to the estimates for the random-intercept model.

The subject-specific or conditional odds ratios from conditional logistic regression represent within-effects, where patients serve as their own controls. As discussed in chapter 5, within-patient estimates cannot be confounded with omitted between-patient covariates and are hence less sensitive to model misspecification than estimates based on the random-intercept model (which makes the strong assumption that the patient-specific intercepts are independent of the covariates). A further advantage of conditional maximum likelihood estimation is that it does not make any assumptions regarding the distribution of the patient-specific effect. Therefore, it is reassuring that the conditional maximum likelihood estimates are fairly similar to the maximum likelihood estimates for the random-intercept model.

If the random-intercept model is correct, the latter estimator is more efficient and tends to yield smaller standard errors leading to smaller p-values, as we can see for the treatment by time interaction. Here the conditional logistic regression method is inefficient because, as noted in the output, 179 subjects whose responses were all 0 or all 1 cannot contribute to the analysis. This is because the conditional probabilities of these response patterns, conditioning on the total response across time, are 1 regardless of the covariates (for example, if the total is zero, all responses must be zero) and the conditional probabilities therefore do not provide any information on covariate effects.

The above model is sometimes referred to as the Chamberlain fixed-effects logit model in econometrics and is used for matched case–control studies in epidemiology. The same trick of conditioning is also used for the Rasch model in psychometrics and the conditional logit model for discrete choice and nominal responses (see section 12.2.2). Unfortunately, there is no counterpart to conditional logistic regression for probit models.

Note that dynamic models with subject-specific effects cannot be estimated consistently by simply including lagged responses in conditional logistic regression. Also, subject-specific predictions are not possible in conditional logistic regression because no inferences are made regarding the subject-specific intercepts.

10.14.2   Generalized estimating equations (GEE)

Generalized estimating equations (GEE), first introduced in section 6.6, can be used to estimate marginal or population-averaged effects. Dependence among the responses of units in a given cluster is taken into account but treated as a nuisance, whereas this dependence is of central interest in multilevel modeling.

The basic idea of GEE is that an algorithm, known as reweighted iterated least squares, for maximum likelihood estimation of single-level generalized linear models requires only the mean structure (expectation of the response variable as a function of the covariates) and the variance function. The algorithm iterates between linearizing the model given current parameter estimates and then updating the parameters using weighted least squares, with weights determined by the variance function. In GEE, this iterative algorithm is extended to two-level data by assuming a within-cluster correlation structure, in addition to the mean structure and variance function, so that the weighted least-squares step becomes a generalized least-squares step (see section 3.10.1), and another step is required for updating the correlation matrix. GEE can be viewed as a special case of generalized methods of moments (GMM) estimation (implemented in Stata's `gmm` command).

In addition to specifying a model for the marginal relationship between the response variable and covariates, it is necessary to choose a structure for the correlations among the observed responses (conditional on covariates). The variance function follows from the Bernoulli distribution. The most common correlation structures are (see section 6.6 for some other correlation structures):

- Independence:
  Same as ordinary logistic regression

- Exchangeable:
  Same correlation for all pairs of units

- Autoregressive lag-1 [AR(1)]:
  Correlation declines exponentially with the time lag—only makes sense for longitudinal data and assumes constant time intervals between occasions (but allows gaps due to missing data).

- Unstructured:
  A different correlation for each pair of responses—only makes sense if units are not exchangeable within clusters, in the sense that the labels i attached to the units mean the same thing across clusters. For instance, it is meaningful in longitudinal data where units are occasions and the first occasion means the same thing across individuals, but not in data on students nested in schools where the numbering of students is arbitrary. In addition, each pair of unit labels i and i' must occur sufficiently often across clusters to estimate the pairwise correlations. Finally, the

number of unique unit labels, say, m, should not be too large because the number of parameters is $m(m-1)/2$.

The reason for specifying a correlation structure is that more efficient estimates (with smaller standard errors) are obtained if the specified correlation structure resembles the true dependence structure. Using ordinary logistic regression is equivalent to assuming an independence structure. GEE is therefore generally more efficient than ordinary logistic regression although the gain in precision can be meagre for balanced data (Lipsitz and Fitzmaurice 2009).

An important feature of GEE (and ordinary logistic regression) is that *marginal effects* can be consistently estimated, even if the dependence among units in clusters is not properly modeled. For this reason, correct specification of the correlation structure is downplayed by using the term "working correlations".

In GEE, the standard errors for the marginal effects are usually based on the robust sandwich estimator, which takes the dependence into account. Use of the sandwich estimator implicitly relies on there being many replications of the responses associated with each distinct combination of covariate values. Otherwise, the estimated standard errors can be biased downward. Furthermore, estimated standard errors based on the sandwich estimator can be very unreliable unless the number of clusters is large, so in this case model-based (nonrobust) standard errors may be preferable. See Lipsitz and Fitzmaurice (2009) for further discussion.

We now use GEE to estimate marginal odds ratios for the toenail data. We request an exchangeable correlation structure (the default) and robust standard errors by using `xtgee` with the `vce(robust)` and `eform` options:

```
. quietly xtset patient

. xtgee outcome treatment month trt_month, link(logit)
> family(binomial) corr(exchangeable) vce(robust) eform

GEE population-averaged model          Number of obs      =        1908
Group variable:               patient  Number of groups   =         294
Link:                           logit  Obs per group: min =           1
Family:                      binomial                 avg =         6.5
Correlation:              exchangeable                 max =           7
                                       Wald chi2(3)       =       63.44
Scale parameter:                    1  Prob > chi2        =      0.0000

                        (Std. Err. adjusted for clustering on patient)
```

outcome	Odds Ratio	Robust Std. Err.	z	P>|z|	[95% Conf. Interval]	
treatment	1.007207	.2618022	0.03	0.978	.6051549	1.676373
month	.8425856	.0253208	-5.70	0.000	.7943911	.893704
trt_month	.9252113	.0501514	-1.43	0.152	.8319576	1.028918
_cons	.5588229	.0963122	-3.38	0.001	.3986309	.7833889

These estimates are given under "GEE" in table 10.2 and can alternatively be obtained using `xtlogit` with the `pa` option.

We can display the fitted working correlation matrix by using `estat wcorrelation`:

```
. estat wcorrelation, format(%4.3f)
Estimated within-patient correlation matrix R:
           c1     c2     c3     c4     c5     c6     c7
    r1 | 1.000
    r2 | 0.422  1.000
    r3 | 0.422  0.422  1.000
    r4 | 0.422  0.422  0.422  1.000
    r5 | 0.422  0.422  0.422  0.422  1.000
    r6 | 0.422  0.422  0.422  0.422  0.422  1.000
    r7 | 0.422  0.422  0.422  0.422  0.422  0.422  1.000
```

A problem with the exchangeable correlation structure is that the true marginal (over the random effects) correlation of the responses is in general not constant but varies according to values of the observed covariates. Using Pearson correlations for dichotomous responses is also somewhat peculiar because the odds ratio is the measure of association in logistic regression.

GEE is an *estimation method* that does not require the specification of a full statistical model. While the mean structure, variance function, and correlation structure are specified, it may not be possible to find a statistical model with such a structure. As we already pointed out, it may not be possible to specify a model for binary responses where the residual Pearson correlation matrix is exchangeable. For this reason, the approach is called an estimating equation approach rather than a modeling approach. This is in stark contrast to multilevel modeling, where statistical models are explicitly specified.

The fact that no full statistical model is specified has three important implications. First, there is no likelihood and therefore likelihood-ratio tests cannot be used. Instead, comparison of nested models typically proceeds by using Wald-tests. Unless the sample size is large, this approach may be problematic because it is known that these tests do not work as well as likelihood-ratio tests in ordinary logistic regression. Second, it is not possible to simulate or predict individual responses based on the estimates from GEE (see section 10.13.2 for prediction and forecasting based on multilevel models). Third, GEE does not share the useful property of ML that estimators are consistent when data are missing at random (MAR). Although GEE produces consistent estimates of marginal effects if the probability of responses being missing is covariate dependent [and for the special case of responses missing completely at random (MCAR)], it produces inconsistent estimates if the probability of a response being missing for a unit depends on observed responses for other units in the same cluster. Such missingness is likely to occur in longitudinal data where dropout could depend on a subjects' previous responses (see sections 5.8.1 and 13.12).

10.15   Summary and further reading

We have described various approaches to modeling clustered dichotomous data, focusing on random-intercept models for longitudinal data. Alternatives to multilevel modeling, such as conditional maximum likelihood estimation and generalized estimating equations, have also been briefly discussed. The important distinction between conditional or subject-specific effects and marginal or population-averaged effects has been emphasized.

We have described adaptive quadrature for maximum likelihood estimation and pointed out that you need to make sure that a sufficient number of quadrature points have been used for a given model and application. We have demonstrated the use of a variety of predictions, either cluster-specific predictions, based on empirical Bayes, or population-averaged predictions. Keep in mind that consistent estimation in logistic regression models with random effects in principle requires a completely correct model specification. Diagnostics for generalized linear mixed models are still being developed.

We did not cover random-coefficient models for binary responses in this chapter but have included two exercises (10.3 and 10.8), with solutions provided, involving these models. The issues discussed in chapter 4 regarding linear models with random coefficients are also relevant for other generalized linear mixed models. The syntax for random-coefficient logistic models is analogous to the syntax for linear random-coefficient models except that `xtmixed` is replaced with `xtmelogit` and `gllamm` is used with a different link function and distribution (the syntax for linear random-coefficient models in `gllamm` can be found in the `gllamm` companion). Three-level random-coefficient logistic models for binary responses are discussed in chapter 16. In chapter 11, `gllamm` will be used to fit random-coefficient ordinal logistic regression models; see section 11.7.2.

Dynamic or lagged-response models for binary responses have not been discussed. The reason is that such models, sometimes called transition models in this context, can suffer from similar kinds of endogeneity problems as those discussed for dynamic models with random intercepts in chapter 5 of volume I. However, these problems are not as straightforward to address for binary responses (but see Wooldridge [2005]).

We have discussed the most common link functions for dichotomous responses, namely, logit and probit links. A third link that is sometimes used is the complementary log-log link, which is introduced in section 14.6. Dichotomous responses are sometimes aggregated into counts, giving the number of successes y_i in n_i trials for unit i. In this situation, it is usually assumed that y_i has a binomial(n_i, π_i) distribution. `xtmelogit` can then be used as for dichotomous responses but with the `binomial()` option to specify the variable containing the values n_i. Similarly, `gllamm` can be used with the binomial distribution and any of the link functions together with the `denom()` option to specify the variable containing n_i.

Good introductions to single-level logistic regression include Collett (2003a), Long (1997), and Hosmer and Lemeshow (2000). Logistic and other types of regression using Stata are discussed by Long and Freese (2006), primarily with examples from social science, and by Vittinghoff et al. (2005), with examples from medicine.

Generalized linear mixed models are described in the books by McCulloch, Searle, and Neuhaus (2008), Skrondal and Rabe-Hesketh (2004), Molenberghs and Verbeke (2005), and Hedeker and Gibbons (2006). See also Goldstein (2011), Raudenbush and Bryk (2002), and volume 3 of the anthology by Skrondal and Rabe-Hesketh (2010). Several examples with dichotomous responses are discussed in Skrondal and Rabe-Hesketh (2004, chap. 9). Guo and Zhao (2000) is a good introductory paper on multilevel modeling of binary data with applications in social science. We also recommend the book chapter by Rabe-Hesketh and Skrondal (2009), the article by Agresti et al. (2000), and the encyclopedia entry by Hedeker (2005) for overviews of generalized linear mixed models. Detailed accounts of generalized estimating equations are given in Hardin and Hilbe (2003), Diggle et al. (2002), and Lipsitz and Fitzmaurice (2009).

Exercises 10.1, 10.2, 10.3, and 10.6 are on longitudinal or panel data. There are also exercises on cross-sectional datasets on students nested in schools (10.7 and 10.8), cows nested in herds (10.5), questions nested in respondents (10.4) and wine bottles nested in judges (10.9). Exercise 10.2 involves GEE, whereas exercises 10.4 and 10.6 involve conditional logistic regression. The latter exercise also asks you to perform a Hausman test. Exercises 10.3 and 10.8 consider random-coefficient models for dichotomous responses (solutions are provided for both exercises). Exercise 10.4 introduces the idea of item-response theory, and exercise 10.8 shows how `gllamm` can be used to fit multilevel models with survey weights.

10.16   Exercises

10.1 Toenail data

1. Fit the probit version of the random-intercept model in (10.6) with `gllamm`. How many quadrature points appear to be needed using adaptive quadrature?

2. Estimate the residual intraclass correlation for the latent responses.

3. Obtain empirical Bayes predictions using both the random-intercept logit and probit models and estimate the approximate constant of proportionality between these.

4. ❖ By considering the residual standard deviations of the latent response for the logit and probit models, work out what you think the constant of proportionality should be for the logit- and probit-based empirical Bayes predictions. How does this compare with the constant estimated in step 3?

10.2 Ohio wheeze data

In this exercise, we use data from the Six Cities Study (Ware et al. 1984), previously analyzed by Fitzmaurice (1998), among others. The dataset includes 537 children from Steubenville, Ohio, who were examined annually four times from age 7 to age 10 to ascertain their wheezing status. The smoking status of the mother was also determined at the beginning of the study to investigate whether maternal smoking increases the risk of wheezing in children. The mother's smoking status is treated as time constant, although it may have changed for some mothers over time.

The dataset `wheeze.dta` has the following variables:

- `id`: child identifier (j)
- `age`: number of years since ninth birthday (x_{2ij})
- `smoking`: mother smokes regularly (1: yes; 0: no) (x_{3j})
- `y`: wheeze status (1: yes; 0: no) (y_{ij})

1. Fit the following transition model considered by Fitzmaurice (1998):

$$\text{logit}\{\Pr(y_{ij}=1|\mathbf{x}_{ij}, y_{i-1,j})\} = \beta_1 + \beta_2 x_{2ij} + \beta_3 x_{3j} + \gamma y_{i-1,j}, \quad i = 2, 3, 4$$

   where x_{2ij} is `age` and x_{3j} is `smoking`. (The lagged responses can be obtained using `by id (age), sort: generate lag = y[_n-1]`. Alternatively, use the time-series operator `L.`; see table 5.3 on page 275.)

2. Fit the following random-intercept model considered by Fitzmaurice (1998):

$$\text{logit}\{\Pr(y_{ij}=1|\mathbf{x}_{ij}, \zeta_j)\} = \beta_1 + \beta_2 x_{2ij} + \beta_3 x_{3j} + \zeta_j, \quad i = 1, 2, 3, 4$$

   It is assumed that $\zeta_j \sim N(0, \psi)$, and that ζ_j is independent across children and independent of $\mathbf{x}_{ij}$.

3. Use GEE to fit the marginal model

$$\text{logit}\{\Pr(y_{ij}=1|\mathbf{x}_{ij})\} = \beta_1 + \beta_2 x_{2ij} + \beta_3 x_{3j}, \quad i = 1, 2, 3, 4$$

   specifying an unstructured correlation matrix (`xtset` the data using `xtset id age`). Try some other correlation structures and compare the fit (using `estat wcorrelation`) to the unstructured version.

4. Interpret the estimated effects of mother's smoking status for the models in steps 1, 2, and 3.

10.3 Vaginal-bleeding data `Solutions`

Fitzmaurice, Laird, and Ware (2011) analyzed data from a trial reported by Machin et al. (1988). Women were randomized to receive an injection of either 100 mg or 150 mg of the long-lasting injectable contraception depot medroxyprogesterone acetate (DMPA) at the start of the trial and at three successive 90-day intervals. In addition, the women were followed up 90 days after the final injection. Throughout the study, each woman completed a menstrual diary that recorded any vaginal bleeding pattern disturbances. The diary data were used to determine whether a woman experienced amenorrhea, defined as the absence of menstrual bleeding for at least 80 consecutive days.

The response variable for each of the four 90-day intervals is whether the woman experienced amenorrhea during the interval. Data are available on 1,151 women for the first interval, but there was considerable dropout after that.

The dataset `amenorrhea.dta` has the following variables:

- `dose`: high dose (1: yes; 0: no)
- `y1–y4`: responses for time intervals 1–4 (1: amenorrhea; 0: no amenorrhea)
- `wt2`: number of women with the same dose level and response pattern

1. Produce an identifier variable for women, and reshape the data to long form, stacking the responses `y1–y4` into one variable and creating a new variable, `occasion`, taking the values 1–4 for each woman.

2. Fit the following model considered by Fitzmaurice, Laird, and Ware (2011):

$$\text{logit}\{\Pr(y_{ij} = 1|x_j, t_{ij}, \zeta_j)\} = \beta_1 + \beta_2 t_{ij} + \beta_3 t_{ij}^2 + \beta_4 x_j t_{ij} + \beta_5 x_j t_{ij}^2 + \zeta_j$$

   where $t_{ij} = 1, 2, 3, 4$ is the time interval and x_j is `dose`. It is assumed that $\zeta_j \sim N(0, \psi)$, and that ζ_j is independent across women and independent of x_j and t_{ij}. Use `gllamm` with the `weight(wt)` option to specify that `wt2` are level-2 weights.

3. Write down the above model but with a random slope of t_{ij}, and fit the model. (See section 11.7.2 for an example of a random-coefficient model fit in `gllamm`.)

4. Interpret the estimated coefficients.

5. Plot marginal predicted probabilities as a function of time, separately for women in the two treatment groups.

10.4 Verbal-aggression data

De Boeck and Wilson (2004) discuss a dataset from Vansteelandt (2000) where 316 participants were asked to imagine the following four frustrating situations where either another or oneself is to blame:

1. Bus: A bus fails to stop for me (another to blame)

2. Train: I miss a train because a clerk gave me faulty information (another to blame)

3. Store: The grocery store closes just as I am about to enter (self to blame)

4. Operator: The operator disconnects me when I have used up my last 10 cents for a call (self to blame)

For each situation, the participant was asked if it was true (yes, perhaps, or no) that

1. I would (want to) curse
2. I would (want to) scold
3. I would (want to) shout

For each of the three behaviors above, the words "want to" were both included and omitted, yielding six statements with a 3×2 factorial design (3 behaviors in 2 modes) combined with the four situations. Thus there were 24 items in total.

The dataset `aggression.dta` contains the following variables:

- `person`: subject identifier
- `item`: item (or question) identifier
- `description`: item description
  (situation: bus/train/store/operator; behavior: curse/scold/shout; mode: do/want)
- `i1`–`i24`: dummy variables for the items, for example, `i5` equals 1 when `item` equals 5 and 0 otherwise
- `y`: ordinal response (0: no; 1: perhaps; 2: yes)
- Person characteristics:
    - `anger`: trait anger score (STAXI, Spielberger [1988]) (w_{1j})
    - `gender`: dummy variable for being male (1: male; 0: female) (w_{2j})
- Item characteristics:
    - `do_want`: dummy variable for mode being "do" (that is, omitting words "want to") versus "want" (x_{2ij})
    - `other_self`: dummy variable for others to blame versus self to blame (x_{3ij})
    - `blame`: variable equal to 0.5 for blaming behaviors curse and scold and -1 for shout (x_{4ij})
    - `express`: variable equal to 0.5 for expressive behaviors curse and shout and -1 for scold (x_{5ij})

1. Recode the ordinal response variable `y` so that either a "2" or a "1" for the original variable becomes a "1" for the recoded variable.
2. De Boeck and Wilson (2004, sec. 2.5) consider the following "explanatory item-response model" for the dichotomous response

$$\text{logit}\{\Pr(y_{ij}=1|\mathbf{x}_{ij},\zeta_j)\} \;\; = \;\; \beta_1 + \beta_2 x_{2ij} + \beta_3 x_{3ij} + \beta_4 x_{4ij} + \beta_5 x_{5ij} + \zeta_j$$

where $\zeta_j \sim N(0,\psi)$ can be interpreted as the latent trait "verbal aggressiveness". Fit this model using `xtlogit`, and interpret the estimated coefficients. In De Boeck and Wilson (2004), the first five terms have minus signs, so their estimated coefficients have the opposite sign.

3. De Boeck and Wilson (2004, sec. 2.6) extend the above model by including a latent regression, allowing verbal aggressiveness (now denoted η_j instead of ζ_j) to depend on the person characteristics w_{1j} and w_{2j}:

$$\text{logit}\{\Pr(y_{ij}=1|\mathbf{x}_{ij},\eta_j)\} = \beta_1 + \beta_2 x_{2ij} + \beta_3 x_{3ij} + \beta_4 x_{4ij} + \beta_5 x_{5ij} + \eta_j$$

$$\eta_j = \gamma_1 w_{1j} + \gamma_2 w_{2j} + \zeta_j$$

Substitute the level-2 model for η_j into the level-1 model for the item responses, and fit the model using `xtlogit`.

4. Use `xtlogit` to fit the "descriptive item-response model", usually called a one-parameter logistic item response (IRT) model or *Rasch model*, considered by De Boeck and Wilson (2004, sec. 2.3):

$$\text{logit}\{\Pr(y_{ij}=1|d_{1i},\ldots,d_{24,i},\zeta_j)\} = \sum_{m=1}^{24} \beta_m d_{mi} + \zeta_j$$

where d_{mi} is a dummy variable for item i, with $d_{mi} = 1$ if $m = i$ and 0 otherwise. In De Boeck and Wilson (2004), the first term has a minus sign, so their β_m coefficients have the opposite sign; see also their page 53.

5. The model above is known as a one-parameter item-response model because there is one parameter β_m for each item. The negative of these item-specific parameters $-\beta_m$ can be interpreted as "difficulties"; the larger $-\beta_m$, the larger the latent trait (here verbal aggressiveness, but often ability) has to be to yield a given probability (for example, 0.5) of a 1 response.

Sort the items in increasing order of the estimated difficulties. For the least and most difficult items, look up the variable `description`, and discuss whether it makes sense that these items are particularly easy and hard to endorse (requiring little and a lot of verbal aggressiveness), respectively.

6. Replace the random intercepts ζ_j with fixed parameters α_j. Set the difficulty of the first item to zero for identification and fit the model by conditional maximum likelihood. Verify that differences between estimated difficulties for the items are similar as in step 4.

7. ❖ Fit the model in step 4 using `gllamm` or `xtmelogit` (this will take longer than `xtlogit`) and obtain empirical Bayes (also called EAP) or empirical Bayes modal (also called MAP) predictions (depending on whether you fit the model in `gllamm` or `xtmelogit`, respectively) and ML estimates of the latent trait. Also obtain standard errors (for ML, this means saving `_se[_cons]` in addition to `_b[_cons]` by adding `mlse = _se[_cons]` in the `statsby` command). Does there appear to be much shrinkage? Calculate the total score (sum of item responses) for each person and plot curves of the different kinds of standard errors with total score on the x axis. Comment on what you find.

See also exercise 11.2 for further analyses of these data.

10.5 Dairy-cow data

Dohoo et al. (2001) and Dohoo, Martin, and Stryhn (2010) analyzed data on dairy cows from Reunion Island. One outcome considered was the "risk" of conception at the first insemination attempt (first service) since the previous calving. This outcome was available for several lactations (calvings) per cow.

The variables in the dataset `dairy.dta` used here are

- `cow`: cow identifier
- `herd`: herd identifier
- `region`: geographic region
- `fscr`: first service conception risk (dummy variable for cow becoming pregnant)
- `lncfs`: log of time interval (in log days) between calving and first service (insemination attempt)
- `ai`: dummy variable for artificial insemination being used (versus natural) at first service
- `heifer`: dummy variable for being a young cow that has calved only once

1. Fit a two-level random-intercept logistic regression model for the response variable `fscr`, an indicator for conception at the first insemination attempt (first service). Include a random intercept for cow and the covariates `lncfs`, `ai`, and `heifer`. (Use either `xtlogit`, `xtmelogit`, or `gllamm`.)
2. Obtain estimated odds ratios with 95% confidence intervals for the covariates and interpret them.
3. Obtain the estimated residual intraclass correlation between the latent responses for two observations on the same cow. Is there much variability in the cows' fertility?
4. Obtain the estimated median odds ratio for two randomly chosen cows with the same covariates, comparing the cow that has the larger random intercept with the cow that has the smaller random intercept.

See also exercises 8.8 and 16.1.

10.6 Union membership data

Vella and Verbeek (1998) analyzed panel data on 545 young males taken from the U.S. National Longitudinal Survey (Youth Sample) for the period 1980–1987. In this exercise, we will focus on modeling whether the men were members of unions or not.

The dataset `wagepan.dta` was provided by Wooldridge (2010) and was previously used in exercise 3.6 and *Introduction to models for longitudinal and panel data (part III)*. The subset of variables considered here is

- nr: person identifier (j)
- year: 1980–1987 (i)
- union: dummy variable for being a member of a union (that is, wage being set in collective bargaining agreement) (y_{ij})
- educ: years of schooling (x_{2j})
- black: dummy variable for being black (x_{3j})
- hisp: dummy variable for being Hispanic (x_{4j})
- exper: labor market experience, defined as age$-6-$educ (x_{5ij})
- married: dummy variable for being married (x_{6ij})
- rur: dummy variable for living in a rural area (x_{7ij})
- nrtheast: dummy variable for living in Northeast (x_{8ij})
- nrthcen: dummy variable for living in Northern Central (x_{9ij})
- south: dummy variable for living in South $(x_{10,ij})$

You can use the **describe** command to get a description of the other variables in the file.

1. Use maximum likelihood to fit the random-intercept logistic regression model

$$\text{logit}\{\Pr(y_{ij} = 1 | \mathbf{x}_{ij}, \zeta_j)\} = \beta_1 + \beta_2 x_{2j} + \cdots + \beta_{11} x_{10,ij} + \zeta_j$$

   where $\zeta_j \sim N(0, \psi)$, and ζ_j is assumed to be independent across persons and independent of $\mathbf{x}_{ij}$. Use **xtlogit** because it is considerably faster than the other commands here.

2. Interpret the estimated effects of the covariates from step 1 in terms of odds ratios, and report the estimated residual intraclass correlation of the latent responses.

3. Fit the marginal model

$$\text{logit}\{\Pr(y_{ij} = 1 | \mathbf{x}_{ij})\} = \beta_1 + \beta_2 x_{2j} + \cdots + \beta_{11} x_{10,ij}$$

   using GEE with an exchangeable working correlation structure.

4. Interpret the estimated effects of the covariates from step 3 in terms of odds ratios, and compare these estimates with those from step 1. Why are the estimates different?

5. Explore the within and between variability of the response variable and covariates listed above. For which of the covariates is it impossible to estimate an effect using a fixed-effects approach? Are there any covariates whose effects you would expect to be imprecisely estimated when using a fixed-effects approach?

6. Use conditional maximum likelihood to fit the fixed-intercept logistic regression model

$$\text{logit}\{\text{Pr}(y_{ij} = 1|\mathbf{x}_{ij})\} \;=\; \beta_1 + \beta_2 x_{2j} + \cdots + \beta_{11} x_{10,ij} + \alpha_j$$

where the α_j are unknown person-specific parameters.

7. Interpret the estimated effects of the covariates from step 6 in terms of odds ratios, and compare these estimates with those from step 1. Why are the estimates different?

8. Perform a Hausman test to assess the validity of the random-intercept model. What do you conclude?

9. Fit the probit versions of the random-intercept model from step 1 using `xtprobit`. Which type of model do you find easiest to interpret?

10.7 School retention in Thailand data

A national survey of primary education was conducted in Thailand in 1988. The data were previously analyzed by Raudenbush and Bhumirat (1992) and are distributed with the HLM software (Raudenbush et al. 2004). Here we will model the probability that a child repeats a grade any time during primary school.

The dataset `thailand.dta` has the following variables:

- `rep`: dummy variable for child having repeated a grade during primary school (y_{ij})
- `schoolid`: school identifier (j)
- `pped`: dummy variable for child having preprimary experience (x_{2ij})
- `male`: dummy variable for child being male (x_{3ij})
- `mses`: school mean socioeconomic status (SES) (x_{4j})
- `wt1`: number of children in the school having a given set of values of `rep`, `pped`, and `male` (level-1 frequency weights)

1. Fit the model

$$\text{logit}\{\text{Pr}(y_{ij} = 1|\mathbf{x}_{ij}, \zeta_j)\} \;=\; \beta_1 + \beta_2 x_{2ij} + \beta_3 x_{3ij} + \beta_4 x_{4j} + \zeta_j$$

where $\zeta_j \sim N(0, \psi)$, and ζ_j is independent across schools and independent of the covariates $\mathbf{x}_{ij}$. Use `gllamm` with the `weight(wt)` option to specify that each row in the data represents `wt1` children (level-1 units).

2. Obtain and interpret the estimated odds ratios and the estimated residual intraschool correlation of the latent responses.

3. Use `gllapred` to obtain empirical Bayes predictions of the probability of repeating a grade. These probabilities will be specific to the schools, as well as dependent on the student-level predictors.

   a. List the values of `male`, `pped`, `rep`, `wt1`, and the predicted probabilities for the school with `schoolid` equal to 10104. Explain why the predicted probabilities are greater than 0 although none of the children in the sample from that school have been retained. For comparison, list the same variables for the school with `schoolid` equal to 10105.

b. Produce box plots of the predicted probabilities for each school by `male` and `pped` (for instance, using `by(male)` and `over(pped)`). To ensure that each school contributes no more than four probabilities to the graph (one for each combination of the student-level covariates), use only responses where `rep` is 0 (that is, `if rep==0`). Do the schools appear to be variable in their retention probabilities?

10.8 PISA data [Solutions]

Here we consider data from the 2000 Program for International Student Assessment (PISA) conducted by the Organization for Economic Cooperation and Development (OECD 2000) that are made available with permission from Mariann Lemke. The survey assessed educational attainment of 15-year-olds in 43 countries in various areas, with an emphasis on reading. Following Rabe-Hesketh and Skrondal (2006), we will analyze reading proficiency, treated as dichotomous (1: proficient; 0: not proficient), for the U.S. sample.

The variables in the dataset `pisaUSA2000.dta` are

- `id_school`: school identifier
- `pass_read`: dummy variable for being proficient in reading
- `female`: dummy variable for student being female
- `isei`: international socioeconomic index
- `high_school`: dummy variable for highest education level by either parent being high school
- `college`: dummy variable for highest education level by either parent being college
- `test_lang`: dummy variable for test language (English) being spoken at home
- `one_for`: dummy variable for one parent being foreign born
- `both_for`: dummy variable for both parents being foreign born
- `w_fstuwt`: student-level or level-1 survey weights
- `wnrschbq`: school-level or level-2 survey weights

1. Fit a logistic regression model with `pass_read` as the response variable and the variables `female` to `both_for` above as covariates and with a random intercept for schools using `gllamm`. (Use the default eight quadrature points.)
2. Fit the model from step 1 with the school mean of `isei` as an additional covariate. (Use the estimates from step 1 as starting values.)
3. Interpret the estimated coefficients of `isei` and school mean `isei` and comment on the change in the other parameter estimates due to adding school mean `isei`.
4. From the estimates in step 2, obtain an estimate of the between-school effect of socioeconomic status.
5. Obtain robust standard errors using the command `gllamm, robust`, and compare them with the model-based standard errors.

6. Add a random coefficient of `isei`, and compare the random-intercept and random-coefficient models using a likelihood-ratio test. Use the estimates from step 2 (or step 5) as starting values, adding zeros for the two additional parameters as shown in section 11.7.2.

7. ❖ In this survey, schools were sampled with unequal probabilities, π_j, and given that a school was sampled, students were sampled from the school with unequal probabilities $\pi_{i|j}$. The reciprocals of these probabilities are given as school- and student-level survey weights, `wnrschbg` ($w_j = 1/\pi_j$) and `w_fstuwt` ($w_{i|j} = 1/\pi_{i|j}$), respectively. As discussed in Rabe-Hesketh and Skrondal (2006), incorporating survey weights in multilevel models using a so-called *pseudolikelihood* approach can lead to biased estimates, particularly if the level-1 weights $w_{i|j}$ are different from 1 and if the cluster sizes are small. Neither of these issues arises here, so implement pseudomaximum likelihood estimation as follows:

   a. Rescale the student-level weights by dividing them by their cluster means [this is scaling method 2 in Rabe-Hesketh and Skrondal (2006)].

   b. Rename the level-2 weights and rescaled level-1 weights to `wt2` and `wt1`, respectively.

   c. Run the `gllamm` command from step 2 above with the additional option `pweight(wt)`. (Only the stub of the weight variables is specified; `gllamm` will look for the level-1 weights under `wt1` and the level-2 weights under `wt2`.) Use the estimates from step 2 as starting values.

   d. Compare the estimates with those from step 2. Robust standard errors are computed by `gllamm` because model-based standard errors are not appropriate with survey weights.

10.9 Wine-tasting data

Tutz and Hennevogl (1996) and Fahrmeir and Tutz (2001) analyzed data on the bitterness of white wines from Randall (1989).

The dataset `wine.dta` has the following variables:

- `bitter`: dummy variable for bottle being classified as bitter (y_{ij})
- `judge`: judge identifier (j)
- `temp`: temperature (low=1; high=0) x_{2ij}
- `contact`: skin contact when pressing the grapes (yes=1; no=0) x_{3ij}
- `repl`: replication

Interest concerns whether conditions that can be controlled while pressing the grapes, such as temperature and skin contact, influence the bitterness. For each combination of temperature and skin contact, two bottles of white wine were randomly chosen. The bitterness of each bottle was rated by the same nine judges, who were selected and trained for the ability to detect bitterness. Here we consider the binary response "bitter" or "nonbitter".

To allow the judgment of bitterness to vary between judges, a random-intercept logistic model is specified

$$\ln\left\{\frac{\Pr(y_{ij}=1|x_{2ij},x_{3ij},\zeta_j)}{\Pr(y_{ij}=0|x_{2ij},x_{3ij},\zeta_j)}\right\} = \beta_1 + \beta_2 x_{2ij} + \beta_3 x_{3ij} + \zeta_j$$

where $\zeta_j \sim N(0,\psi)$. The random intercepts are assumed to be independent across judges and independent of the covariates x_{2ij} and x_{3ij}. Maximum likelihood estimates and estimated standard errors for the model are given in table 10.3 below.

Table 10.3: Maximum likelihood estimates for bitterness model

	Est	(SE)
Fixed part		
β_1	-1.50	(0.90)
β_2	4.26	(1.30)
β_3	2.63	(1.00)
Random part		
ψ	2.80	
Log likelihood	-25.86	

1. Interpret the estimated effects of the covariates as odds ratios.
2. State the expression for the residual intraclass correlation of the latent responses for the above model and estimate this intraclass correlation.
3. Consider two bottles characterized by the same covariates and judged by two randomly chosen judges. Estimate the median odds ratio comparing the judge who has the larger random intercept with the judge who has the smaller random intercept.
4. ❖ Based on the estimates given in table 10.3, provide an approximate estimate of ψ if a probit model is used instead of a logit model. Assume that the estimated residual intraclass correlation of the latent responses is the same as for the logit model.
5. ❖ Based on the estimates given in the table, provide approximate estimates for the marginal effects of x_{2ij} and x_{3ij} in an ordinary logistic regression model (without any random effects).

See also exercise 11.8 for further analysis of these data.

10.10 ❖ Random-intercept probit model

In a hypothetical study, an ordinary probit model was fit for students clustered in schools. The response was whether students gave the right answer to a question, and the single covariate was socioeconomic status (SES). The intercept and regression coefficient of SES were estimated as $\widehat{\beta}_1 = 0.2$ and $\widehat{\beta}_2 = 1.6$, respectively. The analysis was then repeated, this time including a normally distributed random intercept for school with variance estimated as $\widehat{\psi} = 0.15$.

1. Using a latent-response formulation for the random-intercept probit model, derive the marginal probability that $y_{ij} = 1$ given SES. See page 512 and remember to replace ϵ_{ij} with $\zeta_j + \epsilon_{ij}$.

2. Obtain the values of the estimated school-specific regression coefficients for the random-intercept probit model.

3. Obtain the estimated residual intraclass correlation for the latent responses.

11 Ordinal responses

11.1 Introduction

An ordinal variable is a categorical variable with ordered categories. An example from sociology would be the response to an attitude statement such as "Murderers should be executed" in one of the ordered categories "disagree strongly", "disagree", "agree", or "agree strongly". Such questions asking about level of agreement with an attitude statement are called Likert-scale items. A medical example of an ordinal response is diagnosis of multiple sclerosis using the ordered categories "definitely not", "unlikely", "doubtful", "possible", "probable", and "certain". In education, level of proficiency ("below basic", "basic", "proficient", and "advanced") may be of interest.

In this chapter, we generalize the multilevel models for dichotomous responses that were discussed in the previous chapter to handle ordinal responses. Chapter 10 should therefore be read before embarking on this chapter. We start by introducing single-level cumulative logit and probit models for ordinal responses before extending these models to the multilevel and longitudinal setting by including random effects in the linear predictor.

Many of the issues that were discussed in the previous chapter—such as the distinction between conditional and marginal effects and the use of numerical integration for maximum likelihood estimation and empirical Bayes prediction—persist for multilevel modeling of ordinal responses. Indeed, the main difference between the logistic models considered in this and the previous chapter is that odds ratios must now be interpreted in terms of the odds of exceeding a given category.

11.2 Single-level cumulative models for ordinal responses

For simplicity, we start by introducing ordinal regression models for a single covariate x_i in the single-level case where units have index i. Just as for the dichotomous responses discussed in section 10.2, we can specify models for ordinal responses by using either a generalized linear model formulation or a latent-response formulation.

11.2.1 Generalized linear model formulation

Consider an ordinal response variable y_i with S ordinal categories denoted s ($s = 1, \ldots, S$). One way of specifying regression models for ordinal responses y_i is to let the cumulative probability that a response is in a higher category than s, given a covariate x_i, be structured as

$$\Pr(y_i > s|x_i) \;=\; F(\beta_2 x_i - \kappa_s) \qquad s=1,\ldots,S-1 \qquad (11.1)$$

where $F(\cdot)$ is a cumulative distribution function (CDF). $F(\cdot)$ is a standard normal CDF for the *ordinal probit model* and a logistic CDF for the *ordinal logit model*.

$F(\cdot)$ can be viewed as an inverse link function, denoted $h(\cdot)$ or $g^{-1}(\cdot)$, and the category-specific linear predictor ν_{is} is

$$\nu_{is} \;=\; \beta_2 x_i - \kappa_s$$

Here the κ_s are category-specific parameters, often called thresholds (the Greek letter κ is pronounced "kappa"). Notice that we have omitted the intercept β_1 because it is not identified if all the $S-1$ thresholds κ_s are free parameters (see section 11.2.4). The covariate effect β_2 is constant across categories, a property sometimes referred to as the *parallel-regressions assumption* because the linear predictors for different categories s are parallel.

It follows from the cumulative probabilities in (11.1) that the probability for a specific category s can be obtained as

$$
\begin{aligned}
\Pr(y_i = s|x_i) &= \Pr(y_i > s-1|x_i) - \Pr(y_i > s|x_i)\\
&= F(\beta_2 x_i - \kappa_{s-1}) - F(\beta_2 x_i - \kappa_s)
\end{aligned}
$$

For this reason, cumulative models for ordinal responses are sometimes called *difference models*.

Having specified the link function and linear predictor, the final ingredient for a generalized linear model is the conditional distribution of the responses. For ordinal responses, this is a multinomial distribution with category-specific probabilities, as given above. A simple example of a process following a multinomial distribution is rolling a fair die, where the probabilities for all categories $s=1,\ldots,6$ are equal to $1/6$.

Cumulative models for ordinal responses differ from generalized linear models for other response types in one important way: the inverse-link function of the linear predictor $g^{-1}(\nu_{is})$ does not represent the expectation or mean of the response (given the covariates) but represents a cumulative probability. Perhaps for this reason, Stata's `glm` command cannot be used for these models; instead, these models are fit using the specialty commands `ologit` and `oprobit`.

11.2.2 Latent-response formulation

Cumulative models for ordinal responses can alternatively be viewed as linear regression models for latent (unobserved) continuous responses y_i^*,

$$y_i^* \;=\; \beta_2 x_i + \epsilon_i$$

where the conditional distribution of ϵ_i, given the observed covariate x_i, is standard normal for the ordinal probit model and standard logistic for the ordinal logit model.

Observed ordinal responses y_i are generated from the latent continuous responses y_i^* via a threshold model:

$$y_i = \begin{cases} 1 & \text{if} & y_i^* \leq \kappa_1 \\ 2 & \text{if} & \kappa_1 < y_i^* \leq \kappa_2 \\ \vdots & \vdots & \vdots \\ S & \text{if} & \kappa_{S-1} < y_i^* \end{cases}$$

This cutting up of the latent-response scale is shown for $S = 5$ categories in figure 11.1.

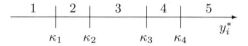

Figure 11.1: Illustration of threshold model for $S = 5$ categories

When there are no covariates and $y_i^* \sim N(0, 1)$, we obtain an ordinal probit model as shown for three categories ($S = 3$) in figure 11.2.

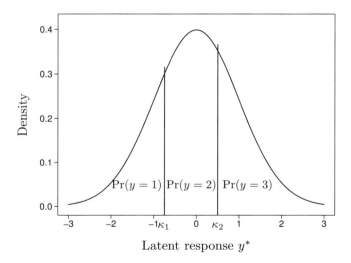

Figure 11.2: Illustration of three-category ordinal probit model without covariates

The observed response is 1 if the latent response y_i^* is less than κ_1 with corresponding probability $\Pr(y_i = 1)$ given by the area under the (standard) normal density to the left of κ_1. The response is 2 if y_i^* is larger than κ_1 but less than κ_2 with $\Pr(y_i = 2)$ given by the area under the density between these thresholds. Finally, the response is 3 if y_i^* is larger than κ_2 with $\Pr(y_i=3)$ given by the area under the density to the right of κ_2. Similar graphs are also shown in figure 11.6.

The latent-response and generalized linear model formulations are equivalent for ordinal responses, just as previously shown for dichotomous responses in section 10.2.2. This can be seen by considering the cumulative probabilities in (11.1)

$$
\begin{aligned}
\Pr(y_i > s|x_i) \;\; &= \;\; \Pr(y_i^* > \kappa_s|x_i) = \Pr(\beta_2 x_i + \epsilon_i > \kappa_s|x_i) \\
&= \;\; \Pr(-\epsilon_i \leq \beta_2 x_i - \kappa_s|x_i) = F(\beta_2 x_i - \kappa_s)
\end{aligned}
$$

Figure 11.3 illustrates the equivalence of the formulations for an ordinal logit model with one covariate x_i. In the lower part of the figure, we present the (solid) regression line for the regression of a latent response y_i^* on a covariate x_i

$$
E(y_i^*|x_i) = \beta_2 x_i
$$

For selected values of x_i, we have plotted the conditional logistic densities of the latent responses y_i^*. The thresholds κ_1 and κ_2 are represented as dashed and dotted lines, respectively. For a given value of x_i, the probability of observing a category above 1 is given by the gray (combined light and dark) area under the corresponding logistic density above κ_1. The probability of a category above 2 is the dark gray area under the logistic density above κ_2. These probabilities are plotted in the upper part of the figure as dashed and dotted lines, respectively. The light gray area is the conditional probability that the response is 2. This probability increases as response 1 (white area) becomes less likely and decreases as response 3 (dark gray) becomes more likely.

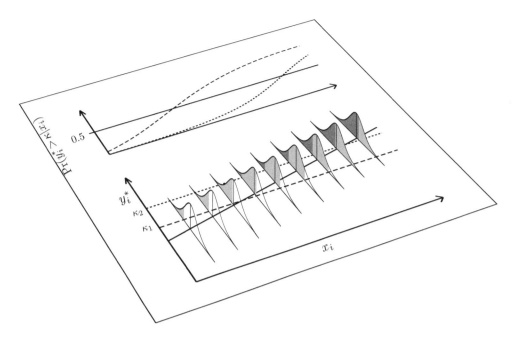

Figure 11.3: Illustration of equivalence of latent-response and generalized linear model formulation for ordinal logistic regression

The top panel of figure 11.4 shows the same cumulative probability curves as the upper part of figure 11.3:

$$\begin{array}{rcl}
\Pr(y_i > 1 | x_i) & = & \Pr(y_i = 2 | x_i) + \Pr(y_i = 3 | x_i) \\
\Pr(y_i > 2 | x_i) & = & \Pr(y_i = 3 | x_i)
\end{array}$$

The corresponding probabilities of individual categories s are given by

$$\begin{array}{rcl}
\Pr(y_i = 3 | x_i) & = & \Pr(y_i > 2 | x_i) \\
\Pr(y_i = 2 | x_i) & = & \Pr(y_i > 1 | x_i) - \Pr(y_i > 2 | x_i) \\
\Pr(y_i = 1 | x_i) & = & 1 - \Pr(y_i > 1 | x_i)
\end{array}$$

and are shown in the bottom panel of figure 11.4. The probabilities of the lowest and highest response categories 1 and S, which are categories 1 and 3 in the current example, are monotonic (strictly increasing or decreasing) functions of the covariates in cumulative models for ordinal responses. In contrast, the probability of an intermediate category is a unimodal or single-peaked function, as shown for $\Pr(y_i = 2 | x_i)$ in the bottom panel (and as we saw when looking at the light gray area in figure 11.3).

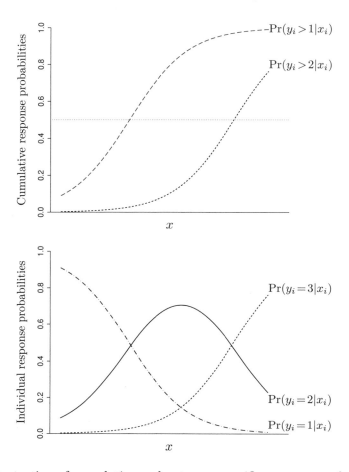

Figure 11.4: Illustration of cumulative and category-specific response probabilities

11.2.3   Proportional odds

If we specify a logit link or equivalently assume a standard logistic distribution for the residual ϵ_i in the latent-response model, the cumulative probabilities become

$$\Pr(y_i > s | x_i) \;=\; \frac{\exp(\beta_2 x_i - \kappa_s)}{1 + \exp(\beta_2 x_i - \kappa_s)}$$

and the corresponding log odds are

$$\ln\left\{\frac{\Pr(y_i > s | x_i)}{1 - \Pr(y_i > s | x_i)}\right\} \;=\; \beta_2 x_i - \kappa_s$$

It follows that the ratio of the odds that the response exceeds s for two covariate values, $x_i = a$ and $x_i = b$, becomes

$$\frac{\Pr(y_i > s | x_i = a)/\{1 - \Pr(y_i > s | x_i = a)\}}{\Pr(y_i > s | x_i = b)/\{1 - \Pr(y_i > s | x_i = b)\}} \;=\; \exp\{\beta_2(a-b)\} \;=\; \exp(\beta_2)^{(a-b)}$$

We see that for a unit increase in the covariate, $a - b = 1$, the odds ratio is $\exp(\beta_2)$. Indeed, for dummy variables the only possible increase is a unit increase from 0 to 1. In contrast, for a continuous covariate, we can also consider, for instance, a 10-unit increase. Then we must raise $\exp(\beta_2)$ to the 10th power because the model for the odds is multiplicative.

The category-specific parameter κ_s cancels out from the odds ratio, so the odds ratio is the same whatever category s is considered. For instance, for a four-category response, the odds ratio of being above 1 (that is, 2, 3, or 4) versus 1 is the same as the odds ratio of being above 2 (that is, 3 or 4) versus being at or below 2 (that is, 1 or 2). The corresponding odds are shown in the first two rows of figure 11.5 where the categories corresponding to 'success' (numerator of odds) are enclosed in thick frames and the categories corresponding to 'failure' (denominator of odds) are enclosed in thin frames. The third row corresponds to the odds of being above 3 (that is, 4) versus at or below 3 (that is, 1, 2, or 3). The odds themselves are not the same across the three rows but the *ratios* of the odds when a covariate increases by a unit are the same. This property, called *proportional odds*, makes it evident that these models are highly structured. It hence comes as no surprise that several models have been suggested that relax this property (see section 11.12).

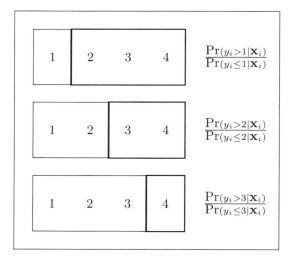

Figure 11.5: Relevant odds $\Pr(y_i > s|\mathbf{x}_i)/\Pr(y_i \leq s|\mathbf{x}_i)$ for $s = 1, 2, 3$ in a proportional odds model with four categories. Odds is a ratio of probabilities of events. Events included in numerator probability are in thick frames and events included in denominator probability are in thin frames [adapted from Brendan Halpin's web notes on "Models for ordered categories (ii)"].

If the proportional odds property holds, we can merge or collapse categories—or in the extreme case, dichotomize an ordinal variable—without changing the regression coefficient β_2. For instance, in the case of three categories, we could redefine the categories as $1 = \{2, 3\}$ and $0 = \{1\}$ and fit a binary logistic regression model. The interpretation of the odds ratios for this model would then correspond to one of the interpretations of the odds ratios for the original proportional odds model: the odds ratios for categories 2 or 3 versus 1. However, if the proportional odds assumption is violated, the estimates may be quite different.

11.2.4 ❖ Identification

Consider now for simplicity a model for y_i^* without covariates, but with an intercept β_1 and a variance parameter θ,

$$y_i^* = \beta_1 + \epsilon_i, \qquad \epsilon_i \sim N(0, \theta)$$

which can alternatively be written as

$$y_i^* \sim N(\beta_1, \theta)$$

It follows from this model that the probability of observing a category larger than s becomes

$$\Pr(y_i > s) \;=\; \Pr(y_i^* > \kappa_s) \;=\; \Pr\left(\frac{y_i^* - \beta_1}{\sqrt{\theta}} > \frac{\kappa_s - \beta_1}{\sqrt{\theta}}\right) \;=\; \Phi\left(\frac{\beta_1 - \kappa_s}{\sqrt{\theta}}\right)$$

where $\Phi(\cdot)$ is the cumulative standard normal density function.

We see from this expression that the cumulative probabilities $\Pr(y_i > s)$ do not change if we add some constant a to β_1 (or to the latent response y_i^*) and to all the thresholds κ_s, because the constant cancels out in the numerator $\beta_1 - \kappa_s$. Similarly, we see that the cumulative probabilities do not change if we multiply y_i^* by a constant b, so that β_1 and $\sqrt{\theta}$ are multiplied by b, as long as we also multiply κ_s by the same constant b, because we are in effect multiplying both the numerator and the denominator by the same constant. It follows that the location β_1 and scale $\sqrt{\theta}$ of the latent responses y_i^* are not identified if the thresholds are free parameters and thus cannot be estimated from the data.

This invariance is also illustrated in figure 11.6. It is evident from the left panel (top and bottom) that response probabilities are invariant to translation of the latent response y_i^*; the effect of adding 2 to y_i^* (or to the intercept β_1) can be counteracted by adding 2 to the thresholds κ_s. From the right panel (top and bottom), we see that the response probabilities are invariant to rescaling of y_i^*; the effect of multiplying y_i^* by 2 (or multiplying β_1 and $\sqrt{\theta}$ by 2) can be counteracted by multiplying the thresholds κ_s by 2. Thus the location and scale of the latent responses y_i^* are identified only relative to the location and spacing of the thresholds κ_s.

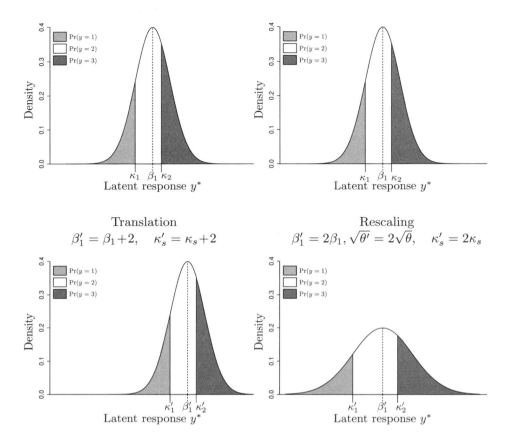

Figure 11.6: Illustration of scale and translation invariance in cumulative probit model

To identify ordinal regression models, the location of the latent response is usually fixed by setting the intercept β_1 to zero, which is what we have done so far in this chapter. The scale is usually fixed by setting $\theta = 1$ for probit models and $\theta = \pi^2/3$ for logit models. These two conventions are most common for ordinal models and are used by Stata's `ologit` and `oprobit` commands.

An alternative to fixing the location of y_i^* by setting $\beta_1 = 0$ is to fix the location of the thresholds, typically by setting $\kappa_1 = 0$ to identify the intercept β_1. This parameterization is common for dichotomous responses and is used in Stata's `logit`, `probit`, and `glm` commands. Less commonly, we could fix both the location and the scale of the thresholds by setting $\kappa_1 = 0$ and $\kappa_2 = 1$ so that both β_1 and θ can be estimated.

11.3 Are antipsychotic drugs effective for patients with schizophrenia?

Schizophrenia is a mental illness characterized by impairments in the perception or expression of reality, most commonly in terms of auditory hallucinations and paranoid or bizarre delusions. According to the U.S. National Institute of Mental Health (NIMH), the prevalence of schizophrenia is about 1% in the U.S. population aged 18 and older.

We now use data from the NIMH Schizophrenia Collaborative Study on treatment-related changes in overall severity of schizophrenia. The data have previously been analyzed by Gibbons and Hedeker (1994), Hedeker and Gibbons (1996, 2006), and Gibbons et al. (1988).

Patients were randomly assigned to receive one of four treatments: placebo (control), chlorpromazine, fluphenazine, or thioridazine. In the present analysis, we will not distinguish between the three antipsychotic drugs; instead, we call the combined group the treatment group. The outcome considered is "severity of illness" as measured by item 79 of the Inpatient Multidimensional Psychiatric Scale (IMPS) of Lorr and Klett (1966). The patients were examined weekly for up to six weeks.

The dataset `schiz.dta` has the following variables:

- `id`: patient identifier
- `week`: week of assessment since randomization $(0, 1, \ldots, 6)$
- `imps`: item 79 of IMPS (-9 represents missing)
- `treatment`: dummy variable for being in treatment group (1: treatment [drug]; 0: control)

We read in the schizophrenia data by typing

```
. use http://www.stata-press.com/data/mlmus3/schiz
```

Following Hedeker and Gibbons (1996), we recode item 79 of the IMPS into an ordinal severity of illness variable, `impso`, with four categories (1: normal or borderline mentally ill; 2: mildly or moderately ill; 3: markedly ill; 4: severely or among the most extremely ill). This can be accomplished using the `recode` command:

```
. generate impso = imps
. recode impso -9=. 1/2.4=1 2.5/4.4=2 4.5/5.4=3 5.5/7=4
```

11.4 Longitudinal data structure and graphs

Before we start modeling the severity of schizophrenia over time, let us look at the dataset, using both descriptive statistics and graphs. This may provide insights that can be used in subsequent model specification (see also *Introduction to models for longitudinal and panel data (part III)*).

11.4.1 Longitudinal data structure

We first `xtset` the data:

```
. xtset id week
       panel variable:  id (unbalanced)
        time variable:  week, 0 to 6, but with gaps
                delta:  1 unit
```

We then use the `xtdescribe` command to describe the participation pattern in the dataset:

```
. xtdescribe if impso<.
         id:  1103, 1104, ..., 9316                          n =         437
       week:  0, 1, ..., 6                                   T =           7
             Delta(week) = 1 unit
             Span(week)  = 7 periods
             (id*week uniquely identifies each observation)
Distribution of T_i:    min      5%     25%     50%     75%     95%     max
                          2       2       4       4       4       4       5

       Freq.   Percent    Cum. | Pattern
     ---------------------------+----------
         308     70.48   70.48 | 11.1..1
          41      9.38   79.86 | 11.1...
          37      8.47   88.33 | 11.....
           8      1.83   90.16 | 11....1
           8      1.83   91.99 | 111....
           6      1.37   93.36 | 11.1.1.
           5      1.14   94.51 | 1..1..1
           5      1.14   95.65 | 11.11..
           3      0.69   96.34 | .1.1..1
          16      3.66  100.00 | (other patterns)
     ---------------------------+----------
         437    100.00         | XXXXXXX
```

All patterns shown have missing assessments for at least three occasions. The predominant pattern, "11.1..1" (308 patients), has assessments only at weeks 0, 1, 3, and 6. We can tabulate the number of observations in each week for the treatment and control groups by using the `table` command:

```
. table week treatment, contents(count impso) col

          |       treatment
    week  |      0        1    Total
    ------+-------------------------
       0  |    107      327      434
       1  |    105      321      426
       2  |      5        9       14
       3  |     87      287      374
       4  |      2        9       11
       5  |      2        7        9
       6  |     70      265      335
```

We see that few assessments took place in weeks 2, 4, and 5.

11.4.2   Plotting cumulative proportions

When considering cumulative models, it is natural to inspect the cumulative proportions in the dataset. We therefore calculate the proportions of patients having responses above 1, 2, and 3 at each occasion and by treatment group. These proportions can be calculated conveniently using Stata's `egen` command with the `mean()` function:

```
. egen propg1 = mean(impso>1), by(week treatment)
. egen propg2 = mean(impso>2), by(week treatment)
. egen propg3 = mean(impso>3), by(week treatment)
```

Because only a few assessments were made in weeks 2, 4, and 5, we will plot cumulative proportions only for weeks 0, 1, 3, and 6. To simplify later commands, we define a dummy variable, `nonrare`, for these weeks by using `egen` with the `anymatch()` function:

```
. egen nonrare = anymatch(week), values(0,1,3,6)
```

We then define a label for `treatment`,

```
. label define t 0 "Control" 1 "Treatment", modify
. label values treatment t
```

and produce a graph for the cumulative sample proportions against week,

```
. sort treatment id week
. twoway (line propg1 week, sort)
> (line propg2 week, sort lpatt(vshortdash))
> (line propg3 week, sort lpatt(dash)) if nonrare==1, by(treatment)
> legend(order(1 "Prop(y>1)" 2 "Prop(y>2)" 3 "Prop(y>3)")) xtitle("Week")
```

The resulting graph, shown in figure 11.7, suggests that antipsychotic drugs have a beneficial effect because the proportions in the higher, more severe categories decline more rapidly in the treatment group. We observe that the cumulative proportions also decline somewhat in the control group.

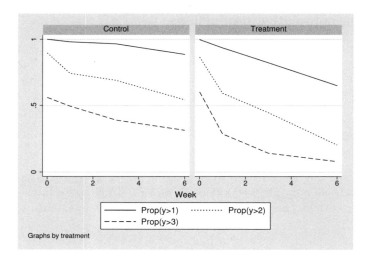

Figure 11.7: Cumulative sample proportions versus week

11.4.3 Plotting cumulative sample logits and transforming the time scale

In ordinal logistic regression, the cumulative log odds or logits, and not the cumulative probabilities, are specified as linear functions of covariates. Before specifying a model for the time trend, it is therefore useful to construct cumulative sample logits (logits of sample proportions),

```
. generate logodds1 = ln(propg1/(1-propg1))
. generate logodds2 = ln(propg2/(1-propg2))
. generate logodds3 = ln(propg3/(1-propg3))
```

and plot these against week, giving the graphs in figure 11.8.

```
. twoway (line logodds1 week, sort)
> (line logodds2 week, sort lpatt(vshortdash))
> (line logodds3 week, sort lpatt(dash)) if nonrare==1, by(treatment)
> legend(order(1 "Log Odds(y>1)" 2 "Log Odds(y>2)" 3 "Log Odds(y>3)"))
> xtitle("Week")
```

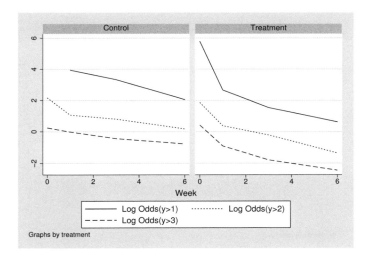

Figure 11.8: Cumulative sample logits versus week

The missing value for the log odds of being in categories above 1 in week 0 for the control group is due to the corresponding proportion being equal to 1 (see figure 11.7), giving an odds of ∞.

The cumulative sample logits in figure 11.8 would be poorly approximated by a linear function of week in the treatment group. One way to address this problem would be to depart from a linear trend for the cumulative logits and instead consider some polynomial function, such as a quadratic or cubic. Another approach, adopted by Hedeker and Gibbons (1996) and followed here, is to retain a linear trend for the logits but transform the time scale by taking the square root of week:

```
. generate weeksqrt = sqrt(week)
```

Plots for the cumulative sample logits against the transformed time variable, weeksqrt, are produced by the command

```
. twoway (line logodds1 weeksqrt, sort)
> (line logodds2 weeksqrt, sort lpatt(vshortdash))
> (line logodds3 weeksqrt, sort lpatt(dash)) if nonrare==1, by(treatment)
> legend(order(1 "Log Odds(y>1)" 2 "Log Odds(y>2)" 3 "Log Odds(y>3)"))
> xtitle("Square root of week")
```

and presented in figure 11.9.

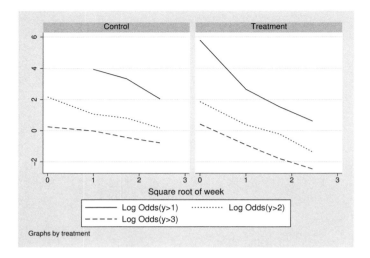

Figure 11.9: Cumulative sample logits versus square root of week

As desired, the curves generally appear to be more linear after the square-root transformation of time.

11.5   A single-level proportional odds model

We start modeling the schizophrenia data by fitting a single-level ordinal logistic regression model.

11.5.1   Model specification

Based on figure 11.9, it seems reasonable to assume a linear relationship between the
cumulative logits and the square root of time, allowing the intercept and the slope to
differ between the treatment and control groups,

$$\begin{aligned}
\text{logit}\{\Pr(y_{ij} > s | x_{2ij}, x_{3j})\} &= \beta_2 x_{2ij} + \beta_3 x_{3j} + \beta_4 x_{2ij} x_{3j} - \kappa_s \\
&= (\beta_2 + \beta_4 x_{3j}) x_{2ij} + \beta_3 x_{3j} - \kappa_s \qquad (11.2)
\end{aligned}$$

where x_{2ij} represents `weeksqrt` and x_{3j} represents `treatment`. Here β_2 is the slope
of transformed time for the control group, β_3 is the difference between treatment and
control groups at week 0, and β_4 is the difference in the slopes of transformed time
between treatment and control groups. From the figure, we would expect β_4 (which we
can interpret as the treatment effect) to be negative and the corresponding odds ratio,
given by $\exp(\beta_4)$, to be less than 1.

To interpret $\exp(\beta_4)$, consider the odds ratios for a unit increase in time x_{2ij} in the two treatment groups:

$$\exp(\beta_2 + \beta_4 x_{3j}) = \begin{cases} \exp(\beta_2) & \text{if } x_{3j} = 0 \quad \text{(control group)} \\ \exp(\beta_2)\exp(\beta_4) & \text{if } x_{3j} = 1 \quad \text{(treatment group)} \end{cases}$$

Thus $\exp(\beta_4)$ is the odds ratio for transformed time for the treatment group divided by the odds ratio for transformed time for the control group.

11.5.2 Estimation using Stata

We start by constructing an interaction term, `interact`, by taking the product of `weeksqrt` and `treatment`:

```
. generate interact = weeksqrt*treatment
```

The Stata commands for fitting the single-level proportional odds model (11.2) using the `ologit` command are

```
. ologit impso weeksqrt treatment interact, vce(cluster id) or
Ordered logistic regression                    Number of obs   =       1603
                                               Wald chi2(3)    =     440.17
                                               Prob > chi2     =     0.0000
Log pseudolikelihood = -1878.0969              Pseudo R2       =     0.1177
                             (Std. Err. adjusted for 437 clusters in id)
```

		Robust				
impso	Odds Ratio	Std. Err.	z	P>|z|	[95% Conf. Interval]	
weeksqrt	.5847056	.0591797	-5.30	0.000	.4794958	.7130004
treatment	.9993959	.2042595	-0.00	0.998	.6695244	1.491793
interact	.4719089	.0568135	-6.24	0.000	.3727189	.5974961
/cut1	-3.807279	.1956796			-4.190804	-3.423754
/cut2	-1.760167	.1811041			-2.115125	-1.40521
/cut3	-.4221112	.1795596			-.7740415	-.0701808

Here we have used the `vce(cluster id)` option to obtain robust standard errors (based on the sandwich estimator), taking into account the clustered nature of the data, and the `or` option to obtain estimated odds ratios $\exp(\widehat{\beta})$ with corresponding 95% confidence intervals. The output looks just like output from binary logistic regression, except for the additional estimates of the thresholds $\kappa_1 - \kappa_3$, labeled `/cut1`–`/cut3`. Maximum likelihood estimates for the proportional odds model are presented under the heading "POM" in table 11.1.

Table 11.1: Maximum likelihood estimates and 95% CIs for proportional odds model (POM), random-intercept proportional odds model (RI-POM), and random-coefficient proportional odds model (RC-POM)

	POM		RI-POM		RC-POM	
	Est	(95% CI)†	Est	(95% CI)	Est	(95% CI)
Fixed part: Odds ratios						
$\exp(\beta_2)$ [weeksqrt]	0.58	(0.48, 0.71)	0.46	(0.36, 0.60)	0.41	(0.27, 0.63)
$\exp(\beta_3)$ [treatment]	1.00	(0.67, 1.49)	0.94	(0.51, 1.75)	1.06	(0.49, 2.28)
$\exp(\beta_4)$ [interact]	0.47	(0.37, 0.60)	0.30	(0.22, 0.40)	0.18	(0.11, 0.30)
Fixed part: Thresholds						
κ_1	−3.81		−5.86		−7.32	
κ_2	−1.76		−2.83		−3.42	
κ_3	−0.42		−0.71		−0.81	
Random part: Variances and covariance						
ψ_{11}			3.77		6.99	
ψ_{22}					2.01	
ψ_{21}					−1.51	
Log likelihood	−1,878.10		−1,701.38		−1,662.76	

†Based on the sandwich estimator

The odds ratios for this model represent marginal or population-averaged effects. All of these odds ratios are ratios of the odds of more severe versus less severe illness regardless of where we cut the ordinal scale to define more versus less. We see from table 11.1 that the odds ratio of more severe illness per unit of time (on the transformed scale) is estimated as 0.58 (95% CI from 0.48 to 0.71) for the control group. The corresponding odds ratio for the treatment group is estimated as $0.58 \times 0.47 = 0.28$. To obtain a confidence interval for the latter odds ratio, we can use the lincom command with the or (or eform) option:

```
. lincom weeksqrt+interact, or
 ( 1)  [impso]weeksqrt + [impso]interact = 0
```

impso	Odds Ratio	Std. Err.	z	P>|z|	[95% Conf. Interval]	
(1)	.2759278	.0179484	-19.80	0.000	.2428997	.3134469

The 95% confidence interval for the odds ratio per square root of week ranges from 0.24 to 0.31 in the treatment group.

We could have used factor-variable notation in the olgit and lincom commands, with the following syntax:

```
ologit impso i.treatment##c.weeksqrt, or
lincom c.weeksqrt+1.treatment#c.weeksqrt, or
```

We can easily assess the model fit by comparing model-implied probabilities with sample proportions. Model-implied probabilities for the individual response categories can be obtained by using the `predict` command with the `pr` option:

```
. predict prob1-prob4, pr
```

The corresponding cumulative probabilities of exceeding categories 3, 2, and 1 are obtained by the following commands:

```
. generate probg3 = prob4
. generate probg2 = prob3 + probg3
. generate probg1 = prob2 + probg2
```

We now plot these marginal probabilities together with the corresponding sample proportions against week:

```
. twoway (line propg1 week if nonrare==1, sort lpatt(solid))
> (line propg2 week if nonrare==1, sort lpatt(vshortdash))
> (line propg3 week if nonrare==1, sort lpatt(dash))
> (line probg1 week, sort lpatt(solid))
> (line probg2 week, sort lpatt(vshortdash))
> (line probg3 week, sort lpatt(dash)), by(treatment)
> legend(order(1 "Prob(y>1)" 2 "Prob(y>2)" 3 "Prob(y>3)")) xtitle("Week")
```

The resulting graphs, shown in figure 11.10, suggest that the sample proportions are well recovered by the model.

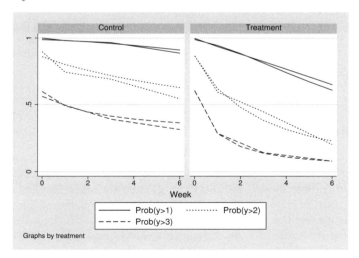

Figure 11.10: Cumulative sample proportions and predicted cumulative probabilities from ordinal logistic regression versus week

Fitting an ordinary proportional odds model and specifying robust standard errors for clustered data can be viewed as GEE (generalized estimating equations) with an independence working correlation matrix. The dependence is treated as a nuisance. (Stata's xtgee command cannot be used for ordinal responses.)

11.6 A random-intercept proportional odds model

11.6.1 Model specification

We now model the longitudinal dependence by including a patient-specific random intercept ζ_{1j} in the proportional odds model:

$$\text{logit}\{\Pr(y_{ij} > s|\mathbf{x}_{ij}, \zeta_{1j})\} = \beta_2 x_{2ij} + \beta_3 x_{3j} + \beta_4 x_{2j} x_{3ij} + \zeta_{1j} - \kappa_s \quad (11.3)$$

The overall level or intercept of the cumulative logits is ζ_{1j} and hence varies over patients j. As usual, we assume that the $\zeta_{1j} \sim N(0, \psi_{11})$ are independently distributed across patients and that ζ_{1j} is independent of the covariates $\mathbf{x}_{ij}$. (We denoted the random intercept ζ_{1j} instead of ζ_j and the variance ψ_{11} instead of ψ because we will add a random coefficient later.)

The model can alternatively be written in terms of a latent response y_{ij}^*,

$$y_{ij}^* = \beta_2 x_{2ij} + \beta_3 x_{3j} + \beta_4 x_{2j} x_{3ij} + \zeta_{1j} + \epsilon_{ij}$$

where the ϵ_{ij} have standard logistic distributions, are independent across patients and occasions, and are independent of $\mathbf{x}_{ij}$. The ordinal severity of illness variable y_{ij} is related to the latent response via the threshold model:

$$y_{ij} = \begin{cases} 1 & \text{if} & y_{ij}^* \leq \kappa_1 \\ 2 & \text{if} & \kappa_1 < y_{ij}^* \leq \kappa_2 \\ 3 & \text{if} & \kappa_2 < y_{ij}^* \leq \kappa_3 \\ 4 & \text{if} & \kappa_3 < y_{ij}^* \end{cases}$$

11.6.2 Estimation using Stata

At the time of writing this book, there is no official Stata command for fitting multilevel models for ordinal responses, so we will use the user-written command gllamm. Ordinal logit and probit models can be fit in gllamm by using the link(ologit) and link(oprobit) options, respectively. To fit the random-intercept proportional odds model in (11.3), we use the following options: i(id) to specify that the responses are nested in patients with identifier id, adapt to request adaptive quadrature, and eform to obtain exponentiated estimates or odds ratios. The gllamm command is

```
. gllamm impso weeksqrt treatment interact, i(id) link(ologit) adapt eform
number of level 1 units = 1603
number of level 2 units = 437

Condition Number = 15.409532

gllamm model

log likelihood = -1701.3807
```

	exp(b)	Std. Err.	z	P>|z|	[95% Conf. Interval]	
impso						
weeksqrt	.4649524	.0608031	-5.86	0.000	.3598276	.6007898
treatment	.9439398	.2962805	-0.18	0.854	.5102372	1.74629
interact	.2993646	.0457031	-7.90	0.000	.2219474	.4037856
_cut11						
_cons	-5.858454	.331792	-17.66	0.000	-6.508754	-5.208153
_cut12						
_cons	-2.825669	.2900513	-9.74	0.000	-3.394159	-2.257179
_cut13						
_cons	-.7077079	.2750904	-2.57	0.010	-1.246875	-.1685405

```
Variances and covariances of random effects
------------------------------------------------------------------------------

***level 2 (id)

    var(1): 3.7733419 (.46496883)
------------------------------------------------------------------------------
. estimates store model1
```

The estimated odds ratios with 95% confidence intervals for the random-intercept proportional odds model were presented under "RI-POM" in table 11.1. As discussed in section 10.8 for binary responses, these estimates of conditional or subject-specific effects are further from 1 than the marginal or population-averaged counterparts from model (11.2). The subject-specific odds ratio of transformed time is estimated as 0.46 in the control group and as 0.14 ($= 0.46495 \times 0.29936$) in the treatment group.

11.6.3 Measures of dependence and heterogeneity

Residual intraclass correlation of latent responses

The random-intercept variance is estimated as $\widehat{\psi}_{11} = 3.773$, implying an estimated residual intraclass correlation for the latent responses y_{ij}^* of

$$\widehat{\rho} = \frac{\widehat{\psi}_{11}}{\widehat{\psi}_{11} + \pi^2/3} = \frac{3.773}{3.773 + \pi^2/3} = 0.53$$

Median odds ratio

The median odds ratio proposed by Larsen et al. (2000) and Larsen and Merlo (2005), and previously discussed in section 10.9.2, can be used as a measure of heterogeneity in ordinal random-intercept models.

The median odds ratio, OR_{median}, is given by

$$OR_{median} = \exp\{\sqrt{2\psi_{11}}\,\Phi^{-1}(3/4)\}$$

Plugging in $\widehat{\psi}_{11}$, we obtain $\widehat{OR}_{median}$:

```
. display exp(sqrt(2*3.7733416)*invnormal(3/4))
6.3783288
```

Hence, if two patients are chosen at random at a given time point from the same treatment group, the odds ratio comparing the subject with the larger odds to the subject with the smaller odds will exceed 6.38 half the time, which is a large odds ratio.

11.7 A random-coefficient proportional odds model

11.7.1 Model specification

To allow the slope of the time variable, weeksqrt, to vary randomly between patients within the two groups, we now include a random slope ζ_{2j} in addition to the random intercept ζ_{1j}:

$$
\begin{aligned}
\text{logit}\{&\Pr(y_{ij} > s|\mathbf{x}_{ij}, \zeta_{1j}, \zeta_{2j})\} \\
&= \beta_2 x_{2ij} + \beta_3 x_{3j} + \beta_4 x_{2j}x_{3ij} + \zeta_{1j} + \zeta_{2j}x_{2ij} - \kappa_s \\
&= \zeta_{1j} + (\beta_2 + \zeta_{2j})x_{2ij} + \beta_3 x_{3j} + \beta_4 x_{2j}x_{3ij} - \kappa_s \quad (11.4)
\end{aligned}
$$

In this model, not only the intercept but also the slope $\beta_2 + \zeta_{2j}$ of weeksqrt (x_{2ij}) varies over patients j. We assume that the random intercept and slope have a bivariate normal distribution with zero mean and covariance matrix

$$
\Psi = \begin{bmatrix} \psi_{11} & \psi_{12} \\ \psi_{21} & \psi_{22} \end{bmatrix}, \qquad \psi_{21} = \psi_{12}
$$

Both the random intercepts and the random slopes are assumed to be independent across patients and independent of the covariates $\mathbf{x}_{ij}$.

The random-coefficient proportional odds model can also be specified using a latent-response formulation, as shown for the random-intercept proportional odds model in section 11.6.1.

11.7.2 Estimation using gllamm

To fit the random-coefficient proportional odds model using gllamm, we must first define equations for the intercept and slope by using the eq command. We need one equation for each random effect, specifying after the equation name and colon the variable that multiplies the random effect. For the random intercept, this variable is just 1, so we first create a variable, cons, equal to 1 and define an equation for the random intercept.

```
. generate cons = 1
. eq inter: cons
```

The random slope is multiplied by weeksqrt, and the corresponding equation is defined using

```
. eq slope: weeksqrt
```

We are now ready to fit the random-coefficient model using gllamm with the following options: nrf(2) to specify the number of random effects as 2 and eqs() to specify the equations defined above:

```
. gllamm impso weeksqrt treatment interact, i(id) nrf(2) eqs(inter slope)
> link(ologit) adapt eform
number of level 1 units = 1603
number of level 2 units = 437

Condition Number = 13.096305

gllamm model

log likelihood = -1662.76
```

impso	exp(b)	Std. Err.	z	P>|z|	[95% Conf. Interval]	
impso						
weeksqrt	.413742	.0901228	-4.05	0.000	.2699714	.6340762
treatment	1.058823	.4152368	0.15	0.884	.4909184	2.283693
interact	.1837951	.0463311	-6.72	0.000	.1121407	.3012346
_cut11						
_cons	-7.317879	.4733159	-15.46	0.000	-8.245561	-6.390197
_cut12						
_cons	-3.417252	.3869461	-8.83	0.000	-4.175652	-2.658851
_cut13						
_cons	-.8120713	.3526985	-2.30	0.021	-1.503348	-.1207948

```
Variances and covariances of random effects
------------------------------------------------------------------------------

***level 2 (id)

    var(1): 6.9861284 (1.3149731)
    cov(2,1): -1.5061266 (.53161343) cor(2,1): -.40212569

    var(2): 2.0079957 (.41890607)
------------------------------------------------------------------------------
. estimates store model2
```

Instead of fitting the model from scratch, it is often faster to use estimates for the random-intercept model as starting values for the random-coefficient model. To illustrate this, we first retrieve the random-intercept estimates,

```
. estimates restore model1
```

so that we can store the estimates in a matrix a by typing

```
. matrix a = e(b)
```

Although we could think of the parameter estimates as a vector, Stata stores them in a row matrix that we have accessed as e(b). We can refer to the elements as, for instance, a[1,3] (for the third element). However, the random-coefficient model requires two additional parameters, which happen to be the last two parameters in the row matrix. We arbitrarily specify values 0.1 and 0 for these and correspondingly add these columns to row matrix a:

```
. matrix a = (a,.1,0)
```

We could then fit the random-coefficient model, using the from(a) and copy options to specify that the augmented matrix a contains the starting values in the correct order:

```
gllamm impso weeksqrt treatment interact, i(id) nrf(2) eqs(inter slope)
    link(ologit) adapt from(a) copy eform
```

We can compare the random-coefficient proportional odds model (11.4) with the random-intercept proportional odds model (11.3) by using a likelihood-ratio test:

```
. lrtest model1 model2
(log likelihoods of null models cannot be compared)
Likelihood-ratio test                          LR chi2(2)  =      77.24
(Assumption: model1 nested in model2)          Prob > chi2 =     0.0000
```

Although the test is conservative (because the null hypothesis is on the boundary of the parameter space; see section 2.6.2 for how to obtain the correct asymptotic p-value), we get a p-value that is close to zero. The random-intercept model is thus rejected in favor of the random-coefficient model.

The maximum likelihood estimates for the random-coefficient proportional odds model were presented under "RC-POM" in table 11.1. In this model, the patient-specific odds ratio per unit of time (in square root of week) is estimated as 0.41 in the control group and as 0.07 (=0.41375×0.183792) in the treatment group. The random-intercept variance is estimated as $\widehat{\psi}_{11} = 6.99$ and the random-slope variance as $\widehat{\psi}_{22} = 2.01$. The covariance between the random intercept and slope is estimated as $\widehat{\psi}_{21} = -1.51$, corresponding to a correlation of -0.40. This means that patients having severe schizophrenia at the onset of the study (week=0) tend to have a greater decline in severity than those with less severe schizophrenia in both the control and the treatment groups.

11.8   Different kinds of predicted probabilities

11.8.1   Predicted population-averaged or marginal probabilities

As shown for binary logistic regression models in section 10.13.1, we can obtain the population-averaged or marginal probabilities implied by the fitted random-coefficient proportional odds model by using `gllapred` with the `mu` and `marginal` options. Because there are several response categories in ordinal models, what `gllapred` actually provides are cumulative probabilities that the response is above category s,

$$\widehat{\Pr}(y_{ij} > s | \mathbf{x}_{ij})$$

The category s is specified in the `above()` option of `gllapred`:

```
. gllapred mprobg1, mu marginal above(1)
(mu will be stored in mprobg1)

. gllapred mprobg2, mu marginal above(2)
(mu will be stored in mprobg2)

. gllapred mprobg3, mu marginal above(3)
(mu will be stored in mprobg3)
```

These marginal cumulative probabilities can be plotted together with the cumulative sample proportions against `week` (we revert to `week` because we find it easier to interpret this natural time scale than `weeksqrt`).

```
. twoway (line propg1 week if nonrare==1, sort lpatt(solid))
> (line propg2 week if nonrare==1, sort lpatt(vshortdash))
> (line propg3 week if nonrare==1, sort lpatt(dash))
> (line mprobg1 week, sort lpatt(solid))
> (line mprobg2 week, sort lpatt(vshortdash))
> (line mprobg3 week, sort lpatt(dash)), by(treatment)
> legend(order(1 "Prob(y>1)" 2 "Prob(y>2)" 3 "Prob(y>3)")) xtitle("Week")
```

The resulting graphs are shown in figure 11.11.

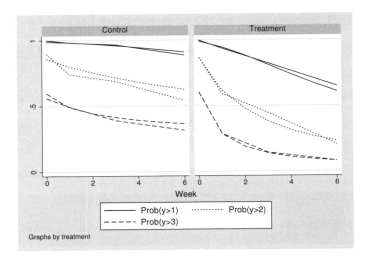

Figure 11.11: Cumulative sample proportions and cumulative predicted marginal prob-abilities from random-coefficient proportional odds model versus week

In addition to modeling the dependence of repeated measurements within patients, we see that the random-coefficient proportional odds model fits the cumulative proportions equally as well as the single-level proportional odds model in figure 11.10.

The corresponding marginal probabilities for the individual response categories s can be obtained using the relationship

$$\widehat{\Pr}(y_{ij} = s | \mathbf{x}_{ij}) = \widehat{\Pr}(y_{ij} > s-1 | \mathbf{x}_{ij}) - \widehat{\Pr}(y_{ij} > s | \mathbf{x}_{ij})$$

where $\widehat{\Pr}(y_{ij} > 0 | \mathbf{x}_{ij}) = 1$. The Stata commands for calculating these predicted proba-bilities are

```
. generate pr1 = 1 - mprobg1
. generate pr2 = mprobg1 - mprobg2
. generate pr3 = mprobg2 - mprobg3
. generate pr4 = mprobg3
```

Graphs of the marginal probabilities for each category against week are produced by the command

```
. twoway (line pr1 week, sort lpatt(solid))
> (line pr2 week, sort lpatt(vshortdash))
> (line pr3 week, sort lpatt(dash_dot))
> (line pr4 week, sort lpatt(dash)), by(treatment)
> legend(order(1 "Prob(y=1)" 2 "Prob(y=2)" 3 "Prob(y=3)" 4 "Prob(y=4)"))
> xtitle("Week")
```

giving the graph in figure 11.12.

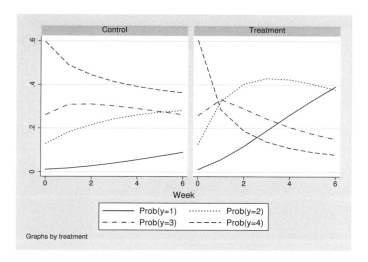

Figure 11.12: Marginal category probabilities from random-coefficient proportional odds model versus week

In the treatment group, the predicted marginal probability of being at most borderline ill (category 1) increases from nearly 0 to about 0.4, whereas the predicted marginal probability of being severely ill or among the most extremely ill (category 4) decreases from about 0.6 to about 0.1. These improvements are much less marked in the control group.

We could also make a graph that is similar to a stacked bar chart, but with a continuous x axis. In a stacked bar chart, the lowest bar represents $\Pr(y = 1)$, one bar up represents $\Pr(y = 2)$, etc., so the height of the top of the kth bar represents the sum of the probabilities up to and including k, $\Pr(y \leq k)$. We first calculate these heights:

```
. generate pr12 = 1 - mprobg2
. generate pr123 = 1 - mprobg3
. generate pr1234 = 1
```

Instead of making bars, we produce area graphs with the command

```
. twoway (area pr1 week, sort fintensity(inten10))
> (rarea pr12 pr1 week, sort fintensity(inten50))
> (rarea pr123 pr12 week, sort fintensity(inten70))
> (rarea pr1234 pr123 week, sort fintensity(inten90)), by(treatment)
> legend(order(1 "Prob(y=1)" 2 "Prob(y=2)" 3 "Prob(y=3)" 4 "Prob(y=4)"))
> xtitle("Week")
```

giving the graph shown in figure 11.13.

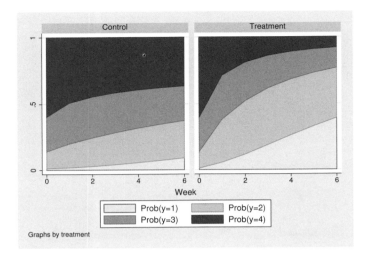

Figure 11.13: Area graph analogous to stacked bar chart for marginal predicted probabilities from random-coefficient proportional odds model

The combined area in the three lightest shades represents the marginal predicted probability that the severity is 3 (markedly ill) or less. The area for any given shade represents the probability that the severity is in the corresponding category. Perhaps the most striking feature is how much more the probability of the highest severity decreases over time in the treatment group compared with the control group.

11.8.2 Predicted subject-specific probabilities: Posterior mean

As shown for binary logistic regression in section 10.13.2, we can obtain empirical Bayes predictions of subject-specific probabilities for subjects in the sample. The corresponding posterior mean cumulative probabilities are

$$\widetilde{\Pr}(y_{ij} > s|\mathbf{x}_{ij}) \equiv \int \widehat{\Pr}(y_{ij} > s|\mathbf{x}_{ij}, \zeta_j) \times \text{Posterior}(\zeta_j|y_{1j}, \ldots, y_{n_jj}, \mathbf{X}_j) \, d\zeta_j$$

and the posterior mean category probabilities can be obtained as the differences

$$\widetilde{\Pr}(y_{ij} = s|\mathbf{x}_{ij}) = \widetilde{\Pr}(y_{ij} > s - 1|\mathbf{x}_{ij}) - \widetilde{\Pr}(y_{ij} > s|\mathbf{x}_{ij})$$

with $\widetilde{\Pr}(y_{ij} > 0|\mathbf{x}_{ij}) = 1$.

To make predictions for weeks 0 through 6 for everyone even if the response variable was not observed, we first use the `fillin` command to produce rows of data for all combinations of the values of `id` and `week` that are missing (see also section 10.13.2):

```
. fillin id week
```

This produces missing values in the new rows of data for all variables except `id` and `week`. To make predictions, none of the variables used for estimation in `gllamm` can be missing except the response variable. We therefore replace the missing values, which occur whenever the dummy variable `_fillin` produced by the `fillin` command is 1:

```
. replace weeksqrt = sqrt(week) if _fillin==1
. egen trt = mean(treatment), by(id)
. replace treatment = trt if _fillin==1
(1456 real changes made)
. replace interact = weeksqrt*treatment if _fillin==1
(1456 real changes made)
. replace cons = 1 if _fillin==1
(1456 real changes made)
```

The posterior mean cumulative probabilities can now be obtained by using the `mu`, `above()`, and `fsample` options (but not the `marginal` option):

```
. gllapred pprobg1, mu above(1) fsample
(mu will be stored in pprobg1)
Non-adaptive log-likelihood:  -1662.1717
-1662.7601 -1662.7600 -1662.7600
log-likelihood:-1662.76
. gllapred pprobg2, mu above(2) fsample
(mu will be stored in pprobg2)
Non-adaptive log-likelihood:  -1662.1717
-1662.7601 -1662.7600 -1662.7600
log-likelihood:-1662.76
. gllapred pprobg3, mu above(3) fsample
(mu will be stored in pprobg3)
Non-adaptive log-likelihood:  -1662.1717
-1662.7601 -1662.7600 -1662.7600
log-likelihood:-1662.76
```

Before plotting the predictions, we define a new identifier, `newid`, that numbers patients within each group from 1:

```
. generate week0 = week==0
. sort treatment id week
. by treatment: generate newid = sum(week0)
```

We now use this identifier to plot the curves for twelve subjects in each treatment group:

```
. twoway (line pprobg1 week, sort lpatt(solid))
> (line pprobg2 week, sort lpatt(vshortdash))
> (line pprobg3 week, sort lpatt(dash))
> (scatter pprobg1 week if _fillin==1, msym(o))
> (scatter pprobg2 week if _fillin==1, msym(o))
> (scatter pprobg3 week if _fillin==1, msym(o) mcol(black))
> if newid<13 & treatment==0, by(newid)
> legend(order(1 "Prob(y>1)" 2 "Prob(y>2)" 3 "Prob(y>3)") rows(1)) xtitle("Week")
. twoway (line pprobg1 week, sort lpatt(solid))
> (line pprobg2 week, sort lpatt(vshortdash))
> (line pprobg3 week, sort lpatt(dash))
> (scatter pprobg1 week if _fillin==1, msym(o))
> (scatter pprobg2 week if _fillin==1, msym(o))
> (scatter pprobg3 week if _fillin==1, msym(o) mcol(black))
> if newid<13 & treatment==1, by(newid)
> legend(order(1 "Prob(y>1)" 2 "Prob(y>2)" 3 "Prob(y>3)") rows(1)) xtitle("Week")
```

Here predictions for occasions where the response variable is missing are plotted as dots. The resulting graphs, shown in figure 11.14, illustrate the considerable variability in the course of schizophrenia over time between patients within the treatment groups.

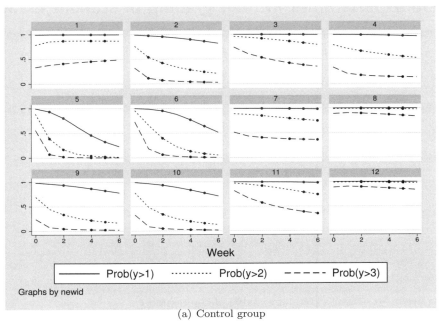

(a) Control group

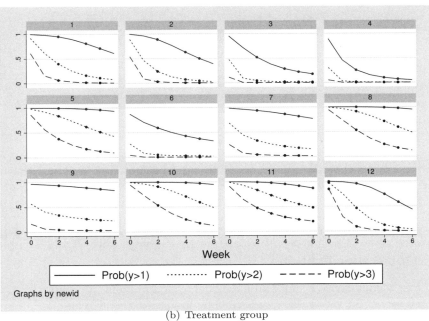

(b) Treatment group

Figure 11.14: Posterior mean cumulative probabilities for 12 patients in control group and 12 patients in treatment group versus week

11.9 Do experts differ in their grading of student essays?

We now consider data from Johnson and Albert (1999) on grades assigned to 198 essays by five experts. The grades are on a 10-point scale, with 10 being "excellent".

The variables in the dataset `essays.dta` used here are

- `essay`: identifier for essays (j)
- `grade`: grade on 10-point scale
- `grader`: identifier of expert grader (i)

We read in the data by typing

```
. use http://www.stata-press.com/data/mlmus3/essays, clear
```

The data can be considered as interrater reliability data. They are similar to the peak-expiratory-flow data considered in chapter 2 except that we have five experts instead of two methods and no replicate measurements for a given expert. However, the grades are on an ordinal scale instead of an interval scale.

It is useful to obtain a frequency table of the response:

```
. tabulate grade
```

grade	Freq.	Percent	Cum.
1	106	10.71	10.71
2	145	14.65	25.35
3	115	11.62	36.97
4	127	12.83	49.80
5	135	13.64	63.43
6	100	10.10	73.54
7	103	10.40	83.94
8	159	16.06	100.00
Total	990	100.00	

This output shows that the top grades of 9 and 10 were never used and that the remaining eight grades all occur at similar frequencies.

11.10 A random-intercept probit model with grader bias

11.10.1 Model specification

We consider a cumulative ordinal probit model with a random intercept, $\zeta_j \sim N(0, \psi)$. The initial model for the grade y_{ij} assigned by grader i to essay j is

$$\Pr(y_{ij} > s \mid \zeta_j) = \Phi(\zeta_j - \kappa_s)$$

where $\Phi(\cdot)$ is the standard normal cumulative density function.

This model can also be written using the latent-response formulation, with the latent-response model and the threshold model specified as

$$y_{ij}^* = \zeta_j + \epsilon_{ij}, \qquad \zeta_j \sim N(0, \psi), \quad \epsilon_{ij}|\zeta_j \sim N(0, \theta) \qquad (11.5)$$

$$y_{ij} = s \qquad \text{if } \kappa_{s-1} < y_{ij}^* \le \kappa_s, \quad s = 1, \ldots, S \qquad (11.6)$$

respectively. When the threshold model is written in this compact form, we must also state that $\kappa_0 = -\infty$ and $\kappa_S = \infty$. This is a classical test-theory model for y_{ij}^*, where ζ_j represents the truth and ϵ_{ij} represents measurement error. The model assumes that all graders i measure the same truth ζ_j with the same measurement error variance θ and assign grades using the same thresholds κ_s ($s = 1, \ldots, S-1$).

At the least, we should allow for grader bias by including grader-specific intercepts β_i. However, one of the intercepts must be set to zero to identify all thresholds. Retaining the threshold model (11.6), we extend the latent-response model (11.5) to

$$y_{ij}^* = \zeta_j + \beta_2 x_{2i} + \beta_3 x_{3i} + \beta_4 x_{4i} + \beta_5 x_{5i} + \epsilon_{ij} \qquad (11.7)$$

where $\mathbf{x}_i = (x_{2i}, x_{3i}, x_{4i}, x_{5i})'$ are dummy variables for graders 2–5. The corresponding regression coefficients (β_2, β_3, β_4, and β_5) represent how much more generous or lenient graders 2–5 are than the first grader.

11.10.2   Estimation using gllamm

We start by creating dummy variables for the graders (called **grad1**–**grad5**) by typing

```
. quietly tabulate grader, generate(grad)
```

The random-intercept probit model with grader-specific bias can then be fit by maximum likelihood using **gllamm** with the **link(oprobit)** option:

```
. gllamm grade grad2-grad5, i(essay) link(oprobit) adapt

number of level 1 units = 990
number of level 2 units = 198

log likelihood = -1808.0929
```

	Coef.	Std. Err.	z	P>|z|	[95% Conf.	Interval]
grade						
grad2	-1.127551	.1123199	-10.04	0.000	-1.347694	-.9074085
grad3	-.6301196	.1102999	-5.71	0.000	-.8463033	-.4139358
grad4	-.673948	.1087182	-6.20	0.000	-.8870318	-.4608642
grad5	-1.283113	.1128503	-11.37	0.000	-1.504295	-1.06193
_cut11						
_cons	-2.760615	.145463	-18.98	0.000	-3.045717	-2.475512
_cut12						
_cons	-1.831267	.1308097	-14.00	0.000	-2.087649	-1.574885
_cut13						
_cons	-1.275763	.1246812	-10.23	0.000	-1.520133	-1.031392
_cut14						
_cons	-.7281207	.1203823	-6.05	0.000	-.9640658	-.4921757
_cut15						
_cons	-.1590002	.1183832	-1.34	0.179	-.3910269	.0730266
_cut16						
_cons	.3013533	.118667	2.54	0.011	.0687702	.5339364
_cut17						
_cons	.8585616	.1224719	7.01	0.000	.6185211	1.098602

```
Variances and covariances of random effects
------------------------------------------------------------------------------

***level 2 (essay)
    var(1): 1.4066991 (.18918365)
------------------------------------------------------------------------------
. estimates store model1
```

We see that grader 1 is the most generous because the estimated coefficients of the dummies for graders 2–5 are all negative. These graders are all significantly more stringent than grader 1 at, say, the 5% level.

11.11 ❖ Including grader-specific measurement error variances

11.11.1 Model specification

Although the above model accommodates grader bias, it is relatively restrictive because it still assumes that all graders i have the same measurement error variance θ. We can relax this homoskedasticity assumption by retaining the previous models (11.6) and (11.7) except that we now also allow each grader to have a grader-specific residual variance or measurement error variance θ_i:

$$\epsilon_{ij}|\mathbf{x}_i, \zeta_j \ \sim \ N(0, \theta_i)$$

In gllamm, this can be accomplished by specifying a linear model for the log standard deviation of the measurement errors:

$$\ln(\sqrt{\theta_i}) = \ln(\theta_i)/2 \ = \ \delta_2 x_{2i} + \delta_3 x_{3i} + \delta_4 x_{4i} + \delta_5 x_{5i} \tag{11.8}$$

In this model for level-1 heteroskedasticity, we have again omitted the dummy variable for the first grader, which amounts to setting the standard deviation of the measurement error for this grader to 1 because $\exp(0) = 1$. A constraint like this is necessary to identify the model because all thresholds κ_s $(s = 1, \dots, S)$ are freely estimated (see section 11.2.4). In terms of the above parameterization, the measurement error variance θ_i for grader i (apart from grader 1) becomes $\exp(2\delta_i)$.

In this model, each grader has his or her own mean and variance,

$$y_{ij}^*|\zeta_j \sim N(\zeta_j + \beta_i, \theta_i), \quad \beta_1 = 0$$

but applies the same thresholds to the latent responses to generate the observed ratings. The cumulative probabilities are

$$\Pr(y_{ij} > s \,|\, \zeta_j) \ = \ \Pr(y_{ij}^* > \kappa_s \,|\, \zeta_j) \ = \ \Pr \left(\frac{y_{ij}^* - \zeta_j - \beta_i}{\sqrt{\theta_i}} > \frac{\kappa_s - \zeta_j - \beta_i}{\sqrt{\theta_i}} \right)$$

$$= \ \Phi \left(\frac{\zeta_j + \beta_i - \kappa_s}{\sqrt{\theta_i}} \right)$$

This model can be thought of as a generalized linear model with a *scaled probit link*, where the scale parameter $\sqrt{\theta_i}$ differs between graders, i.

11.11.2   Estimation using gllamm

To specify model (11.8) for the log standard deviations of the measurement errors in gllamm, we need to first define a corresponding equation:

```
. eq het: grad2-grad5
```

(gllamm will not add an intercept to this equation for heteroskedasticity). Then we pass this equation on to gllamm using the s() option, change the link to soprobit (for scaled ordinal probit), and use the previous estimates as starting values:

```
. matrix a = e(b)
. gllamm grade grad2-grad5, i(essay) link(soprobit) s(het) from(a) adapt
number of level 1 units = 990
number of level 2 units = 198

log likelihood = -1767.6284
```

	Coef.	Std. Err.	z	P>|z|	[95% Conf. Interval]	
grade						
grad2	-1.372239	.1625285	-8.44	0.000	-1.690789	-1.053689
grad3	-.6924764	.183522	-3.77	0.000	-1.052173	-.33278
grad4	-.8134375	.1183584	-6.87	0.000	-1.045416	-.5814594
grad5	-1.552205	.1632283	-9.51	0.000	-1.872127	-1.232284
_cut11						
_cons	-3.443449	.3145287	-10.95	0.000	-4.059914	-2.826984
_cut12						
_cons	-2.248858	.2234508	-10.06	0.000	-2.686814	-1.810902
_cut13						
_cons	-1.547513	.1784254	-8.67	0.000	-1.897221	-1.197806
_cut14						
_cons	-.8606409	.1468759	-5.86	0.000	-1.148512	-.5727694
_cut15						
_cons	-.1507223	.1338984	-1.13	0.260	-.4131584	.1117138
_cut16						
_cons	.420217	.1403799	2.99	0.003	.1450775	.6953565
_cut17						
_cons	1.120534	.1679034	6.67	0.000	.7914498	1.449619

```
Variance at level 1
-----------------------------------------------------------------------------

    equation for log standard deviation:

    grad2: .24143083 (.11558242)
    grad3: .72935227 (.1063642)
    grad4: -.09545464 (.13094183)
    grad5: .04306859 (.12477309)

Variances and covariances of random effects
-----------------------------------------------------------------------------


***level 2 (essay)

    var(1): 2.0587449 (.41200888)
-----------------------------------------------------------------------------
```

We see that grader 3 has the largest estimated measurement error variance of $\exp(0.729)^2 = 4.30$.

We can use a likelihood-ratio test to test the null hypothesis that the measurement error variances are identical for the graders, H_0: $\theta_i = \theta$, against the alternative that the measurement error variances are different for at least two graders:

```
. estimates store model2
. lrtest model1 model2
Likelihood-ratio test                              LR chi2(4)  =     80.93
(Assumption: model1 nested in model2)              Prob > chi2 =    0.0000
```

There is strong evidence to suggest that the graders differ in their measurement error variances.

11.12 ❖ Including grader-specific thresholds

11.12.1 Model specification

Model (11.7), which allows for grader bias by including grader-specific intercepts β_i, can equivalently be specified by omitting the β_i from the latent regression model and instead defining grader-specific thresholds

$$\kappa_{si} = \kappa_s - \beta_i$$

Here all the thresholds of a given grader are translated relative to the thresholds of another grader.

A final model extension would be to allow different graders to apply a different set of thresholds that are not just shifted or translated by a constant between graders:

$$\kappa_{si} = \alpha_{s1} + \alpha_{s2}x_{2i} + \cdots + \alpha_{s5}x_{5i} \tag{11.9}$$

The model now becomes

$$\Pr(y_{ij} > s \mid \zeta_j) = \Phi\left(\frac{\zeta_j - \kappa_{si}}{\sqrt{\theta_i}}\right) = \Phi\left(\frac{\zeta_j}{\sqrt{\theta_i}} - \frac{\kappa_{si}}{\sqrt{\theta_i}}\right)$$

Here $\sqrt{\theta_i}$ is identified because it determines the effect of ζ_j on the response probabilities for the individual raters. In fact, $1/\sqrt{\theta_i}$ can be interpreted as a discrimination parameter or factor loading, and the model is equivalent to a specific item-response model (see exercise 11.2) called Samejima's graded-response model (see Embretson and Reise [2000] for a discussion of this model). Because the thresholds are free parameters, dividing them by $\sqrt{\theta_i}$ does not modify the threshold model (the ratios $\kappa_{si}/\sqrt{\theta_i}$ can take the same set of values as the original thresholds κ_{si}).

11.12.2 Estimation using gllamm

In gllamm, we can specify model (11.9) for the thresholds by first defining an equation:

```
. eq thr: grad2-grad5
```

The intercept α_{s1} will be added by gllamm to the equation for the thresholds and does not need to be specified. This equation is then passed to gllamm using the thresh() option (beware that it takes a long time to fit the model).

```
. gllamm grade, i(essay) link(soprobit) s(het) thresh(thr) from(a) adapt skip
number of level 1 units = 990
number of level 2 units = 198

log likelihood = -1746.814
```

grade	Coef.	Std. Err.	z	P>|z|	[95% Conf. Interval]	
_cut11						
grad2	1.728073	.2924749	5.91	0.000	1.154833	2.301314
grad3	1.061355	.3765653	2.82	0.005	.3233002	1.799409
grad4	.6380823	.3413275	1.87	0.062	-.0309073	1.307072
grad5	1.277385	.3128488	4.08	0.000	.6642124	1.890557
_cons	-3.170991	.3363843	-9.43	0.000	-3.830292	-2.51169
_cut12						
grad2	1.654293	.2124865	7.79	0.000	1.237827	2.070759
grad3	1.237556	.2612418	4.74	0.000	.7255312	1.74958
grad4	.9826652	.2164376	4.54	0.000	.5584553	1.406875
grad5	1.53219	.21934	6.99	0.000	1.102292	1.962089
_cons	-2.269541	.240086	-9.45	0.000	-2.740101	-1.798981
_cut13						
grad2	1.668863	.1944094	8.58	0.000	1.287828	2.049899
grad3	1.088168	.2214152	4.91	0.000	.6542024	1.522134
grad4	.8398826	.1800113	4.67	0.000	.4870669	1.192698
grad5	1.738926	.2082086	8.35	0.000	1.330847	2.147009
_cons	-1.658147	.192857	-8.60	0.000	-2.03614	-1.280162
_cut14						
grad2	1.236493	.1748537	7.07	0.000	.8937858	1.5792
grad3	.7111893	.1962876	3.62	0.000	.3264727	1.095906
grad4	.8942423	.1603884	5.58	0.000	.5798868	1.208598
grad5	1.601494	.2126157	7.53	0.000	1.184775	2.018213
_cons	-.8794996	.156047	-5.64	0.000	-1.185346	-.573653
_cut15						
grad2	1.054329	.1857238	5.68	0.000	.6903165	1.418341
grad3	.4492806	.1922589	2.34	0.019	.07246	.8261011
grad4	.8514869	.1742843	4.89	0.000	.509896	1.193078
grad5	1.664346	.2603403	6.39	0.000	1.154089	2.174604
_cons	-.1793809	.1425642	-1.26	0.208	-.4588016	.1000399
_cut16						
grad2	.8328646	.2000713	4.16	0.000	.440732	1.224997
grad3	.3282962	.2088497	1.57	0.116	-.0810417	.7376341
grad4	.7664851	.1991761	3.85	0.000	.3761071	1.156863
grad5	1.409385	.2812254	5.01	0.000	.8581937	1.960577
_cons	.4440509	.1446484	3.07	0.002	.1605452	.7275565
_cut17						
grad2	.466475	.2244542	2.08	0.038	.0265529	.9063971
grad3	.1412568	.2524279	0.56	0.576	-.3534927	.6360063
grad4	.5492535	.2395074	2.29	0.022	.0798276	1.018679
grad5	1.233665	.3404629	3.62	0.000	.5663696	1.90096
_cons	1.233925	.1691786	7.29	0.000	.9023413	1.565509

```
Variance at level 1
----------------------------------------------------------------------------

     equation for log standard deviation:

     grad2: -.20109559 (.16420779)
     grad3: .45238552 (.15515942)
     grad4: -.13881034 (.1708705)
     grad5: .03002795 (.16735946)

Variances and covariances of random effects
----------------------------------------------------------------------------


***level 2 (essay)

     var(1): 1.676465 (.40806705)
----------------------------------------------------------------------------
```

Here the coefficient [_cut16]_cons is the estimate of $\kappa_{61} = \alpha_{61}$, the sixth threshold for grader 1, whereas [_cut16]grad2 is the estimate of α_{62}, the difference between the sixth thresholds for graders 2 and 1.

A likelihood-ratio test again suggests that the more elaborate model is preferred:

```
. estimates store model3

. lrtest model2 model3
Likelihood-ratio test                              LR chi2(24) =     41.63
(Assumption: model2 nested in model3)              Prob > chi2 =     0.0142
```

However, it may not be necessary to relax the parallel-regressions assumption for all graders. For instance, we can consider the null hypothesis that all thresholds for grader 2 are simply translated by a constant relative to those of grader 1:

$$H_0\text{: } \alpha_{12} = \alpha_{22} = \cdots = \alpha_{72}$$

A multivariate Wald test of this null hypothesis can be performed using the **test** command,

```
. test [_cut11=_cut12=_cut13=_cut14=_cut15=_cut16=_cut17]: grad2
 ( 1)   [_cut11]grad2 - [_cut12]grad2 = 0
 ( 2)   [_cut11]grad2 - [_cut13]grad2 = 0
 ( 3)   [_cut11]grad2 - [_cut14]grad2 = 0
 ( 4)   [_cut11]grad2 - [_cut15]grad2 = 0
 ( 5)   [_cut11]grad2 - [_cut16]grad2 = 0
 ( 6)   [_cut11]grad2 - [_cut17]grad2 = 0
          chi2(  6) =     25.60
        Prob > chi2 =      0.0003
```

and similarly for other pairs of graders, keeping in mind that we may have to correct for multiple testing (for instance, using the Bonferroni method by dividing the significance level of each test by the number of tests).

The **thresh()** option can be used more generally to relax the parallel-regressions assumption of constant covariate effects across categories s. As explained in section 11.2.3, this assumption corresponds to the proportional odds assumption in ordinal logit models. To relax the parallel-regressions assumption for a covariate x_{2i}, we can simply move it to the threshold model (11.9) to estimate $S - 1$ threshold parameters α_{s2} $(s = 1, \ldots, S - 1)$ instead of one regression parameter β_2. We cannot estimate both β_2 and the α_{s2} because such a model would not be identified.

Maximum likelihood estimates for the sequence of models developed in this section are given in table 11.2, using the notation for the final model. For "Model 1", the measurement error variances are set to 1, and the thresholds of all graders are set to be equal; that is, in models (11.8) and (11.9), the constraints

$$\delta_i = 0 \quad \text{and} \quad \alpha_{si} = 0, \quad i = 2, 3, 4, 5$$

are in place, but the intercepts β_i for raters 2–5 are free parameters. "Model 2" is the same except that the constraints for the δ_i are relaxed. In "Model 1" and "Model 2", $\alpha_{s1} \equiv \kappa_s$. Finally, in "Model 3" the constraints for both the δ_i and the α_{si} are relaxed, but the intercepts are set to zero: $\beta_i = 0$.

To facilitate comparison of estimates between models, we also report estimates of the "reduced-form" thresholds:

$$\frac{\kappa_{si}}{\sqrt{\theta_i}} = \begin{cases} \frac{\alpha_{s1} - \beta_i}{\exp(\delta_i)} & \text{for } i = 1 \\ \frac{\alpha_{s1} + \alpha_{si} - \beta_i}{\exp(\delta_i)} & \text{for } i > 1 \end{cases}$$

Table 11.2: Maximum likelihood estimates for essay grading data (for models 1 and 2, $\alpha_{s1} \equiv \kappa_s$)

Grader i	Model 1					Model 2					Model 3				
	1	2	3	4	5	1	2	3	4	5	1	2	3	4	5
β_i	0	−1.1	−0.6	−0.7	−1.3	0	−1.4	−0.7	−0.8	−1.6	0	0	0	0	0
α_{1i}	−2.8	0	0	0	0	−3.4	0	0	0	0	−3.2	1.7	1.1	0.6	1.3
α_{2i}	−1.8	0	0	0	0	−2.2	0	0	0	0	−2.3	1.7	1.2	1.0	1.5
α_{3i}	−1.3	0	0	0	0	−1.5	0	0	0	0	−1.7	1.7	1.1	0.8	1.7
α_{4i}	−0.7	0	0	0	0	−0.9	0	0	0	0	−0.9	1.2	0.7	0.9	1.6
α_{5i}	−0.2	0	0	0	0	−0.2	0	0	0	0	−0.2	1.1	0.4	0.9	1.7
α_{6i}	0.3	0	0	0	0	0.4	0	0	0	0	0.4	0.8	0.3	0.8	1.4
α_{7i}	0.9	0	0	0	0	1.1	0	0	0	0	1.2	0.5	0.1	0.5	1.2
δ_i	0	0	0	0	0	0	0.2	0.7	−0.1	0.0	0	−0.2	0.5	−0.1	0.0
$\frac{\alpha_{1i}-\beta_i}{\exp(\delta_i)}$	−2.8	−1.6	−2.1	−2.1	−1.5	−3.4	−1.6	−1.3	−2.9	−1.8	−3.2	−1.8	−1.3	−2.9	−1.8
$\frac{\alpha_{1i}+\alpha_{2i}-\beta_i}{\exp(\delta_i)}$	−1.8	−0.7	−1.2	−1.2	−0.5	−2.4	−0.7	−0.8	−1.6	−0.7	−2.3	−0.8	−0.7	−1.5	−0.7
$\frac{\alpha_{1i}+\alpha_{3i}-\beta_i}{\exp(\delta_i)}$	−1.3	−0.1	−0.6	−0.6	0.0	−1.5	−0.1	−0.4	−0.8	0.0	−1.7	0.0	−0.4	−0.9	0.0
$\frac{\alpha_{1i}+\alpha_{4i}-\beta_i}{\exp(\delta_i)}$	−0.7	0.4	−0.1	−0.1	0.6	−0.9	0.4	−0.1	−0.1	0.7	−0.9	0.4	−0.1	0.0	0.7
$\frac{\alpha_{1i}+\alpha_{5i}-\beta_i}{\exp(\delta_i)}$	−0.2	1.0	0.5	0.5	1.1	−0.2	1.0	0.3	0.7	1.3	−0.2	1.1	0.2	0.8	1.4
$\frac{\alpha_{1i}+\alpha_{6i}-\beta_i}{\exp(\delta_i)}$	0.3	1.4	0.9	1.0	1.6	0.4	1.4	0.5	1.4	1.9	0.4	1.6	0.5	1.4	1.8
$\frac{\alpha_{1i}+\alpha_{7i}-\beta_i}{\exp(\delta_i)}$	0.9	2.0	1.5	1.5	2.1	1.1	2.0	0.9	2.1	2.6	1.2	2.1	0.9	2.0	2.4
ψ	1.41					2.06					1.68				
Log likelihood	−1808.09					−1767.63					−1746.81				

11.13 ❖ Other link functions

In this chapter, we have considered cumulative logit and probit models for ordinal responses. In both models, the link function (logit or probit) is applied to the probability of being above a given category s. The linear predictor then includes a category-specific intercept $-\kappa_s$ but the coefficients of covariates are assumed constant across categories. With two covariates and no random effects, the linear predictor for category s can be written as

$$\nu_{is} = \beta_2 x_{2i} + \beta_3 x_{3i} - \kappa_s \tag{11.10}$$

This parallel-regressions assumption (that only the intercept differs between categories) translates into the proportional odds assumption if a logit link is used.

Cumulative complementary log-log model

In addition to logit and probit links, we could also apply a complementary log-log link to the probability of being above category s. The link function can be written as

$$\ln[-\ln\{1 - \Pr(y > s)\}]$$

(where we have omitted the i subscript and the conditioning on covariates, as we will do whenever possible in this section). Cumulative complementary log-log link models can be written as latent-response models if $-\epsilon_i$ has a Gumbel or type I extreme value distribution with variance $\pi^2/6$. This distribution is not symmetric, so unlike the logit and probit counterparts, reversing the ordinal scale will change the log likelihood and not simply reverse the sign of the regression coefficients. An appealing property of all cumulative models is that the interpretation of the regression coefficients does not change when groups of adjacent categories are merged. Without random effects, the model can be fit by using the user-contributed command `oglm` (Williams 2010), and with or without random effects using `gllamm` with the `link(ocll)` option. See sections 14.6 and 14.7 for more information on the complementary log-log link function but for binary responses.

Continuation-ratio logit model

In addition to cumulative logit models, there are several other models for ordinal data based on the logit link. In a continuation-ratio logit model, the linear predictor in (11.10) is set equal to the log odds of stopping at a given category s, given that the individual reached at least category s,

$$
\begin{aligned}
\ln\{\text{Odds}(y = s | y \geq s)\} &= \text{logit}\{\Pr(y = s | y \geq s)\} \\
&= \ln\left\{\frac{\Pr(y = s | y \geq s)}{\Pr(y \neq s | y \geq s)}\right\} = \ln\left\{\frac{\Pr(y = s)}{\Pr(y > s)}\right\} \tag{11.11}
\end{aligned}
$$

This model is often used for discrete-time survival (see chapter 14), such as the number of semesters until completion of a degree. Then $\Pr(y = s | y \geq s)$, the probability of completing the degree in semester s given that the student has studied at least s semesters, is called a discrete-time hazard. If a student has not yet completed a degree during the observation period, his or her response is said to be right-censored, and this is easily handled with the continuation-ratio logit link.

If a complementary log-log link is specified for $\Pr(y = s | y \geq s)$ instead of a logit link, the model is equivalent to an interval-censored survival model where the underlying continuous survival time follows a proportional hazards model (see section 14.6).

The model can also be used for any sequential process where a given category can be reached only after reaching the previous category, such as educational attainment ("no high school", "high school", "some college", "college degree"), or stages of development or disease, such as cancer stages (0, I, II, III, and IV). See exercise 14.8 for an example.

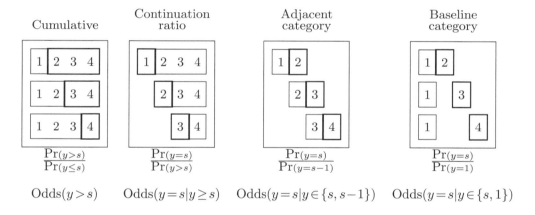

Figure 11.15: Relevant odds for different logit link models for ordinal responses. Events corresponding to "success" are in thick frames, and events corresponding to "failure" are in thin frames—when not all categories are shown, the odds are conditional on the response being in one of the categories shown [adapted from Brendan Halpin's web notes on "Models for ordered categories (ii)"].

The odds involved in the continuation-ratio logit model with four categories are shown in the second column of figure 11.15. The odds are defined as the ratio of two probabilities, as shown in (11.11). The numerator probability is $\Pr(y = s)$, so category s is enclosed in a thick frame for $s = 1, 2, 3$. The corresponding denominator probability is $\Pr(y > s)$, so categories above s are enclosed in a thin frame. The odds are conditional on y being equal to one of the categories that is displayed.

Adjacent-category logit model

Another logit model for ordinal responses is the adjacent-category logit model. Here the linear predictor is equated with

$$\ln\{\text{Odds}(y = s | y \in \{s, s-1\})\} = \text{logit}\{\Pr(y = s | y \in \{s, s-1\})\} = \ln\left\{\frac{\Pr(y = s)}{\Pr(y = s-1)}\right\}$$

A diagram illustrating this model is given in the third row of figure 11.15 for four categories. We see that the probability that $y = s$ (thick frame) is divided by the probability that $y = s - 1$ (thin frame) and that the odds should be interpreted as conditional on y being equal to either s or $s-1$. Again, a parallel-regressions assumption is made by constraining the regression coefficients of covariates to be the same for all s. Zheng and Rabe-Hesketh (2007) show how adjacent-category logit models can be fit in gllamm with the link(mlogit) option.

As discussed for cumulative logit models in section 11.12, the parallel-regressions assumption made by all models discussed so far can be relaxed for some covariates.

Baseline-category logit and stereotype models

The final model displayed in figure 11.15 is the baseline category logit model. In this model, the odds for category s is the ratio of the probability of that category divided by the probability of the baseline category, here category $s = 1$. In this model, it is not assumed that covariates have the same coefficient for each category s. With two covariates, the model can be written as

$$\ln\left\{\frac{\Pr(y_i = s | x_i)}{\Pr(y_i = 1 | x_i)}\right\} = \beta_1^{[s]} + \beta_2^{[s]} x_{2i} + \beta_3^{[s]} x_{3i}$$

This looks like $s - 1$ binary logistic regression models for subsets of the data comprising the category of interest and the baseline category (but some efficiency is gained by fitting the model to all data jointly). Because the model does not impose any constraints on the coefficients, it can be used for nominal or unordered categorical responses (see chapter 12). Baseline category logit models, often simply called multinomial logit models, can be fit in mlogit without random effects and in gllamm with the link(mlogit) option with or without random effects (see chapter 12).

An ordinal version of the multinomial logit model is the stereotype model where category-specific regression coefficients are constrained as follows:

$$\beta_2^{[s]} = \alpha^{[s]}\beta_2 \qquad \beta_3^{[s]} = \alpha^{[s]}\beta_3$$

The ratios of regression coefficients $\beta_k^{[s]}/\beta_k^{[s']} = \alpha^{[s]}/\alpha^{[s']}$ for all pairs of categories s and s' are then the same for all covariates x_{ki}. If the relationship between covariates and the response variable is ordinal, $\alpha^{[1]} \geq \alpha^{[2]} \geq \cdots \geq \alpha^{[S]}$. The Stata command slogit can be used to fit stereotype models without random effects, but at the time of writing this book we are not aware of any Stata command for stereotype models with random effects.

11.14 Summary and further reading

We started by introducing cumulative models for ordinal responses, with special emphasis on the proportional odds model. Both the generalized linear and the latent-response formulations of the model were outlined, and the problem of identification in ordinal regression models was discussed. Finally, we discussed various extensions to cumulative models for ordinal responses. The most important is relaxing the parallel-regressions or proportional odds assumption in cumulative models, as shown in (11.9), using the `thresh()` option in `gllamm` (see also exercise 11.7, for which solutions are available).

Cumulative models are sometimes applied to discrete-time duration or survival data. However, this is appropriate only if there is no left- or right-censoring or if the model is extended to handle censoring (see Rabe-Hesketh, Yang, and Pickles [2001b] and Skrondal and Rabe-Hesketh [2004, sec. 12.3]), which is possible using composite links in `gllamm`. More common approaches to modeling discrete-time survival data are discussed in chapter 14.

Other models for ordinal responses were briefly described in section 11.13, and some of these will be discussed in detail in chapters 12 and 14 on nominal or unordered categorical responses and discrete-time survival, respectively. For introductions to these models (mostly without random effects), see for instance, Fahrmeir and Tutz (2001), Agresti (2002, 2010), Long (1997, chap. 5), and Greenland (1994).

Generalized estimating equations (GEE) can be used for ordinal responses, but this is currently not implemented in Stata. However, GEE with an independence structure is easily implemented by fitting an ordinary cumulative logit model and using robust standard errors for clustered data. It is unfortunately not possible to use conditional maximum likelihood estimation for ordinal models with fixed cluster-specific intercepts, in contrast to the dichotomous case.

Useful books discussing random-effects models for ordinal responses include Agresti (2010), Hedeker and Gibbons (2006), and Johnson and Albert (1999), the latter adopting a Bayesian approach. We also recommend the book chapter by Rabe-Hesketh and Skrondal (2009), the encyclopedia entry by Hedeker (2005), and the review article by Agresti and Natarajan (2001). Skrondal and Rabe-Hesketh (2004, chap. 10) analyze several datasets with ordinal responses by using various approaches, including growth-curve modeling.

The exercises of this chapter are based on longitudinal data on children's respiratory status (exercise 11.1), children's recovery after surgery (exercise 11.7), and adults' attitudes to abortion (exercise 11.5); repeated measures data on verbal aggression (exercise 11.2), essay grades (exercise 11.4), and wine assessments by different raters (exercise 11.8); data from a cluster-randomized trial of classroom interventions to prevent smoking (exercise 11.3); cross-sectional data on first graders' math proficiency (exercise 11.9); and on married spouse's closeness to each other (exercise 11.6). Exercise 11.7, for which solutions are provided, involves relaxing the proportional odds assumption.

11.15   Exercises

11.1 Respiratory-illness data

Koch et al. (1990) analyzed data from a clinical trial comparing two treatments for respiratory illness. In each of two centers, eligible patients were randomized to active treatment or placebo. The respiratory status was determined prior to randomization and during four visits after randomization. The dichotomized response was analyzed by Davis (1991) and Everitt and Pickles (2004), among others. Here we analyze the original ordinal response.

The dataset `respiratory.dta` has the following variables:

- `center`: center (1 or 2)
- `patient`: patient identifier
- `drug`: treatment group (1: active treatment; 0: placebo)
- `male`: dummy variable for patient being male
- `age`: age of patient in years
- `bl`: respiratory status at baseline (0: terrible; 1: poor; 2: fair; 3: good; 4: excellent)
- `v1` to `v4`: respiratory status at visits 1–4 (0: terrible; 1: poor; 2: fair; 3: good; 4: excellent)

1. Reshape the data by stacking the responses for visits 1–4 into a single variable.
2. Fit a proportional odds model with `drug`, `male`, `age`, and `bl` as covariates and a random intercept for patients.
3. Check whether there is a linear trend for the logits over time (after controlling for the patient-specific covariates) and whether the slope differs between treatment groups.
4. For your chosen model, plot the model-implied posterior mean cumulative probabilities over time for some of the patients.

11.2 Verbal-aggression data

Consider the data in `aggression.dta` from Vansteelandt (2000) and De Boeck and Wilson (2004), described in exercise 10.4. Use five-point adaptive quadrature to speed up estimation, which will be slow.

1. Fit the following explanatory item-response model for the original ordinal response y_{ij} (0: no; 1: perhaps; 2: yes):

$$\text{logit}\{\Pr(y_{ij} > s | \mathbf{x}_{ij}, \zeta_j)\} = \beta_2 x_{2ij} + \beta_3 x_{3ij} + \beta_4 x_{4ij} + \beta_5 x_{5ij} + \zeta_j - \kappa_s$$

where $s = 0, 1$ and $\zeta_j \sim N(0, \psi)$ can be interpreted as the latent trait "verbal aggressiveness". Item-response models with cumulative logit or probit links for ordinal data are called graded-response models. Interpret the estimated coefficients.

2. Now extend the above model by including a latent regression, allowing verbal aggressiveness (now denoted η_j instead of ζ_j) to depend on the personal characteristics w_{1j} and w_{2j}:

$$\text{logit}\{\Pr(y_{ij} > s | \mathbf{x}_{ij}, \eta_j)\} = \beta_2 x_{2ij} + \beta_3 x_{3ij} + \beta_4 x_{4ij} + \beta_5 x_{5ij} + \eta_j - \kappa_s$$

$$\eta_j = \gamma_1 w_{1j} + \gamma_2 w_{2j} + \zeta_j$$

Substitute the level-2 model for η_j into the level-1 model for the item responses, and fit the resulting model.

3. ❖ Relax the proportional odds assumption for x_{2ij} (*do* versus *want*). Interpret the estimates, and perform a likelihood-ratio test to assess whether there is any evidence that the proportional odds assumption is violated for this variable.

11.3 Smoking-intervention data

Gibbons and Hedeker (1994) and Hedeker and Gibbons (2006) analyzed data from a subset of the Television School and Family Smoking Prevention and Cessation Project (TVSFP) (see Flay et al. [1988]).

Schools were randomized to one of four conditions defined by different combinations of two factors: 1) TV, a media (television) intervention (1: present; 0: absent), and 2) CC, a social-resistance classroom curriculum (1: present; 0: absent). The outcome measure is the tobacco and health knowledge scale (THKS) score, defined as the number of correct answers to seven items on tobacco and health knowledge. This variable has been collapsed into four ordinal categories.

Students are nested in classes, which are nested in schools. We consider two-level models in this exercise but will revisit the data for three-level modeling in exercise 16.4.

The dataset `tvsfpors.dta` has the following variables:

- `school`: school identifier
- `class`: class identifier
- `thk`: ordinal THKS score postintervention (four categories)
- `prethk`: ordinal THKS score preintervention (four categories)
- `cc`: social-resistance classroom curriculum (dummy variable)
- `tv`: television intervention (dummy variable)

1. Investigate how tobacco and health knowledge is influenced by the interventions by fitting a two-level, random-intercept proportional odds model with students nested in schools. Include the covariates `cc`, `tv`, and their interaction, and control for `prethk`.

   a. Obtain estimated odds ratios for the interventions and interpret these.
   b. Calculate the estimated residual intraclass correlation for the latent responses underlying the observed ordinal response.

2. Fit a two-level, random-intercept proportional odds model with students nested in classes (instead of schools).

   a. Obtain estimated odds ratios for the interventions and interpret these.
   b. Calculate the estimated residual intraclass correlation for the latent responses underlying the observed ordinal response.
   c. Does class or school appear to induce more dependence among students?

See also exercise 16.4 for further analyses of these data.

11.4 Essay-grading data

Here we consider the dataset from Johnson and Albert (1999) that was analyzed in section 11.9.

The dataset `essays.dta` has the following variables:

- `essay`: identifier for essays
- `grade`: grade on 10-point scale
- `grader`: identifier of expert grader
- `wordlength`: average word length
- `sqrtwords`: square root of the number of words in the essay
- `commas`: number of commas $\times$ 100 divided by number of words
- `errors`: percentage of spelling errors
- `prepos`: percentage of prepositions
- `sentlength`: average sentence length

1. For simplicity, collapse the variable `grade` into four categories: {1,2}, {3,4}, {5,6}, and {7,8,9,10}. Fit an ordinal probit model without covariates and with a random intercept for essays. Obtain the estimated intraclass correlation for the latent responses.

2. Include dummy variables for graders 2–5 to allow some graders to be more generous in their grading than others. Does this model fit better?

3. Include the six essay characteristics as further covariates. Interpret the estimated coefficients.

4. Extend the model to investigate whether the graders differ in the importance they attach to the length of the essay (`sqrtwords`). Discuss your findings.

11.5 Attitudes-to-abortion data

Wiggins et al. (1990) analyzed data from the British Social Attitudes (BSA) survey. All adults aged 18 or over and living in private households in Britain were eligible to participate. A multistage sample was drawn, and in this dataset we have identifiers for districts that were drawn at stage 2. Respondents were sampled in stage 3. A subset of the respondents in the 1983 survey were followed up each year until 1986. Here we analyze the subset of respondents with complete data at all four waves that can be downloaded from the webpage of the Centre for Multilevel Modelling at the University of Bristol.

The respondents were asked for each of seven circumstances whether abortion should be allowed by law. The circumstances included "The woman decides on her own that she does not wish to have the child" and "The woman became pregnant as a result of rape". The variables in the dataset `abortion.dta` are

- `district`: district identifier
- `person`: subject identifier
- `year`: year (1–4)
- `score`: number of items (circumstances) where respondents answered "yes" to the question if abortion should be allowed by law (0–7)
- `male`: dummy variable for being male
- `age`: age in years
- `religion`: religion (1: catholic; 2: protestant; 3: other; 4: none)
- `party`: party chosen (1: conservative; 2: labour; 3: liberal; 4: other; 5: none)
- `class`: self-assessed social class (1: middle class; 2: upper working class; 3: lower working class)

1. Recode the variable `score` to merge the relatively rare responses 0, 1, and 2 into one category.

2. Fit an ordinal logistic regression model with the recoded `score` variable as the response and with `male`, `age`, and dummy variables for `religion` and `year` as covariates. Use the `vce(cluster person)` option to obtain appropriate standard errors. Obtain odds ratios and interpret them.

3. Now include a normally distributed random intercept for subjects in the above model. You can use the `nip(5)` option to speed up estimation and get relatively accurate estimates.

   a. Is there any evidence for residual between-subject variability in attitudes to abortion?

   b. Compare the estimated odds ratios for this model with the odds ratios for the model not including a random intercept. Explain why they differ the way they do.

   c. Obtain the estimated residual intraclass correlation for the latent responses.

   d. The model does not take the clustering of subjects within districts into account. The `cluster()` option can be used to obtain standard errors based on the sandwich estimator that take clustering into account. Either include the `cluster(district)` option in the `gllamm` command when fitting the model or fit the model without this option, and then issue the command

   ```
   gllamm, cluster(district)
   ```

11.6 Marriage data

Kenny, Kashy, and Cook (2006) analyzed data from Acitelli (1997) on 148 married couples. Both husbands and wives were asked to rate their closeness to their spouse, their commitment to the marriage, and their satisfaction with the marriage. The length of the marriage at the time of data collection was also recorded.

The variables in the dataset `marriage.dta` are

- `couple`: couple identifier
- `husband`: dummy variable for being the husband
- `close`: closeness, rated from 1 to 4 (from less close to closer)
- `commit`: commitment, rated from 1 to 4
- `satis`: satisfaction, rated from 1 to 4
- `lmarr`: length of marriage in years

1. Fit an ordinal probit model for the variable `close` for both spouses including a random intercept for couples and a dummy variable for the wife. Assume that husbands and wives have the same threshold parameters.

2. Obtain the estimated residual intraclass correlation for the latent responses.

3. Investigate whether length of marriage has an impact on closeness, allowing for different regression coefficients for husbands and wives.

   a. Write down the model using the latent-response formulation.
   b. Fit the model.

4. Interpret the estimated odds ratios.

5. State the following null hypotheses in terms of the model parameters and conduct hypothesis testing (with two-sided alternatives) using Stata:

   a. The coefficient of length of marriage is zero for husbands.
   b. The coefficient of length of marriage is zero for wives.
   c. The coefficients of length of marriage are zero both for husbands and wives (jointly).
   d. The coefficients of length of marriage are equal for husbands and wives.

6. ❖ In the model from step 3, relax the assumption that the residual variance in the latent-response formulation is the same for husbands and wives (see section 11.11 on using the `s()` option in `gllamm`). Use a likelihood-ratio test to compare this model with the model from step 3.

11.7 Recovery after surgery data [Solutions]

Davis (1991, 2002) analyzed data from a clinical trial comparing the effects of different dosages of anesthetic on postsurgical recovery. Sixty young children undergoing outpatient surgery were randomly assigned to four groups of size 15, receiving 15, 20, 25, and 30 milligrams of anesthetic per kilogram of body weight. Recovery was assessed upon admission to the recovery room (0 minutes) and 5,

15, and 30 minutes after admission to the recovery room. The recovery score was an ordinal variable with categories 1 (least favorable) through 6 (most favorable).

The variables in `recovery.dta` are

- `dosage`: dosage of anesthetic in milligram per kilogram bodyweight
- `id`: subject identifier, taking values 1–15 in each dosage group
- `age`: age of child in months
- `duration`: duration of surgery in minutes
- `score1`, `score2`, `score3`, `score4`: recovery scores 0, 5, 15, and 30 minutes after admission to the recovery room

1. Reshape the data to long form, stacking the recovery scores at the four occasions into a single variable and generating an identifier, `occ`, for the four occasions. (You can specify several variables in the `i()` option of the `reshape` command if one variable does not uniquely identify the individuals.) Recode the recovery score to four categories (to simplify some of the commands below) by merging {0,1}, {2,3}, and {4,5} and calling the new categories 1, 2, 3, and 4.

2. Construct a variable, `time`, taking the values 0, 5, 15, and 30 at the four occasions. Fit a random-intercept proportional odds model with dummy variables for the dosage groups, `age`, `duration`, and `time` as covariates. (Make sure there are 60 level-2 clusters.)

3. Compare the model from step 2 with a model including `dosage` as a continuous covariate instead of the dummy variables for dosage groups, using a likelihood-ratio test at the 5% significance level.

4. Extend the model chosen in step 3 to include an interaction between `dosage` and `time`. Test the interaction using a Wald test at the 5% level of significance.

5. For the model selected in step 4, interpret the estimated odds ratios and random-intercept variance.

6. ❖ Extend the model selected in step 4 by relaxing the proportional odds assumption for dosage (see section 11.12 on using the `thresh()` option in `gllamm` to relax proportional odds). Test whether the odds are proportional by using a likelihood-ratio test.

7. For age equal to 37 months, duration equal to 80 minutes, and time in recovery room equal to 15 minutes, produce a graph of predicted marginal probabilities similar to figure 11.13 for the model selected in step 6 or for the model selected in step 4. Also produce a stacked bar chart, treating dosage group as categorical.

11.8 Wine-tasting data

In exercise 10.9, data on judges' bitterness ratings of white wines were analyzed. There we considered a binary response, `bitter`, taking on the values 1: bitter and

0: nonbitter. In this exercise, we will also analyze the original ordinal variable, `bitterness`. Judges were asked to indicate the intensity of bitterness by making a mark on a horizontal line marked "none" at one end and "intense" at the other. The categories were derived by dividing the line into five intervals of equal length and assigning the values 1–5 according to the interval in which the mark was located.

The following model will be used to analyze `bitterness`:

$$\ln \left\{ \frac{\Pr(y_{ij} > s | x_{2ij}, x_{3ij}, \zeta_j)}{\Pr(y_{ij} \leq s | x_{2ij}, x_{3ij}, \zeta_j)} \right\} = \beta_2 x_{2ij} + \beta_3 x_{3ij} + \zeta_j - \kappa_s$$

where x_{2ij} is the temperature, `temp`, and x_{3ij} is a dummy variable, `contact`, for skin contact. It is assumed that $\zeta_j \sim N(0, \psi)$ is independent across judges and independent of x_{2ij} and x_{3ij}.

1. Fit the random-intercept proportional odds model given above and store the estimates.
2. Interpret the estimates.
3. Fit a random-intercept binary logistic regression model with response variable `bitter` that has the same linear predictor as above with $-\kappa_s$ replaced by β_1.
4. Comment on how the estimates from step 3 compare with those for the proportional odds model in step 1.
5. Obtain predicted marginal probabilities that the bitterness ratings are above 3 for the models in steps 1 and 3 (`bitter` is an indicator for `bitterness` being above 3) and use graphs to compare the predictions.

11.9 Early childhood math proficiency data

Here we analyze data from the Early Childhood Longitudinal Study—kindergarten cohort (ECLS-K), made available by O'Connell and McCoach (2008) and analyzed by O'Connell et al. (2008). In the ECLS-K, a random sample of kindergarten children was drawn from a random sample of U.S. schools in 1998 and followed up into eighth grade.

Here we consider the mathematics proficiency for a subsample of the children as they neared the end of first grade. Specifically, the children were native English speakers, were not repeating first grade, and had not been retained in kindergarten. Furthermore, only schools with complete information on school-level variables and with at least five children in the sample were selected. Finally, students with missing data on student-level variables of interest were dropped.

The response variable is math proficiency conceptualized as the highest developmental milestone reached by the child. Five clusters of four test items each were used to assess learning milestones in mathematics. Getting any three items within a cluster correct was viewed as passing the milestone, and it was verified that children who pass a given milestone have also mastered all the previous milestones (only 5% of response patterns did not fit this hierarchical structure). The milestones are (Rock and Pollack 2002)

1 Number and shape: identifying some one-digit numerals, recognizing geometric shapes, and one-to-one counting of up to ten objects.

2 Relative size: reading all single-digit numerals, counting beyond ten, recognizing a sequence of patterns, and using nonstandard units of length to compare objects.

3 Ordinary number sequence: reading two-digit numerals, recognizing the next number in a sequence, identifying the ordinal position of an object, and solving a simple word problem.

4 Addition/subtraction: solving simple addition and subtraction problems.

5 Multiplication/division: solving simple multiplication and division problems and recognizing more complex number patterns.

Mathematics proficiency was rated 0 if the child did not pass proficiency level 1, and 1–5 according to the highest proficiency level passed.

There are two datasets: `ecls_child.dta` contains the student-level data, whereas `ecls_school.dta` contains the school-level data. The variables in these datasets that we will use here are the following:

- `ecls_child.dta`
    - `s4_id`: school identifier (string variable)
    - `profmath`: proficiency level (0 to 5)
    - `numrisks`: number of risk factors out of the following four: living in a single parent household, living in a family that receives welfare payments or food stamps, having a mother with less than a high school education, and having parents whose primary language is not English

- `ecls_school.dta`
    - `s4_id`: school identifier (string variable)
    - `nbhoodcl`: neighborhood climate, a composite of the principal's perception of six specific problems in the vicinity of the school, including extent of litter, drug activity, gang activity, crime, violence, and existence of vacant lots or homes
    - `pubpriv2`: dummy variable for school being private (1: private, 0: public)

1. Use the `merge` command to combine the two datasets.
2. Create a numeric school identifier from the string variable `s4_id`.
3. Summarize the student-level and school-level variables, treating school as the unit of analysis when summarizing school-level variables (see section 3.2.1).
4. Following O'Connell et al. (2008), fit a proportional odds model to the math proficiency data with `numrisks`, `nbhoodcl`, and `pubpriv2` as covariates. Because `profmath` takes the value 0 for only nine children, merge the lowest two categories before fitting the model.
5. Interpret the estimates.

6. Obtain predicted marginal probabilities of proficiency being 4 or 5 for all possible combinations of `numrisks` and `pubpriv2` when `nbhoodcl` takes the value 0. Plot the predicted probabilities versus `numrisks` with separate lines for public and private schools.

See also exercise 14.8, where we consider continuation-ratio logit models for these data.

12 Nominal responses and discrete choice

12.1 Introduction

A nominal variable is a categorical variable with categories that do not have a unique ordering. An example from biomedicine is blood type with categories "O", "A", "B", and "AB". A classical example from economics is the choice of travel mode from the set of alternatives "plane", "train", "bus", and "car". In marketing, the purchasing behavior of consumers is important, for instance, the brand of yogurt that is bought among the alternatives "Yoplait", "Dannon", and "Weight Watchers". In political science, a central nominal variable is the political party voted for in an election, such as "Labour", "Conservatives", or "Liberal Democrats" in the United Kingdom. The last three examples all concern individuals' choices from a discrete set of alternatives or, in other words, *discrete choices*. While it may be possible to order the categories of nominal variables according to some criterion, there is not one unique ordering as for ordinal variables.

Sometimes variables have partially ordered categories, such as Likert-scale items for attitude measurement with a "don't know" category, labeled, for instance, as "disagree strongly", "disagree", "agree", "agree strongly", and "don't know". These kinds of responses are often treated as ordinal, with "don't know" ordered between "disagree" and "agree". However, this may not be appropriate because lack of knowledge does not necessarily imply a neutral attitude. It might instead be preferable to treat such partially ordered responses as nominal.

In this chapter, we generalize the multilevel models for dichotomous responses discussed in chapter 10 to handle nominal responses and discrete choices. Many of the issues discussed in that chapter persist for multilevel modeling of nominal responses. The main difference between the logistic models considered here and those discussed in the chapter on dichotomous responses is that several odds ratios involving different pairs of categories must now be considered.

Models for nominal responses can include covariates characterizing the subjects (or units), the alternatives (or categories), or the combination of subjects and alternatives. For example, for choice of mobile (or cell) phone, the age of the customer characterizes the subject, the price characterizes the alternative, and whether any friends have the phone characterizes the subject and alternative combination.

It is common to distinguish between two kinds of logistic models for nominal responses. Multinomial logit models are for subject-specific covariates only and are popular, for instance, in biostatistics. In contrast, conditional logit models traditionally use covariates that vary between alternatives and possibly between subjects as well and are popular, for instance, in economics. As we will see later, the multinomial logit model can actually be viewed as a special case of the conditional logit model. Both types of models can be specified either directly, writing down the expression for the probability that the response falls in a particular category, or via a utility-maximization formulation where the category with the highest utility or attractiveness is chosen. The utility-maximization formulation was developed in psychometrics and econometrics, and the term "discrete choice" is often used in economics.

We start by describing single-level models for nominal responses, using the direct formulation to introduce multinomial logit models and conditional logit models before discussing the utility-maximization formulation of these models. We then extend these models to the multilevel and longitudinal setting by including random effects in the linear predictor.

12.2   Single-level models for nominal responses

In this section, we consider a nominal response variable y_i with S unordered categories denoted $s = 1, \ldots, S$. Regression models for nominal responses are often called "multinomial" because the conditional distribution of the response given the covariates is a multinomial distribution for the S possible categories of the response. The multinomial distribution was also used for ordinal responses in the previous chapter, and the special case of the binomial distribution (where $S = 2$) was used in chapter 10.

We first discuss multinomial logit models for covariates that vary only over subjects i but do not vary over response categories or alternatives s. Thereafter, we describe conditional logit models for covariates that vary between alternatives or response categories s and possibly between subjects i.

12.2.1   Multinomial logit models

Recall the binary logistic regression model (10.2) for a single covariate x_i that we discussed in chapter 10. There we only specified $\Pr(y_i = 1|x_i)$ because $\Pr(y_i = 0|x_i) = 1 - \Pr(y_i = 1|x_i)$, but these probabilities can be written as

$$\Pr(y_i = 0|x_i) \;=\; \frac{1}{1 + \exp(\beta_1 + \beta_2 x_i)}$$

$$\Pr(y_i = 1|x_i) \;=\; \frac{\exp(\beta_1 + \beta_2 x_i)}{1 + \exp(\beta_1 + \beta_2 x_i)}$$

The denominator for the two probabilities above is the sum of the numerators, guaranteeing that the probabilities sum to one.

For the case of $S = 3$ categories (typically coded 1, 2, and 3 in this context), a natural extension would be to model the probabilities that $y_i = 1$, $y_i = 2$, and $y_i = 3$ as

$$\Pr(y_i = 1|x_i) = \frac{1}{1 + \exp(\beta_1^{[2]} + \beta_2^{[2]} x_i) + \exp(\beta_1^{[3]} + \beta_2^{[3]} x_i)}$$

$$\Pr(y_i = 2|x_i) = \frac{\exp(\beta_1^{[2]} + \beta_2^{[2]} x_i)}{1 + \exp(\beta_1^{[2]} + \beta_2^{[2]} x_i) + \exp(\beta_1^{[3]} + \beta_2^{[3]} x_i)}$$

$$\Pr(y_i = 3|x_i) = \frac{\exp(\beta_1^{[3]} + \beta_2^{[3]} x_i)}{1 + \exp(\beta_1^{[2]} + \beta_2^{[2]} x_i) + \exp(\beta_1^{[3]} + \beta_2^{[3]} x_i)} \tag{12.1}$$

In the binary case, we needed only one intercept and one coefficient, but for three categories, two intercepts and two coefficients are required. We use the superscript [2] for the intercept and coefficient that enter the numerator for category $s = 2$, and we use the superscript [3] for the intercept and coefficient that enter the numerator for category $s = 3$. Again to ensure that the probabilities sum to one, the denominator is equal to the sum of the numerators.

Another way of expressing the three-category model is by defining additional parameters $\beta_1^{[1]}$ and $\beta_2^{[1]}$ for category 1 and setting them to zero, giving the general expression

$$\Pr(y_i = s|x_i) = \frac{\exp(\beta_1^{[s]} + \beta_2^{[s]} x_i)}{\exp(\beta_1^{[1]} + \beta_2^{[1]} x_i) + \exp(\beta_1^{[2]} + \beta_2^{[2]} x_i) + \exp(\beta_1^{[3]} + \beta_2^{[3]} x_i)}$$

$$= \frac{\exp(\beta_1^{[s]} + \beta_2^{[s]} x_i)}{\sum_{c=1}^{3} \exp(\beta_1^{[c]} + \beta_2^{[c]} x_i)}$$

In the sum in the denominator, the index c takes the values 1, 2, and 3 to produce the three required terms. The probabilities are the same as in equation (12.1) because $\exp(\beta_1^{[1]} + \beta_2^{[1]} x_i) = \exp(0 + 0 x_i) = 1$.

Following the same reasoning, we can consider the general case of S categories where the probability that the response is category s becomes

$$\Pr(y_i = s|x_i) = \frac{\exp(\beta_1^{[s]} + \beta_2^{[s]} x_i)}{\sum_{c=1}^{S} \exp(\beta_1^{[c]} + \beta_2^{[c]} x_i)}, \qquad s = 1, \ldots, S \tag{12.2}$$

Because the coding of the categories is arbitrary, we can designate any category r as *baseline category* or *base outcome* by setting the intercept and coefficients for that category to zero; $\beta_1^{[r]} = 0$ and $\beta_2^{[r]} = 0$. However, in practice, we will often have a preference regarding which category we find it most natural to compare the others with.

Category probabilities from (12.2) are shown in figure 12.1 for $S = 4$ categories (for some made-up values for the coefficients).

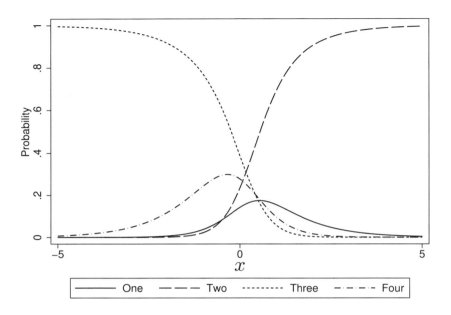

Figure 12.1: Illustration of category probabilities for multinomial logit model with four categories

When there are three or more categories, there will always be one monotonically increasing curve, and one monotonically decreasing curve; the other curve(s) are unimodal with upper and lower tails approaching zero as x becomes extreme. These features of the category probabilities may not be apparent for the range of x observed in a given dataset.

The model can alternatively be expressed as

$$
\ln \overbrace{\left\{ \frac{\Pr(y_i = s|x_i)}{\Pr(y_i = r|x_i)} \right\}}^{\text{Odds}_{(y_i=s \text{ vs. } y_i=r|x_i)}}
$$

$$
= \ln \left[\left\{ \frac{\exp(\beta_1^{[s]}+\beta_2^{[s]}x_i)}{\sum_{c=1}^{S} \exp(\beta_1^{[c]}+\beta_2^{[c]}x_i)} \right\} \Big/ \left\{ \frac{\exp(\beta_1^{[r]}+\beta_2^{[r]}x_i)}{\sum_{c=1}^{S} \exp(\beta_1^{[c]}+\beta_2^{[c]}x_i)} \right\} \right]
$$

$$
= \ln[\exp\{(\beta_1^{[s]} - \underbrace{\beta_1^{[r]}}_{=0}) + (\beta_2^{[s]} - \underbrace{\beta_2^{[r]}}_{=0})x_i\}] \;=\; \beta_1^{[s]} + \beta_2^{[s]}x_i
$$

The coefficients represent log odds-ratios for the odds of each category versus the baseline category, and the model is therefore also called a *baseline category logit model*. The relevant odds for a four-alternative example, $S = 4$, where $r = 1$ is chosen as base outcome are shown in figure 12.2

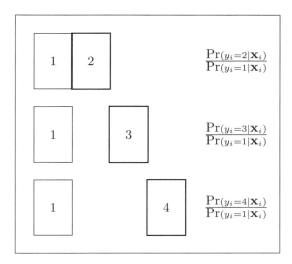

Figure 12.2: Relevant odds $\Pr(y_i = s|\mathbf{x}_i)/\Pr(y_i = 1|\mathbf{x}_i)$ for $(s = 2, 3, 4)$ in a baseline category logit model with four categories. Odds is the ratio of probabilities of events; events included in the numerator probability are in thick frames, and events included in the denominator probability are in thin frames [adapted from Brendan Halpin's web notes on "Models for ordered categories (ii)"].

A one-unit increase in the covariate x_i from some value a to $a + 1$ corresponds to the odds ratio $\mathrm{OR}_{s:r}$ for category s versus the base outcome r.

$$\mathrm{OR}_{s:r} \equiv \frac{\mathrm{Odds}(y_i = s \text{ vs. } y_i = r|x_i = a+1)}{\mathrm{Odds}(y_i = s \text{ vs. } y_i = r|x_i = a)} = \frac{\exp(\beta_1^{[s]} + \beta_2^{[s]}(a+1))}{\exp(\beta_1^{[s]} + \beta_2^{[s]}a)} = \exp(\beta_2^{[s]})$$

The odds ratio for category s versus a category other than the base outcome r, say, category t, can simply be obtained as

$$\mathrm{OR}_{s:t} = \exp(\beta_2^{[s]})/\exp(\beta_2^{[t]}) = \exp(\beta_2^{[s]} - \beta_2^{[t]})$$

because

$$\frac{\Pr(y_i = s|x_i)}{\Pr(y_i = t|x_i)} = \frac{\Pr(y_i = s|x_i)}{\Pr(y_i = r|x_i)} \bigg/ \frac{\Pr(y_i = t|x_i)}{\Pr(y_i = r|x_i)}$$

To illustrate regression modeling with a nominal response, we will use data from Greene (2012) on choice of transport between Melbourne and Sydney in Australia. The dataset `travel1.dta` contains the travelers' chosen mode of transport and two explanatory variables:

- `traveler`: identifier for traveler
- `alt`: chosen mode of transport
  (1: Air; 2: Train; 3: Bus; 4: Car)
- `Hinc`: household income of traveler in Australian \$1,000 (x_{2i})
- `Psize`: number of people traveling together (x_{3i})

We first read in the dataset,

```
. use http://www.stata-press.com/data/mlmus3/travel1
```

before listing the variables for the first 10 travelers:

```
. list traveler alt Hinc Psize in 1/10, noobs clean
    traveler   alt   Hinc   Psize
           1     4     35       1
           2     4     30       2
           3     4     40       1
           4     4     70       3
           5     4     45       2
           6     2     20       1
           7     1     45       1
           8     4     12       1
           9     4     40       1
          10     4     70       2
```

We see that eight of these individuals drive (alternative or category 4 is car), traveler 6 uses the train, and traveler 7 flies. Note that the covariates `Hinc` and `Psize` are characteristics of the travelers, not specific to the mode of transport for a traveler (a covariate that could depend on the mode of transport would be the cost of travel).

The response variable has four unordered categories $s = 1, \ldots, 4$, and we will use Car (category 4) as the base outcome r henceforth, because we want the parameters to refer to comparisons with this alternative. We specify the following multinomial logit model:

$$\Pr(y_i = s \mid x_{2i}, x_{3i}) = \frac{\exp(\beta_1^{[s]} + \beta_2^{[s]} x_{2i} + \beta_3^{[s]} x_{3i})}{\sum_{c=1}^{4} \exp(\beta_1^{[c]} + \beta_2^{[c]} x_{2i} + \beta_3^{[c]} x_{3i})}, \qquad s = 1, \ldots, 4$$

where $\beta_1^{[4]} = \beta_2^{[4]} = \beta_3^{[4]} = 0$.

The model can be fit by maximum likelihood using the `mlogit` command (for "multinomial logit"). We use the `baseoutcome(4)` option to designate Car as the base outcome (Stata otherwise uses the most frequent category, in this case Train, as base outcome).

```
. mlogit alt Hinc Psize, baseoutcome(4)
```

Multinomial logistic regression					Number of obs	=	210
					LR chi2(6)	=	60.84
					Prob > chi2	=	0.0000
Log likelihood = -253.34085					Pseudo R2	=	0.1072

alt	Coef.	Std. Err.	z	P>|z|	[95% Conf.	Interval]
1						
Hinc	.0035438	.0103047	0.34	0.731	-.0166531	.0237407
Psize	-.6005541	.1992005	-3.01	0.003	-.9909798	-.2101284
_cons	.9434923	.549847	1.72	0.086	-.1341881	2.021173
2						
Hinc	-.0573078	.0118416	-4.84	0.000	-.0805169	-.0340987
Psize	-.3098126	.1955598	-1.58	0.113	-.6931028	.0734775
_cons	2.493848	.5357211	4.66	0.000	1.443854	3.543842
3						
Hinc	-.0303253	.0132228	-2.29	0.022	-.0562415	-.004409
Psize	-.940414	.3244532	-2.90	0.004	-1.576331	-.3044974
_cons	1.977971	.671715	2.94	0.003	.6614334	3.294508
4	(base outcome)					

For each category of the response, apart from the base outcome, the output gives the estimated intercept $\widehat{\beta}_1^{[s]}$ and coefficients $\widehat{\beta}_2^{[s]}$ and $\widehat{\beta}_3^{[s]}$. It also shows the estimated standard errors; Wald statistics z with corresponding p-values, P>|z|; and 95% confidence intervals for the parameters. In the top panel where alt is 1, estimates are given for alternative 1, Air versus the base outcome 4, which is Car. In the panel labeled 2, estimates for Train versus Car are given, and the panel labeled 3 contains estimates for Bus versus Car. The final panel, labeled 4 for the base outcome (which the other alternatives are compared with), is empty. The estimated coefficients and their standard errors are also given under "Only traveler-specific covariates" in table 12.1.

Table 12.1: Estimates for nominal regression models for choice of transport

| | Multinomial logit | | | Conditional logit | | | | | |
| | Only traveler-specific covariates | | | Only alternative-specific covar. | | | Both traveler & alt.-specific covar. | | |
	Est (SE)	OR	(95% CI)	Est (SE)	OR	(95% CI)	Est (SE)	OR	(95% CI)
Traveler-specific covariates									
Air									
$\beta_1^{[1]}$ [_cons]	0.94 (0.55)						7.87 (0.99)		
$\beta_2^{[1]}$ [Hinc]	0.00 (0.01)	1.00	(0.98, 1.02)				0.00 (0.01)	1.00	(0.98, 1.03)
$\beta_3^{[1]}$ [Psize]	−0.60 (0.55)	0.55	(0.37, 0.81)				−1.03 (0.27)	0.36	(0.21, 0.60)
Train									
$\beta_1^{[2]}$ [_cons]	2.49 (0.54)						5.56 (0.70)		
$\beta_2^{[2]}$ [Hinc]	−0.06 (0.01)	0.94	(0.92, 0.97)				−0.06 (0.01)	0.95	(0.92, 0.97)
$\beta_3^{[2]}$ [Psize]	−0.31 (0.20)	0.73	(0.50, 1.08)				−0.30 (0.23)	1.35	(0.87, 2.11)
Bus									
$\beta_1^{[3]}$ [_cons]	1.98 (0.67)						4.43 (0.78)		
$\beta_2^{[3]}$ [Hinc]	−0.03 (0.01)	0.97	(0.95, 1.00)				−0.02 (0.02)	0.98	(0.95, 1.01)
$\beta_3^{[3]}$ [Psize]	−0.94 (0.32)	0.39	(0.21, 0.74)				−0.03 (0.33)	0.97	(0.50, 1.87)
Alternative-specific covariates									
β_2 [CC]				−0.01 (0.00)	0.99	(0.98,1.00)	−0.02 (0.01)	0.98	(0.97,0.99)
β_3 [Ttime]				−0.01 (0.00)	0.99	(0.98,0.99)	−0.10 (0.11)	0.90	(0.88,0.92)
Log likelihood	−253.34			−270.11[†]			−177.45[†]		

[†]Log conditional likelihood

Rather than considering changes in log odds, most people find it more informative to interpret odds ratios, comparing the odds for an alternative versus the base outcome (here Car) for a unit difference in a covariate. We use "odds" to refer to the probability of a category s divided by the probability of another category r. Strictly speaking, it is the conditional odds given that the category is r or s. In contrast, Stata's definition of "odds" is the probability of a category s divided by the probability of the complement (all other categories than s), and what we have called "odds ratios" (ORs) are referred to as "relative-risk ratios" (RRRs). Display 12.1 shows that the odds ratios can be interpreted as relative-risk ratios (this can be skipped if you like).

$$
=
\overbrace{\frac{\text{Odds}(y_i = s \text{ vs. } y_i = r | x_i = a+1)}{\text{Odds}(y_i = s \text{ vs. } y_i = r | x_i = a)}}^{\text{Odds ratio used in this book}} = \frac{\Pr(y_i = s | x_i = a+1)/\Pr(y_i = r | x_i = a+1)}{\Pr(y_i = s | x_i = a)/\Pr(y_i = r | x_i = a)}
$$

$$
= \underbrace{\frac{\text{Risk ratio}(y_i = s | x_i = a+1 \text{ vs. } x_i = a)}{\text{Risk ratio}(y_i = r | x_i = a+1 \text{ vs. } x_i = a)}}_{\text{Stata's relative-risk ratio}} = \frac{\Pr(y_i = s | x_i = a+1)/\Pr(y_i = s | x_i = a)}{\Pr(y_i = r | x_i = a+1)/\Pr(y_i = r | x_i = a)}
$$

Display 12.1: Odds ratio used in this book equals Stata's relative-risk ratio

Estimated odds ratios can be obtained by using the `mlogit` command but now with the `rrr` option (for "relative-risk ratio"). We also define value labels for `alt` to make them appear in the output instead of the numerical values of `alt`:

```
. label define modelabels 1 "Air" 2 "Train" 3 "Bus" 4 "Car"
. label values alt modelabels
```

We fit the model using `mlogit`:

```
. mlogit alt Hinc Psize, baseoutcome(4) rrr
Multinomial logistic regression              Number of obs   =         210
                                             LR chi2(6)      =       60.84
                                             Prob > chi2     =      0.0000
Log likelihood = -253.34085                  Pseudo R2       =      0.1072
```

alt	RRR	Std. Err.	z	P>|z|	[95% Conf. Interval]	
Air						
Hinc	1.00355	.0103413	0.34	0.731	.9834848	1.024025
Psize	.5485076	.109263	-3.01	0.003	.3712128	.8104802
_cons	2.568937	1.412523	1.72	0.086	.8744256	7.547171
Train						
Hinc	.9443033	.0111821	-4.84	0.000	.9226393	.9664761
Psize	.7335844	.1434596	-1.58	0.113	.5000222	1.076244
_cons	12.10778	6.486392	4.66	0.000	4.236994	34.5996
Bus						
Hinc	.9701299	.0128279	-2.29	0.022	.9453108	.9956007
Psize	.3904662	.126688	-2.90	0.004	.2067323	.7374939
_cons	7.22806	4.855196	2.94	0.003	1.937568	26.96414
Car	(base outcome)					

These estimated odds ratios with their 95% confidence intervals were also reported in table 12.1 under "Only traveler-specific covariates". It is often useful to express odds ratios as percentage changes using $100\%(\text{OR}-1)$. We see that household income does not appear to affect the odds of traveling by air compared with going by car (the base outcome) because the estimated odds ratio of 1.004 is practically 1 with a 95% confidence interval including 1, controlling for the number of people traveling together. However, for given a number of travelers, the odds of taking the train compared with driving is reduced by an estimated 6% $[-6\% = 100\%(0.9443033-1)]$ per \$1,000 increase in income, and the odds of taking the bus compared with driving is reduced by an estimated 3% $[-3\% = 100\%(0.9701299-1)]$ per \$1,000. According to the fitted model, increasing the number of people traveling together reduces the odds of flying, taking the train, and taking the bus versus driving considerably, for given household income. Specifically, the odds of flying compared with driving is reduced by an estimated 45% $[-45\% = 100\%(0.5485076-1)]$, the odds of taking the train compared with driving is reduced by an estimated 27% $[-27\% = 100\%(0.7335844-1)]$, and the odds of taking the bus compared with driving is reduced by an estimated 61% $[-61\% = 100\%(0.3904662-1)]$ per person traveling.

12.2.2 Conditional logit models

In conditional logit models, there are covariates that characterize the alternatives instead of characterizing the units as in multinomial logit models. The covariates are often called the attributes of the different alternatives. An example would be the cost of traveling by Air, Train, Bus, or Car. A change in an attribute, such as cost, is assumed to have the *same* effect β for all alternatives, so there are no alternative-specific coefficients for attributes. This makes it possible to make predictions of the marginal probability (for example, market share) for a hypothetical alternative (for example, product), because we can use the estimated β to obtain predictions for hypothetical alternatives that are not in the data.

Much of the theory and practice of conditional logit modeling was developed by the econometrician Daniel McFadden to forecast usage of the Bay Area Rapid Transit (BART) system, a subway (underground) system in the San Francisco Bay area of the U.S.A., before it was open for service.

Classical conditional logit models

We first consider the classical conditional logit models for a single covariate or attribute $x_i^{[s]}$ that varies between alternatives or response categories s and possibly between units i. In contrast to the multinomial logit model discussed in section 12.2.1, where the covariate x_i had alternative-specific effects $\beta^{[s]}$, the covariate $x_i^{[s]}$ has a coefficient β that does not vary over alternatives s. Moreover, there are no covariates x_i included that vary only over units and there are no alternative-specific intercepts $\beta^{[s]}$.

The standard model for such data is the *conditional logit model*:

$$\Pr(y_i = s | x_i^{[1]}, \ldots, x_i^{[S]}) \ = \ \frac{\exp(\beta x_i^{[s]})}{\sum_{c=1}^{S} \exp(\beta x_i^{[c]})} \tag{12.3}$$

The probabilities have the same form as (12.2) except that the regression parameter β does not vary over alternatives while the covariate does. Because β is constant across alternatives, there is no base outcome as in the multinomial logit model.

It follows from the conditional logit model that the log of the odds of choosing alternative s versus t is

$$\ln\left\{ \frac{\Pr(y_i = s | x_i^{[1]}, \ldots, x_i^{[S]})}{\Pr(y_i = t | x_i^{[1]}, \ldots, x_i^{[S]})} \right\} \ = \ \ln\left\{ \frac{\exp(\beta x_i^{[s]}) / \sum_{c=1}^{S} \exp(\beta x_i^{[c]})}{\exp(\beta x_i^{[t]}) / \sum_{c=1}^{S} \exp(\beta x_i^{[c]})} \right\}$$

$$= \ \beta(x_i^{[s]} - x_i^{[t]}) \tag{12.4}$$

and depends only on the difference in the attribute for these alternatives, for instance, the difference in the cost of traveling by Air and Train. We see that only parameters for attributes that vary over alternatives $x_i^{[s]}$ can be estimated in (12.3), because the parameters would cancel out for covariates x_i that vary only over units, $\beta(x_i - x_i) = 0$.

A one-unit increase in the difference of the attribute for two alternatives s and t from $x_i^{[s]} - x_i^{[t]} = a$ to $x_i^{[s]} - x_i^{[t]} = a + 1$ corresponds to the odds ratio

$$\text{OR} \ \equiv \ \frac{\text{Odds}(y_i = s \text{ vs. } y_i = t | x_i^{[s]} - x_i^{[t]} = a + 1)}{\text{Odds}(y_i = s \text{ vs. } y_i = t | x_i^{[s]} - x_i^{[t]} = a)} \ = \ \frac{\exp\{\beta(a + 1)\}}{\exp\{\beta(a)\}} \ = \ \exp(\beta)$$

The exponential of the regression coefficient, $\exp(\beta)$, can actually be interpreted in three different ways. First, as shown above, it represents the ratio of the odds of alternative s versus t per unit increase in the difference $x_i^{[s]} - x_i^{[t]}$ in the attribute, for example, the price difference. Second, if $x_i^{[t]}$ stays constant, $\exp(\beta)$ is the odds ratio comparing s with t per unit change in the attribute $x_i^{[s]}$. Third, $\exp(\beta)$ is the ratio of the odds of choosing an alternative s versus choosing *any* of the other alternatives per unit increase in $x_i^{[s]}$ when the attributes of all other alternatives remain the same. To see this, we note that

$$\ln\left\{\frac{\Pr(y_i = s | x_i^{[1]}, \ldots, x_i^{[S]})}{\sum_{t \neq s} \Pr(y_i = t | x_i^{[1]}, \ldots, x_i^{[S]})}\right\} = \beta(x_i^{[s]} - \sum_{t \neq s} x_i^{[t]})$$

so that $\exp(\beta)$ is the odds ratio per unit increase in the difference $x_i^{[s]} - \sum_{t \neq s} x_i^{[t]}$. As pointed out earlier, only the third interpretation would be consistent with Stata's use of the term "odds ratio".

The conditional logit model (12.3) is often used for making predictions regarding marginal (not conditional on the attributes $x_i^{[s]}$) probabilities $\Pr(y_i^{[s]})$ for nominal responses. It is particularly useful for forecasting the market share for a new alternative or product $S + 1$. Such predictions can be accomplished by specifying the attributes $\mathbf{x}^{[S+1]}$ of the new product, such as the price, keeping the attributes for the existing alternatives at their sample values, and using the fitted conditional logit model to obtain predicted probabilities for the new product for each unit,

$$\widehat{\Pr}(y_i = S + 1 | x_i^{[1]}, \ldots, x_i^{[S]}, x_i^{[S+1]}) = \frac{\exp(\widehat{\beta} x_i^{[S+1]})}{\sum_{c=1}^{S+1} \exp(\widehat{\beta} x_i^{[c]})}$$

The market share can then be estimated as the marginal or mean probability of choosing the new product in the sample (of size N),

$$\widehat{\Pr}(y_i = S + 1) = \frac{1}{N} \sum_{i=1}^{N} \widehat{\Pr}(y_i = S + 1 | x_i^{[1]}, \ldots, x_i^{[S]}, x_i^{[S+1]})$$

A potential problem with using conditional logit models for prediction is the independence of irrelevant alternatives (IIA) property of these models, an issue that will be discussed in section 12.3.

To illustrate conditional logit modeling, we use the dataset **travel2.dta**, which contains a different set of covariates for the same Australian travelers as before. Here we consider the following variables:

- **traveler**: identifier for traveler

- **alt**: mode of transport
  (1: Air; 2: Train; 3: Bus; 4: Car)

- **Choice**: dummy variable for chosen alternative for traveler

- **GC**: generalized cost in Australian \$ $(x_{2i}^{[s]})$, defined as
  (in-vehicle cost) + (time spent traveling) × (wagelike measure)

- **Ttime**: time spent in terminal in minutes $(x_{3i}^{[s]})$

We read in the data by typing

```
. use http://www.stata-press.com/data/mlmus3/travel2
```

In contrast to the dataset used in the previous section, **travel2.dta** is "expanded" in the sense that it contains a row (record) for each possible alternative for each traveler, in this case, four records per traveler.

We take a look at the variables for the first two travelers contained in the first eight records of the dataset:

```
. list traveler alt Choice GC Ttime in 1/8, sepby(traveler) noobs
```

traveler	alt	Choice	GC	Ttime
1	1	0	70	69
1	2	0	71	34
1	3	0	70	35
1	4	1	30	0
2	1	0	68	64
2	2	0	84	44
2	3	0	85	53
2	4	1	50	0

The variable **alt** no longer represents the alternative being chosen, as it did in the dataset **travel1.dta**, but instead represents all alternatives s ($s = 1, 2, 3, 4$) a traveler i could choose among. For each traveler, the binary variable **Choice** is 1 for the alternative chosen and 0 for the three alternatives not chosen. Also note that there are not four real responses per traveler because only one alternative is chosen by each traveler and there is hence only one piece of information per traveler. As observed earlier when inspecting the travel data, the first two travelers both chose to drive (**Choice** is 1 when **alt** is 4 for Car). The covariates **GC** and **Ttime** vary over mode of transport for each traveler and between travelers for a given mode of transport. For travelers 1 and 2, we see that the cost, **GC**, happens to be lowest for the chosen alternative. **Ttime** takes on the value 0 for Car because there is no waiting time in a terminal when driving.

An expanded dataset not only makes it straightforward to include variables that vary over alternatives but also allows different travelers to choose from different alternative sets. For instance, if air traffic controllers had been on strike when traveler 2 made his trip, the alternative Air would no longer be available to him and line 5 could have been deleted from the dataset.

We now consider estimation of the conditional logit model (12.3). We use $v_i^{[s]}$ to denote the binary choice indicator `Choice` for alternative s for traveler i. Note that this notation departs from the usual convention of placing all indices in the subscript, here i and s, that identify the rows in the data. Similarly, the value of a covariate x for alternative s and traveler i is denoted $x_i^{[s]}$.

It would be tempting to use a *binary* logistic regression model for the probability that $v_i^{[s]} = 1$. However, such a model would be misspecified because it falsely assumes that the responses $v_i^{[s]}$ are independent given the covariates. The $v_i^{[s]}$ are clearly dependent because a 1 can occur only once for each traveler. So if we know that a traveler is driving, there is a zero probability of traveling by any other mode.

In the expanded dataset with one record per alternative, the conditional logit model actually represents the conditional probability that the choice indicator takes the value 1 for the chosen alternative s, *given* that exactly one alternative is chosen by the traveler, $\sum_{c=1}^{S} v_i^{[c]} = 1$,

$$\Pr(v_i^{[s]} = 1 | x_i^{[1]}, \ldots, x_i^{[S]}, \textstyle\sum_{c=1}^{S} v_i^{[c]} = 1) = \frac{\exp(\beta x_i^{[s]})}{\sum_{c=1}^{S} \exp(\beta x_i^{[c]})} = \Pr(y_i = s | x_i^{[1]}, \ldots, x_i^{[S]})$$

(12.5)

The conditioning on the sum of indicators being 1 is the reason why the model is called a *conditional* logit model. To avoid conditioning on $\sum_{c=1}^{S} v_i^{[c]} = 1$ in all expressions, we write the probability in terms of $y_i = s$ as shown after the second equality, although the response variable is really the binary choice indicator $v_i^{[s]}$. A proof of the first equality in (12.5) is presented in display 12.2 (feel free to skip it).

Imagine making an independent decision for each alternative whether to choose that alternative (without any restriction on the number of alternatives that can be chosen). In this case, a binary logistic regression model would be appropriate for the probability that alternative s is chosen, $v_i^{[s]} = 1$:

$$\Pr(v_i^{[s]} = 1 | x_i^{[s]}) = \frac{\exp(\beta x_i^{[s]})}{1 + \exp(\beta x_i^{[s]})}$$

However, when exactly *one* alternative must be chosen, the sum of the choice indicators $v_i^{[s]}$ must be one. We must therefore consider the conditional probability of choosing alternative s, given that $\sum_{c=1}^{S} v_i^{[c]} = 1$. It follows that (suppressing the conditioning on $x_i^{[1]}, \ldots, x_i^{[S]}$)

1. The probability that $v_i^{[s]} = 1$ and the other choice indicators for unit i are zero is

$$\Pr(v_i^{[s]} = 1, \textstyle\sum_{c=1}^{S} v_i^{[c]} = 1) = \frac{\exp(\beta x_i^{[s]})}{\prod_{c=1}^{S} \{1 + \exp(\beta x_i^{[c]})\}}$$

For instance, for $S = 4$ the probability that $v_i^{[1]} = 0$, $v_i^{[2]} = 1$, $v_i^{[3]} = 0$, and $v_i^{[4]} = 0$ is

$$\frac{1}{1 + \exp(\beta x_i^{[1]})} \times \frac{\exp(\beta x_i^{[2]})}{1 + \exp(\beta x_i^{[2]})} \times \frac{1}{1 + \exp(\beta x_i^{[3]})} \times \frac{1}{1 + \exp(\beta x_i^{[4]})}$$

$$= \frac{\exp(\beta x_i^{[2]})}{\prod_{c=1}^{4} \{1 + \exp(\beta x_i^{[c]})\}}$$

2. The probability that the indicator variable is 1 for exactly one alternative for unit i, $\sum_{c=1}^{S} v_i^{[c]} = 1$, is the sum of the probabilities of all the S ways this can be accomplished,

$$\Pr(\textstyle\sum_{c=1}^{S} v_i^{[c]} = 1) = \frac{\exp(\beta x_i^{[1]})}{\prod_{c=1}^{S}[1 + \exp(\beta x_i^{[c]})]} + \cdots + \frac{\exp(\beta x_i^{[S]})}{\prod_{c=1}^{S}[1 + \exp(\beta x_i^{[c]})]}$$

$$= \frac{\sum_{c=1}^{S} \exp(\beta x_i^{[c]})}{\prod_{c=1}^{S}[1 + \exp(\beta x_i^{[c]})]}$$

3. The required conditional probability that alternative s is chosen by unit i, given that only one alternative is chosen, becomes

$$\Pr(v_i^{[s]} = 1 | \textstyle\sum_{c=1}^{S} v_i^{[c]} = 1) = \frac{\Pr(v_i^{[s]} = 1, \sum_{c=1}^{S} v_i^{[c]} = 1)}{\Pr(\sum_{c=1}^{S} v_i^{[c]} = 1)} = \frac{\exp(\beta x_i^{[s]})}{\sum_{c=1}^{S} \exp(\beta x_i^{[c]})}$$

Display 12.2: Conditional probability of choosing an alternative s given that exactly one alternative is chosen

Returning to the travel-choice application, we specify the following conditional logit model for the alternatives Air, Train, Bus, and Car:

$$\Pr(y_i = s | \mathbf{x}_i^{[1]}, \ldots, \mathbf{x}_i^{[4]}) = \frac{\exp(\beta_2 x_{2i}^{[s]} + \beta_3 x_{3i}^{[s]})}{\sum_{c=1}^{4} \exp(\beta_2 x_{2i}^{[c]} + \beta_3 x_{3i}^{[c]})}, \qquad s = 1, 2, 3, 4 \qquad (12.6)$$

where $\mathbf{x}_i^{[s]} = (x_{2i}^{[s]}, x_{3i}^{[s]})'$ is the vector of attributes, GC and Ttime, for mode or alternative s.

We now fit the model by conditional maximum likelihood using the clogit command (for "conditional logit") that we have previously used for the fixed-effects approach to clustered dichotomous responses in section 10.14.1. Here we use the group(traveler) option to specify that traveler is the unit identifier (so Choice takes the value 1 once for each unique value of traveler)

```
. clogit Choice GC Ttime, group(traveler)
Conditional (fixed-effects) logistic regression    Number of obs   =        840
                                                   LR chi2(2)      =      42.03
                                                   Prob > chi2     =     0.0000
Log likelihood = -270.10821                        Pseudo R2       =     0.0722
```

Choice	Coef.	Std. Err.	z	P>|z|	[95% Conf. Interval]	
GC	-.0106331	.0034624	-3.07	0.002	-.0174192	-.003847
Ttime	-.012981	.0028943	-4.49	0.000	-.0186537	-.0073083

These estimates were reported in table 12.1 under "Only alternative-specific covariates".

Estimated odds ratios with associated 95% confidence intervals are produced by including the or option in the clogit command,

```
. clogit Choice GC Ttime, group(traveler) or
Conditional (fixed-effects) logistic regression    Number of obs   =        840
                                                   LR chi2(2)      =      42.03
                                                   Prob > chi2     =     0.0000
Log likelihood = -270.10821                        Pseudo R2       =     0.0722
```

Choice	Odds Ratio	Std. Err.	z	P>|z|	[95% Conf. Interval]	
GC	.9894232	.0034257	-3.07	0.002	.9827316	.9961604
Ttime	.9871029	.0028569	-4.49	0.000	.9815192	.9927183

and were also reported under "Only alternative-specific covariates" in table 12.1.

According to the fitted model a \$10 increase in the generalized cost for any of the alternatives would decrease the odds of that alternative being chosen by an estimated 10% $[-10\% = 100\%\{\exp(10 \times -0.0106331) - 1\}]$, keeping the cost of the other alternatives constant. Increasing terminal time by 10 minutes for a particular mode of transport would decrease the odds of traveling with that alternative by an estimated 12% $[-12\% = 100\%\{\exp(10 \times -0.012981) - 1\}]$. Inversely, decreasing terminal time by

10 minutes would increase the odds of traveling with that alternative by an estimated 14% [$= 100\%\{\exp(-10 \times -0.012981) - 1\}$].

Conditional logit models also including covariates that vary only over units

We now consider two additional variables from `travel2.dta` that vary only over travelers i:

- `Hinc`: household income in Australian \$1,000 (x_{2i})
- `Psize`: number of people traveling together (x_{3i})

As discussed for the multinomial logit model in section 12.2.1, covariates like x_{2i} that do not vary between alternatives must have alternative-specific coefficients $\beta_2^{[s]}$ because a term like $\beta_2 x_{2i}$ would cancel out of the expression for the probability. With data in expanded form, as here, where s denotes a row in the data, we can include alternative-specific coefficients by using dummy variables for the alternatives. Let $d_1^{[s]}$ be a dummy variable for the first alternative, taking the value 1 if $s = 1$ and 0 otherwise. We need dummy variables $d_1^{[s]}$, $d_2^{[s]}$, and $d_3^{[s]}$ for each alternative apart from the fourth (Car), which is the base outcome. The required term $\beta_2^{[s]} x_{2i}$ is then obtained using

$$\beta_2^{[s]} x_{2i} = \beta_2^{[1]} d_1^{[s]} x_{2i} + \beta_2^{[2]} d_2^{[s]} x_{2i} + \beta_2^{[3]} d_3^{[s]} x_{2i}$$

Only one dummy variable—namely, the dummy variable $d_2^{[s]}$ for alternative 2—takes the value 1 and hence picks out the appropriate coefficient $\beta_2^{[s]}$. We will also include alternative-specific intercepts by using the same dummy variables.

To write down the model compactly, we use the notation $V_i^{[s]}$ for the linear predictor in the numerator of the probability,

$$\Pr(y_i = s | x_{2i}, x_{3i}, \mathbf{x}_{ij}^{[1]}, \ldots, \mathbf{x}_{ij}^{[4]}) = \frac{\exp(V_i^{[s]})}{\sum_{c=1}^{4} \exp(V_i^{[c]})}$$

We can now specify the form of the linear predictor as

$$
\begin{aligned}
V_i^{[s]} = {} & \beta_1^{[1]} d_1^{[s]} + \beta_1^{[2]} d_2^{[s]} + \beta_1^{[3]} d_3^{[s]} \\
& + \beta_2^{[1]} d_1^{[s]} x_{2i} + \beta_2^{[2]} d_2^{[s]} x_{2i} + \beta_2^{[3]} d_3^{[s]} x_{2i} + \beta_3^{[1]} d_1^{[s]} x_{3i} + \beta_3^{[2]} d_2^{[s]} x_{3i} + \beta_3^{[3]} d_3^{[s]} x_{3i} \\
& + \beta_4 x_{4i}^{[s]} + \beta_5 x_{5i}^{[s]}
\end{aligned}
$$

The model includes alternative-specific intercepts ($\beta_1^{[1]}$, $\beta_1^{[2]}$, and $\beta_1^{[3]}$), alternative-specific coefficients for `Hinc` ($\beta_2^{[1]}$, $\beta_2^{[2]}$, and $\beta_2^{[3]}$) and `Psize` ($\beta_3^{[1]}$, $\beta_3^{[2]}$, and $\beta_3^{[3]}$), and coefficients β_4 and β_5 for `GC` and `Ttime`.

Before estimation using `clogit`, we construct dummy variables for the alternatives $d_i^{[s]}$,

```
. tabulate alt, gen(a)
```

and give the dummy variables more descriptive names

```
. rename a1 air
. rename a2 train
. rename a3 bus
. rename a4 car
```

We construct interactions between the dummy variables (apart from the dummy for the base outcome) and the traveler-specific covariates:

```
. generate airXhinc = air*Hinc
. generate trainXhinc = train*Hinc
. generate busXhinc = bus*Hinc
. generate airXpsize = air*Psize
. generate trainXpsize = train*Psize
. generate busXpsize = bus*Psize
```

If there are many covariates, it is faster to use a `foreach` loop, as follows:

```
foreach var of varlist Hinc Psize {
    generate airX'var' = air*'var'
    generate trainX'var' = train*'var'
    generate busX'var' = bus*'var'
}
```

The coefficients of these interactions correspond to the alternative-specific coefficients $\beta_2^{[s]}$ and $\beta_3^{[s]}$ of the covariates `Hinc` and `Psize`. Specifically, the coefficient for `airXhinc` is $\beta_2^{[1]}$, the coefficient for `trainXhinc` is $\beta_2^{[2]}$, and the coefficient for `airXhinc` is $\beta_2^{[3]}$. Likewise, the coefficient for `airXpsize` is $\beta_3^{[1]}$, the coefficient for `trainXpsize` is $\beta_3^{[2]}$, and the coefficient for `airXpsize` is $\beta_3^{[3]}$.

We can now fit the model by conditional maximum likelihood using `clogit`:

```
. clogit Choice airXhinc airXpsize air trainXhinc trainXpsize train
> busXhinc busXpsize bus GC Ttime, group(traveler)
Conditional (fixed-effects) logistic regression   Number of obs   =        840
                                                   LR chi2(11)     =     227.34
                                                   Prob > chi2     =     0.0000
Log likelihood =  -177.4541                        Pseudo R2       =     0.3904
```

Choice	Coef.	Std. Err.	z	P>|z|	[95% Conf. Interval]	
airXhinc	.004071	.0127247	0.32	0.749	-.0208689	.029011
airXpsize	-1.027423	.2656569	-3.87	0.000	-1.548101	-.5067448
air	7.873608	.9868475	7.98	0.000	5.939423	9.807794
trainXhinc	-.0551849	.0144824	-3.81	0.000	-.0835698	-.0268
trainXpsize	.3023954	.2256155	1.34	0.180	-.1398029	.7445937
train	5.559205	.6991387	7.95	0.000	4.188918	6.929492
busXhinc	-.0233237	.0162973	-1.43	0.152	-.0552658	.0086185
busXpsize	-.0300096	.3339774	-0.09	0.928	-.6845933	.624574
bus	4.433192	.7783339	5.70	0.000	2.907685	5.958698
GC	-.019685	.0054015	-3.64	0.000	-.0302717	-.0090983
Ttime	-.1015659	.0112306	-9.04	0.000	-.1235776	-.0795543

These estimates, as well as estimated odds ratios with associated 95% confidence intervals produced by including the `or` option, were reported in table 12.1 on page 636 under "Both traveler and alternative-specific covariates".

According to this model, a $10 increase in the generalized cost for any of the alternatives would decrease the odds of that alternative being chosen by an estimated 18% $[-18\% = 100\%\{\exp(10 \times -0.019685) - 1\}]$, keeping the cost of the other alternatives as well as `Ttime`, `Hinc`, and `Psize` constant. Increasing terminal time by 10 minutes for a particular mode of transport would decrease the odds of traveling with that alternative by an estimated 64% $[-64\% = 100\%\{\exp(10 \times -0.1015659) - 1\}]$, keeping the other covariates constant. A $1,000 increase in household income would hardly change the odds of flying $[0.4\% = 100\%\{\exp(0.004071) - 1\}]$, decrease the odds of taking the train by 5% $[5\% = 100\%\{\exp(-0.0551849) - 1\}]$, and decrease the odds of taking the bus by 2% $[2\% = 100\%\{\exp(-0.0233237) - 1\}]$, keeping the other covariates constant. An extra person in the traveling party would reduce the odds of flying by 64% $[64\% = 100\%\{\exp(-1.027423) - 1\}]$, increase the odds of taking the train by 35% $[35\% = 100\%\{\exp(0.3023954) - 1\}]$, and decrease the odds of taking the bus by 3% $[3\% = 100\%\{\exp(-0.0300096) - 1\}]$, keeping the other covariates constant.

We could alternatively have used the `asclogit` command (for "alternative-specific conditional logit"), which does not require explicit construction of the interactions. However, the setup for `clogit` is similar to the one that we will use for models involving random effects estimated by `gllamm` later in this chapter.

In `asclogit`, we specify the unit identifier by using the `case()` option, the covariates having alternative-specific coefficients by using the `casevars()` option, the variable labeling the alternatives by using the `alternatives()` option, and the base outcome by using the `basealternative()` option. Alternative-specific covariates are listed after the response variable. In the present setting, the `asclogit` command is

```
. asclogit Choice GC Ttime, case(traveler) casevars(Hinc Psize)
> alternatives(alt) basealternative(4)
```

```
Alternative-specific conditional logit          Number of obs     =        840
Case variable: traveler                         Number of cases   =        210

Alternative variable: alt                       Alts per case: min =         4
                                                               avg =       4.0
                                                               max =         4

                                                Wald chi2(8)      =     112.25
Log likelihood =  -177.4541                     Prob > chi2       =     0.0000
```

Choice	Coef.	Std. Err.	z	P>|z|	[95% Conf. Interval]	
alt						
GC	-.019685	.0054015	-3.64	0.000	-.0302717	-.0090983
Ttime	-.1015659	.0112306	-9.04	0.000	-.1235776	-.0795543
1						
Hinc	.004071	.0127247	0.32	0.749	-.0208689	.029011
Psize	-1.027423	.2656569	-3.87	0.000	-1.548101	-.5067448
_cons	7.873608	.9868475	7.98	0.000	5.939423	9.807794
2						
Hinc	-.0551849	.0144824	-3.81	0.000	-.0835698	-.0268
Psize	.3023954	.2256155	1.34	0.180	-.1398029	.7445937
_cons	5.559205	.6991387	7.95	0.000	4.188918	6.929492
3						
Hinc	-.0233237	.0162973	-1.43	0.152	-.0552658	.0086185
Psize	-.0300096	.3339774	-0.09	0.928	-.6845933	.624574
_cons	4.433192	.7783339	5.70	0.000	2.907685	5.958698
4	(base alternative)					

We see that the estimates are identical to those produced by `clogit` with the appropriate interactions included.

12.3 Independence from irrelevant alternatives

An important feature of multinomial logit and conditional logit models is so-called independence of irrelevant alternatives (IIA). This means that the odds comparing an alternative s to another alternative t only depends on the characteristics of these two alternatives and *not* on any third alternative u. This is evident for the conditional logit model, where we see from expression (12.4) that the odds are

$$\frac{\Pr(y_i = s | x_i^{[1]}, \ldots, x_i^{[S]})}{\Pr(y_i = t | x_i^{[1]}, \ldots, x_i^{[S]})} = \exp\{\beta(x_i^{[s]} - x_i^{[t]})\} \qquad (12.7)$$

which clearly depends only on the attributes of alternatives s and t. The odds also remain the same if an alternative is removed from or added to the alternative set because this affects the denominators of both probabilities equally; see (12.4).

The notion of IIA was proposed in decision theory, where it was considered to be a required property of appropriately specified choice probabilities. Indeed, the conditional logit model was originally derived by Luce (1959) from IIA, viewing IIA as a desirable axiom.

Although IIA may be realistic in some choice situations, it is clearly inappropriate in others. A famous example is the red-bus–blue-bus problem of McFadden (1974). Consider a commuter who can travel to work either by car or by taking a blue bus, and for simplicity assume that the probability of each of these alternatives is $1/2$, giving an odds of 1. Now suppose that a red bus is introduced and that the commuter finds the two bus alternatives (which differ only in color) to be equally attractive. The probability of taking the red bus hence equals the probability of taking the blue bus. However, as shown above, the odds of choosing the blue bus compared with the car remains 1 in the conditional logit model, implying that each of the three alternatives is equally likely, with probabilities $1/3$. It is clearly counterintuitive that the probability of driving decreases from $1/2$ to $1/3$ due to the introduction of the red bus. In reality, we would expect the probability of taking a bus to be split evenly at $1/4$ between the old (blue) bus and the new (red) bus, keeping the probability of driving equal to $1/2$ and increasing the odds of car versus blue bus from 1 to 2.

The utility-maximization formulation discussed in the next section is convenient for relaxing the IIA assumption.

12.4   Utility-maximization formulation

Instead of writing down multinomial logit and conditional logit models directly in terms of the probability of a category s, we can alternatively derive and motivate these models from a rational-choice perspective.

Consider the utility or "attractiveness" $U_i^{[s]}$ associated with alternative or category s for a subject i. The rational choice for subject i is then to choose the alternative for which his or her utility is the highest. In other words, alternative s is chosen by subject i if the utility of this alternative is greater than the utility of any other alternative t:

$$U_i^{[s]} > U_i^{[t]} \quad \text{for all } t \neq s$$

We then let the utilities be linear functions of subject-specific covariates x_i and alternative-specific covariates $x_i^{[s]}$,

$$U_i^{[s]} \;=\; \underbrace{\beta_1^{[s]} + \beta_2^{[s]} x_i + \beta_3 x_i^{[s]}}_{V_i^{[s]}} + \epsilon_i^{[s]}$$

where $\epsilon_i^{[s]}$ is an error term that varies over both alternatives and subjects, and

$$V_i^{[s]} \equiv \beta_1^{[s]} + \beta_2^{[s]} x_i + \beta_3 x_i^{[s]}$$

represents the fixed part of the utility and is equal to the linear predictor of model (12.6). This provides an appealing interpretation of the coefficients in regression models for nominal responses as linear effects of covariates on utilities.

For the case of three alternatives, 1, 2, and 3, it follows that the probability of choosing an alternative—say, alternative 2—given the covariates, becomes

$$
\begin{aligned}
\Pr(y_i = 2 | x_i, x_i^{[1]}, x_i^{[2]}, x_i^{[3]}) &= \Pr(U_i^{[2]} > U_i^{[1]}, \; U_i^{[2]} > U_i^{[3]}) \\
&= \Pr(U_i^{[2]} - U_i^{[1]} > 0, \; U_i^{[2]} - U_i^{[3]} > 0) \\
&= \Pr\left(\epsilon_i^{[2]} - \epsilon_i^{[1]} > V_i^{[2]} - V_i^{[1]}, \; \epsilon_i^{[2]} - \epsilon_i^{[3]} > V_i^{[2]} - V_i^{[3]} \right)
\end{aligned}
$$

where

$$V_i^{[2]} - V_i^{[1]} = \beta_1^{[2]} - \beta_1^{[1]} + (\beta_2^{[2]} - \beta_2^{[1]}) x_i + \beta_3 (x_i^{[2]} - x_i^{[1]})$$

and

$$V_i^{[2]} - V_i^{[3]} = \beta_1^{[2]} - \beta_1^{[3]} + (\beta_2^{[2]} - \beta_2^{[3]}) x_i + \beta_3 (x_i^{[2]} - x_i^{[3]})$$

There are analogous expressions for $\Pr(y_i = 1 | x_i, x_i^{[1]}, x_i^{[2]}, x_i^{[3]})$ and $\Pr(y_i = 3 | x_i, x_i^{[1]}, x_i^{[2]}, x_i^{[3]})$.

We can identify the regression coefficients β_3 for alternative-specific covariates but only identify the differences between intercepts $\beta_1^{[2]} - \beta_1^{[1]}$ and $\beta_1^{[2]} - \beta_1^{[3]}$, and differences between regression coefficients for subject-specific covariates $\beta_2^{[2]} - \beta_2^{[1]}$ and $\beta_2^{[2]} - \beta_2^{[3]}$. However, the intercepts and coefficients for subject-specific covariates become identified once we fix the intercepts and coefficients for a base or reference category—say, category 1—to zero; $\beta_1^{[1]} = 0$ and $\beta_2^{[1]} = 0$.

If the error terms $\epsilon_i^{[s]}$ have independent Gumbel or standard extreme value type I distributions (with expectation equal to Euler's constant of about 0.577 and variance of $\pi^2/6$), the differences $\epsilon_i^{[s]} - \epsilon_i^{[t]}$ have standard logistic distributions (with expectation 0 and variance $\pi^2/3$). It can be shown that the resulting choice probability is given by the *conditional logit model* (12.3). Conversely, this particular utility-maximization formulation follows from the expression for the probabilities in (12.3).

Seen from the utility-maximization perspective, IIA can be relaxed by letting the utilities of different alternatives be dependent. The more similar the alternatives are, the more correlated their utilities would be expected to be. Indeed, in the red-bus–blue-bus example, the utilities for the two almost identical types of bus would be almost perfectly correlated. In the multilevel setting, correlation among utilities can be introduced by including random effects in multinomial logit and conditional logit models. This is the topic of the rest of the chapter.

12.5   Does marketing affect choice of yogurt?

Although Americans on average eat only about 7 pounds of yogurt a year compared with about 49 pounds in France, yogurt sales in the U.S. have grown by nearly 10% annually for the past three decades, unlike most other grocery products. It hence comes as no surprise that marketing is viewed as crucial by the competitors in the yogurt market.

In the rest of this chapter, we show how conditional logit models—in particular, multilevel conditional logit models that include various kinds of random effects—can be used to investigate how marketing variables affect the brand choice of yogurt. To this end, we use panel data on choice of brand of yogurt by 100 households in Springfield, Missouri, U.S.A. from the marketing research firm A. C. Nielsen.

The data were collected by optical scanners at supermarket checkouts over a period of about two years and contain information on brand purchases, store environment variables (for example, prices of brands), the marketing environment (for example, newspaper feature advertisements), and the value of any coupons used by the panel members (that is, households). Each participant was provided with an identification card to present at the time of purchase, and all purchases were scanned under the corresponding identification number. These data provide a reasonably complete record of the households' purchases over time.

The data considered here were confined to purchases of either the six-ounce size of Yoplait or Weight Watchers or the comparable eight-ounce size of Dannon (known by its original name, Danone, in Europe) or Hiland. Between 4 and 185 purchases per household were recorded (mean 24), giving a total of 2,412 purchases. The market shares for Yoplait, Dannon, Weight Watchers, and Hiland in these data were 34%, 40%, 23%, and 3%, respectively.

This dataset was previously analyzed by Jain, Vilcassim, and Chintagunta (1994) and Chen and Kuo (2001), and provided by Lynn Kuo. The variables are

- `house`: identifier for household (j)
- `occ`: purchasing occasion (i)
- `brand`: brand (1: Yoplait; 2: Dannon; 3: Hiland; 4: Weight Watchers)
- `choice`: dummy variable for the chosen brand
- `feature`: dummy variable for newspaper feature advertisement for brand ($x_{2i}^{[s]}$)
- `price`: price of brand in U.S.\$/oz ($x_{3i}^{[s]}$). For the brand purchased, the price is the actual price (shelf price net of value of coupons redeemed); for all other brands, the price is the shelf price.

The dataset `yogurt.dta`, which is in expanded form, is read in by typing

```
. use http://www.stata-press.com/data/mlmus3/yogurt, clear
```

Because the market share of Hiland was only 3%, we will consider only those purchasing occasions where either Yoplait, Dannon, or Weight Watchers was chosen. We therefore drop Hiland from all choice sets,

```
. drop if brand==3
(2412 observations deleted)
```

and drop all choice sets where Hiland was chosen (that is, where choice takes the value 0 for all three brands that now remain in the data),

```
. egen sum = total(choice), by(house occ)
. drop if sum==0
(213 observations deleted)
```

We then recode the value of Weight Watchers from 4 to 3 so that brand takes the values 1: Yoplait, 2: Dannon, and 3: Weight Watchers:

```
. recode brand(4=3)
(brand: 2341 changes made)
. label define b 1 "Yoplait" 2 "Dannon" 3 "WeightW", modify
. label values brand b
```

We construct the variable set, which numbers the purchasing occasions or choice sets where Yoplait, Dannon, or Weight Watchers was chosen:

```
. egen set = group(house occ)
```

The variable price is measured in U.S.\$/oz in the dataset. However, it is more convenient to interpret price coefficients in terms of cent/oz, and we therefore construct a new price variable, pricec, and drop the old one:

```
. generate pricec = price*100
. drop price
```

We can now list the variables for household number 24:

```
. sort house occ
. list if house==24, sepby(occ) noobs
```

house	occ	brand	choice	feature	sum	set	pricec
24	1	Yoplait	0	0	1	602	11
24	1	Dannon	1	0	1	602	7.4
24	1	WeightW	0	0	1	602	7.900001
24	2	Yoplait	0	0	1	603	11
24	2	Dannon	1	0	1	603	8.1
24	2	WeightW	0	0	1	603	7.900001
24	3	Yoplait	0	0	1	604	11
24	3	Dannon	1	1	1	604	8.1
24	3	WeightW	0	0	1	604	7.900001
24	4	Yoplait	0	0	1	605	10.3
24	4	Dannon	1	0	1	605	8.1
24	4	WeightW	0	0	1	605	7.900001
24	5	Yoplait	1	0	1	606	10.3
24	5	Dannon	0	0	1	606	9.8
24	5	WeightW	0	0	1	606	7.900001
24	6	Yoplait	0	0	1	607	10.8
24	6	Dannon	1	1	1	607	8.1
24	6	WeightW	0	0	1	607	8.6
24	7	Yoplait	0	0	1	608	11.5
24	7	Dannon	1	0	1	608	8.1
24	7	WeightW	0	0	1	608	8.6

There are three possible choices of yogurt at each purchasing occasion for a household, and the choice set at each occasion is hence of size 3. We see that household 24 makes seven purchases of yogurt during the study period, buying Dannon at all occasions except the fifth, where Yoplait was bought. For this household, the only newspaper feature advertisements coinciding with purchasing occasions were for Dannon at occasions 3 and 6. The prices for the competing brands differ at each occasion and change over time for a given brand.

This dataset contains only covariates that vary over alternatives. There are no covariates that are constant across alternatives at a given occasion for a given household (such as income).

12.6   Single-level conditional logit models

We start by fitting standard conditional logit models to the yogurt data, letting i be the index for occasions, j the index for households (the clusters in this application), and as before letting s designate the choice.

12.6.1 Conditional logit models with alternative-specific intercepts

Following Jain, Vilcassim, and Chintagunta (1994) and Chen and Kuo (2001), we first consider a conditional logit model, using the marketing variables `feature` and `pricec` as covariates.

The probability that alternative s is chosen by household j at the ith purchasing occasion is modeled as

$$\Pr(y_{ij} = s | \mathbf{x}_{ij}^{[1]}, \mathbf{x}_{ij}^{[2]}, \mathbf{x}_{ij}^{[3]}) = \frac{\exp(\beta_1^{[1]} d_1^{[s]} + \beta_1^{[3]} d_3^{[s]} + \beta_2 x_{2ij}^{[s]} + \beta_3 x_{3ij}^{[s]})}{\sum_{c=1}^3 \exp(\beta_1^{[1]} d_1^{[c]} + \beta_1^{[3]} d_3^{[c]} + \beta_2 x_{2ij}^{[c]} + \beta_3 x_{3ij}^{[c]})}, \quad s = 1, 2, 3$$

(12.8)

where $\mathbf{x}_{ij}^{[s]} = (x_{2ij}^{[s]}, x_{3ij}^{[s]})'$ is the vector of covariates, `feature` and `pricec`, for brand s. In contrast to the classical conditional logit model (12.3) specified on page 639, alternative-specific intercepts $\beta_1^{[s]}$ are included in this model by using dummy variables $d_1^{[s]}$ and $d_3^{[s]}$ for alternatives 1 and 3 (Yoplait and Weight Watchers) with alternative 2 (Dannon) as base outcome.

The model is equivalent to a linear model for the utility $U_{ij}^{[s]}$ of alternative s for household j at the ith purchasing occasion,

$$U_{ij}^{[s]} = \beta_1^{[1]} d_1^{[s]} + \beta_1^{[3]} d_3^{[s]} + \beta_2 x_{2ij}^{[s]} + \beta_3 x_{3ij}^{[s]} + \epsilon_{ij}^{[s]}$$

where the $\epsilon_{ij}^{[s]}$ have Gumbel distributions (with variance $\pi^2/6$) that are independent over alternatives, occasions, and households. In this formulation, alternative 1 (Yoplait) is chosen by household j at occasion i if its utility is higher than the utilities of alternatives 2 (Dannon) and 3 (Weight Watchers); $U_{ij}^{[1]} > U_{ij}^{[2]}$ and $U_{ij}^{[1]} > U_{ij}^{[3]}$. Similarly, alternative 2 is chosen if $U_{ij}^{[2]} > U_{ij}^{[1]}$ and $U_{ij}^{[2]} > U_{ij}^{[3]}$, and alternative 3 is chosen if $U_{ij}^{[3]} > U_{ij}^{[1]}$ and $U_{ij}^{[3]} > U_{ij}^{[2]}$ (see also section 12.4).

Before fitting this model using `clogit`, we construct dummy variables for the alternatives,

```
. tabulate brand, generate(br)
      brand |      Freq.     Percent        Cum.
------------+-----------------------------------
    Yoplait |      2,341       33.33       33.33
     Dannon |      2,341       33.33       66.67
    WeightW |      2,341       33.33      100.00
------------+-----------------------------------
      Total |      7,023      100.00
```

and rename these:

```
. rename br1 Yoplait
. rename br2 Dannon
. rename br3 WeightW
```

We can then use `clogit` to obtain the estimates for the conditional logit model noting that Dannon is base outcome because the alternative-specific covariate `Dannon` is omitted:

```
. clogit choice Yoplait WeightW feature pricec, group(set) vce(cluster house)
Conditional (fixed-effects) logistic regression   Number of obs   =      7023
                                                   Wald chi2(4)    =     61.11
                                                   Prob > chi2     =    0.0000
Log pseudolikelihood = -2356.6791                  Pseudo R2       =    0.0837

                               (Std. Err. adjusted for 100 clusters in house)
```

choice	Coef.	Robust Std. Err.	z	P>|z|	[95% Conf. Interval]	
Yoplait	.7463574	.2757005	2.71	0.007	.2059945	1.28672
WeightW	-.6393524	.4462032	-1.43	0.152	-1.513895	.2351897
feature	.3047969	.1651837	1.85	0.065	-.0189571	.6285509
pricec	-.3676407	.0552442	-6.65	0.000	-.4759174	-.259364

These estimated coefficients have marginal or population-averaged interpretations, as discussed for dichotomous responses in chapter 10. Because we have used `clogit` with the `vce(cluster house)` option, the estimated standard errors, test-statistics, and confidence intervals for the marginal effects are "robust", taking the dependence of purchases across occasions within households into account.

The first two coefficients represent the estimated log odds of buying the corresponding brand versus Dannon when the covariates `feature` and `pricec` are zero. According to the estimates for `feature` and `pricec`, the use of newspaper feature advertising for a brand increases the odds of buying that brand by an estimated 36% $[36\% = 100\% \times \{\exp(0.3047969) - 1\}]$, keeping the prices of the brands constant and for given advertising of the other brands. The estimated coefficient of `pricec` corresponds to an odds ratio of $\exp(-0.3676407) = 0.6923659$. It follows that increasing the price of a brand by 1 cent per oz (corresponding to an 8 cent increase for an 8 oz yogurt) reduces the odds of buying that brand by an estimated 31% $[-31\% = 100\% \times (0.6923659 - 1)]$, controlling for the prices of other brands and feature advertising.

As a precursor for our treatment of multilevel modeling of these data, we also estimate the conditional logit model using gllamm:

```
. gllamm brand Yoplait WeightW feature pricec, i(house) link(mlogit)
> expanded(set choice o) noconstant cluster(house) robust init

number of level 1 units = 7023

Condition Number = 9.5142813

gllamm model

log likelihood = -2356.6791

Robust standard errors for clustered data: cluster(house)
```

brand	Coef.	Std. Err.	z	P>|z|	[95% Conf. Interval]	
Yoplait	.7463572	.2757004	2.71	0.007	.2059943	1.28672
WeightW	-.6393524	.4462032	-1.43	0.152	-1.513895	.2351897
feature	.3047968	.1651837	1.85	0.065	-.0189572	.6285508
pricec	-.3676406	.0552442	-6.65	0.000	-.4759173	-.2593639

We store the estimates for later use under the name condlogit:

```
. estimates store condlogit
```

We see that the estimates for the conditional logit model from gllamm are practically identical to those produced by clogit.

The part of the gllamm command preceding the options is the same as for the clogit command used above except that the response variable is brand instead of choice. However, the options (given after the comma) are different. The option i(house) designates household as a cluster variable, but estimation proceeds without considering any random effects here because the init option (for initial values) is also given. The option link(mlogit) specifies a multinomial logit link, and the expanded(set choice o) option means that the data are in the form required by the clogit command (with three records per purchase), where

- set is the identifier for the alternative set
- choice is the indicator for the response category or chosen alternative
- o means that one set of coefficients should be estimated for the covariates (m would mean that a separate set would be estimated for each alternative as in the multinomial logit model)

The options cluster(house) and robust produce robust standard errors, taking the clustering of purchasing occasions within household into account.

We can obtain estimated odds ratios with associated 95% confidence intervals by replaying the results, without reestimation of the model, with the eform option (for exponentiated form):

```
. gllamm, eform

number of level 1 units = 7023

Condition Number = 9.5142813

gllamm model

log likelihood = -2356.6791
```

brand	exp(b)	Std. Err.	z	P>|z|	[95% Conf. Interval]	
Yoplait	2.109302	.1750415	8.99	0.000	1.792674	2.481854
WeightW	.527634	.0287251	-11.74	0.000	.4742335	.5870476
feature	1.356349	.1725318	2.40	0.017	1.057051	1.740392
pricec	.692366	.0175438	-14.51	0.000	.6588206	.7276193

The estimated coefficients and their robust standard errors as well as corresponding odds ratios with 95% confidence intervals are practically identical to the results from clogit and are reported under "Conditional logit" in table 12.2.

Table 12.2: Estimates for nominal regression models for choice of yogurt

| | Single-level | | | | Multilevel | | | | | | | |
| | Conditional logit | | | | Random alt.-specific intercepts | | | | Random coefficients | | | |
	Est	(SE*)	OR	(95% CI*)	Est	(SE)	OR	(95% CI)	Est	(SE)	OR	(95% CI)
Fixed intercepts												
$\beta_1^{[1]}$ [Yoplait]	0.75	(0.28)			0.88	(0.54)			0.99	(0.10)		
$\beta_1^{[2]}$ [Dammon]	0				0				0			
$\beta_1^{[3]}$ [WeightW]	−0.64	(0.45)			−2.55	(0.67)			−0.66	(0.06)		
Fixed coefficients												
β_2 [feature]	0.30	(0.17)	1.36	(1.06, 1.74)	0.81	(0.22)	2.24	(1.47, 3.43)	0.79	(0.31)	2.21	(1.21, 4.02)
β_3 [pricec]	−0.37	(0.06)	0.69	(0.66, 0.73)	−0.46	(0.05)	0.63	(0.58, 0.69)	−0.53	(0.08)	0.59	(0.50, 0.69)
Random intercepts												
$\psi^{[1]}$ [Yoplait]					15.01	(3.62)						
$\psi^{[2]}$ [Dammon]					0							
$\psi^{[3]}$ [WeightW]					17.69	(4.61)						
$\psi^{[21]}$					0							
$\psi^{[31]}$					10.28	(3.34)						
$\psi^{[32]}$					0							
Random coefficients												
ψ_2 [feature]									0.50	(0.10)		
ψ_3 [pricec]									3.24	(1.19)		
ψ_{32}									0.54	(0.27)		
Log likelihood	−2356.68†				−1009.54				−1891.28			

*Robust standard errors accommodating clustering of purchasing occasions within households
†Log conditional likelihood

12.7   Multilevel conditional logit models

In the yogurt application, we have panel data with purchasing occasions i nested in households j. Hence, the assumption of conditional logit models that brand choice is independent over time for households, given newspaper feature advertising and prices, is unrealistic. To accommodate longitudinal dependence, we introduce household-specific random effects. The random effects represent unobserved heterogeneity, such as taste variation in households' preferences for yogurt or household-specific effects of the marketing variables. In contrast to the marginal models discussed in section 12.6, all coefficients in this section have conditional or household-specific interpretations.

12.7.1   Preference heterogeneity: Brand-specific random intercepts

Let $\zeta_{1j}^{[s]}$ be an alternative-specific random intercept for brand s that varies randomly over households j. These intercepts represent the households' unobserved heterogeneity or taste variation for the different brands and have variances denoted $\psi^{[s]}$. In marketing, this kind of heterogeneity is often called *preference heterogeneity*.

As for fixed alternative-specific intercepts, we must designate a base outcome to ensure identification. We continue to use Dannon ($s = 2$) as base category, for which we set $\zeta_{1j}^{[2]} = 0$ or equivalently fix its variance to zero, $\psi^{[2]} = 0$.

The conditional probability of choosing brand s ($s = 1, \ldots, 3$), given the covariates $\mathbf{x}_{ij}^{[s]}$ and the random alternative-specific intercepts $\zeta_{1j}^{[s]}$, is specified as

$$\Pr(y_{ij} = s | \mathbf{x}_{ij}^{[1]}, \mathbf{x}_{ij}^{[2]}, \mathbf{x}_{ij}^{[3]}, \zeta_{1j}^{[1]}, \zeta_{1j}^{[3]})$$

$$= \frac{\exp\left\{ (\beta_1^{[1]} + \zeta_{1j}^{[1]})d_1^{[s]} + (\beta_1^{[3]} + \zeta_{1j}^{[3]})d_3^{[s]} + \beta_2 x_{2ij}^{[s]} + \beta_3 x_{3ij}^{[s]} \right\}}{\sum_{c=1}^3 \exp\left\{ (\beta_1^{[1]} + \zeta_{1j}^{[1]})d_1^{[c]} + (\beta_1^{[3]} + \zeta_{1j}^{[3]})d_3^{[c]} + \beta_2 x_{2ij}^{[c]} + \beta_3 x_{3ij}^{[c]} \right\}} \tag{12.9}$$

Here $\beta_1^{[s]} + \zeta_{1j}^{[s]}$ is the total alternative-specific intercept for brand s, composed of the sum of a fixed alternative-specific intercept $\beta_1^{[s]}$ and a random deviation $\zeta_{1j}^{[s]}$ for household j.

Given the covariates and the random intercepts, the above model is equivalent to a linear model for the utility $U_{ij}^{[s]}$ of alternative s for household j at the ith purchasing occasion:

$$U_{ij}^{[s]} = (\beta_1^{[1]} + \zeta_{1j}^{[1]})d_1^{[s]} + (\beta_1^{[3]} + \zeta_{1j}^{[3]})d_3^{[s]} + \beta_2 x_{2ij}^{[s]} + \beta_3 x_{3ij}^{[s]} + \epsilon_{ij}^{[s]}$$

Here the $\epsilon_{ij}^{[s]}$ have Gumbel distributions (with variance $\pi^2/6$) that are independent over alternatives, occasions, and households.

Compared with the ordinary conditional logit model with alternative-specific intercepts, discussed in section 12.6.1, the new feature of the model is the random alternative-specific intercepts $\zeta_{1j}^{[s]}$ that are shared for different purchasing occasions of a household.

Importantly, the utilities are no longer independent as they were in section 12.4, and the assumption of independence of irrelevant alternatives (IIA) discussed in section 12.3 is hence relaxed.

The covariances $\psi^{[st]}$ between the random intercepts for any pair of alternatives s and t are free parameters except for the covariances involving the intercept for Dannon ($s = 2$) that are all zero because $\zeta_{1j}^{[2]}$ does not vary.

A normal distribution is then specified for the vector of alternative-specific random intercepts,

$$\begin{bmatrix} \zeta_{1j}^{[1]} \\ \zeta_{1j}^{[3]} \end{bmatrix} \sim N\left(\begin{bmatrix} 0 \\ 0 \end{bmatrix}, \begin{bmatrix} \psi^{[1]} & \\ \psi^{[31]} & \psi^{[3]} \end{bmatrix} \right)$$

where the random intercepts are assumed to be independent of the observed covariates for all alternatives ($\mathbf{x}_{ij}^{[1]}$, $\mathbf{x}_{ij}^{[2]}$, and $\mathbf{x}_{ij}^{[3]}$).

The correlations between the utility differences implied by the model are shown in display 12.3 (this can be skipped if you like).

Given the covariates, it follows from model (12.9) that the correlation between the utility differences $u_{ij}^{[1]} - u_{ij}^{[2]}$ and $u_{ij}^{[3]} - u_{ij}^{[2]}$ (where the base category 2 is involved in both differences) becomes

$$\text{Cor}\{(u_{ij}^{[1]} - u_{ij}^{[2]}), (u_{ij}^{[3]} - u_{ij}^{[2]}) \mid \mathbf{x}_{ij}^{[1]}, \mathbf{x}_{ij}^{[2]}, \mathbf{x}_{ij}^{[3]}\} = \frac{\psi^{[31]} + \pi^2/6}{\sqrt{\psi^{[1]} + \pi^2/3}\sqrt{\psi^{[3]} + \pi^2/3}}$$

where $\pi^2/3$ is the variance of the difference between two independent Gumbel distributions with variances $\pi^2/6$. The expression for the correlation between utility differences such as $u_{ij}^{[2]} - u_{ij}^{[1]}$ and $u_{ij}^{[3]} - u_{ij}^{[1]}$ (where the base category 2 is involved in only one of the differences) is more convoluted:

$$\text{Cor}\{(u_{ij}^{[2]} - u_{ij}^{[1]}), (u_{ij}^{[3]} - u_{ij}^{[1]}) \mid \mathbf{x}_{ij}^{[1]}, \mathbf{x}_{ij}^{[2]}, \mathbf{x}_{ij}^{[3]}\} = \frac{\psi^{[1]} - \psi^{[31]} + \pi^2/6}{\sqrt{\psi^{[1]} + \pi^2/3}\sqrt{\psi^{[1]} + \psi^{[3]} - 2\psi^{[31]} + \pi^2/3}}$$

Similarly, the correlation between $u_{ij}^{[1]} - u_{ij}^{[3]}$ and $u_{ij}^{[2]} - u_{ij}^{[3]}$ is

$$\text{Cor}\{(u_{ij}^{[1]} - u_{ij}^{[3]}), (u_{ij}^{[2]} - u_{ij}^{[3]}) \mid \mathbf{x}_{ij}^{[1]}, \mathbf{x}_{ij}^{[2]}, \mathbf{x}_{ij}^{[3]}\} = \frac{\psi^{[1]} - \psi^{[31]} + \pi^2/6}{\sqrt{\psi^{[1]} + \psi^{[3]} - 2\psi^{[31]} + \pi^2/3}\sqrt{\psi^{[3]} + \pi^2/3}}$$

Conditional on both covariates and random intercepts, the correlations between all utility differences (from base category) become $\frac{\pi^2/6}{\sqrt{\pi^2/3}\sqrt{\pi^2/3}} = 0.5$. It also follows that utility *differences* (from base category) have a correlation of 0.5 even if the utilities themselves are uncorrelated. Correlated utilities therefore correspond to correlations among utility differences (from base category) that are different from 0.5.

Display 12.3: Correlations between utility differences in conditional logit model with alternative-specific random intercepts

Before using `gllamm` to fit the model, we use the `eq` command to specify the variables that multiply the random effects. Here the alternative-specific intercepts for Yoplait and Dannon are multiplied by the corresponding dummy variables $d_1^{[s]}$ and $d_3^{[s]}$:

```
. eq Yo: Yoplait
. eq We: WeightW
```

The names `Yo` and `We` are arbitrary; they were chosen to be short but descriptive.

We can then fit the model by maximum likelihood using the following `gllamm` command with the equation names given in the `eqs()` option:

```
. gllamm brand Yoplait WeightW feature pricec, i(house) link(mlogit)
> expanded(set choice o) noconstant nrf(2) eqs(Yo We) adapt

number of level 1 units = 7023
number of level 2 units = 100

Condition Number = 18.432026

gllamm model

log likelihood = -1009.5358
```

brand	Coef.	Std. Err.	z	P>|z|	[95% Conf. Interval]	
Yoplait	.8782923	.5437034	1.62	0.106	-.1873469	1.943931
WeightW	-2.549511	.6673952	-3.82	0.000	-3.857582	-1.241441
feature	.8085093	.2160702	3.74	0.000	.3850194	1.231999
pricec	-.4606993	.0450446	-10.23	0.000	-.5489851	-.3724136

```
Variances and covariances of random effects
------------------------------------------------------------------------------

***level 2 (house)

    var(1): 15.007983 (3.6153107)
    cov(2,1): 10.282717 (3.341922) cor(2,1): .63115183

    var(2): 17.685843 (4.6099075)
------------------------------------------------------------------------------
. estimates store rint
```

This `gllamm` command adds some new options to the one used earlier. Specifically, `nrf(2)` stands for two random effects, and `eqs(Yo We)` is used to specify that the variables defined above with the `eq` commands have random coefficients (in this case, a bivariate normal distribution is assumed for the alternative-specific intercepts).

We can use a likelihood-ratio test to compare the fitted model (12.9) with the standard conditional logit model (12.8) that did not include random alternative-specific intercepts. This can be accomplished using the command

```
. lrtest rint condlogit
Likelihood-ratio test                                    LR chi2(3)  =   2694.29
(Assumption: condlogit nested in rint)                   Prob > chi2 =    0.0000
```

The asymptotic null distribution is not simply chi-squared with 3 degrees of freedom (for two variances and one covariance) in this case because the null hypothesis is on the border of the parameter space (because variances cannot be negative). As mentioned in section 4.6, we do not know the distribution for testing the null hypothesis that several variances are zero if the corresponding random effects are correlated. Although the naïve p-value is too large, it is still less than 0.05, so we know that we can reject the standard conditional logit model in favor of the conditional logit model with random alternative-specific intercepts at the 5% significance level.

Estimated variances and covariances of the random alternative-specific intercepts are given under **Variances and covariances of random effects** in the output. The index for these parameters runs from 1 to 2 (because there are two random intercepts) and hence differs from the one used in our model specification (where the base outcome is 2). Hence, 1 refers to Yoplait as before, but 2 now refers to Weight Watchers.

The estimated coefficients in the upper part of the output represent conditional or household-specific effects on the log odds for the brands compared with the base outcome Dannon. Because odds ratios are usually preferred for interpretation, we obtain estimated odds ratios with confidence intervals without refitting the model as

```
. gllamm, eform
number of level 1 units = 7023
number of level 2 units = 100

Condition Number = 18.432026

gllamm model

log likelihood = -1009.5358
```

brand	exp(b)	Std. Err.	z	P>|z|	[95% Conf. Interval]	
Yoplait	2.406786	1.308578	1.62	0.106	.8291561	6.986163
WeightW	.0781198	.0521368	-3.82	0.000	.021119	.2889676
feature	2.244559	.4849824	3.74	0.000	1.469643	3.428076
pricec	.6308423	.028416	-10.23	0.000	.5775356	.6890692

(*Estimates for random part not shown*)

The estimated coefficients and their standard errors as well as corresponding odds ratios with 95% confidence intervals were reported under "Random alt.-specific intercepts" in table 12.2 on page 658. The estimated coefficient of **feature** corresponds to a household-specific odds ratio of $\exp(0.8085093) = 2.244559$. According to the model, the use of newspaper feature advertising for a brand hence increases the odds of buying that brand by an estimated 124% $[124\% = 100\% \times (2.244559 - 1)]$ for a given household, keeping the prices of the brands and the advertising of the other brands constant.

The estimated coefficient of `pricec` corresponds to a household-specific odds ratio of $\exp(-0.4606993) = 0.6308423$. It follows that increasing the price of a brand by 1 cent per oz (corresponding to an 8 cent increase for an 8 oz yogurt) reduces the odds of buying that brand by an estimated 37% $[-37\% = 100\% \times (0.6308423 - 1)]$ for a given household, controlling for the prices of other brands and feature advertising.

Comparing these conditional or household-specific estimates for the odds ratios with the marginal or population-averaged estimates presented for the conditional logit model under "Single-level" in the table, we see that the latter are closer to 1 as expected. For instance, the estimated odds ratio for feature advertising is about 1.36 for the conditional logit model and 2.24 for the model with random alternative-specific intercepts.

12.7.2 Response heterogeneity: Marketing variables with random coefficients

It may very well be that the effects of the marketing variables vary over households. For instance, some households may be more or less susceptible to newspaper feature advertisements than others and more or less sensitive to changes in the prices of the yogurt brands.

We now consider a model where the coefficients for the marketing variables are allowed to vary randomly over households j but not over alternatives s. In contrast to the previous model, the alternative-specific intercepts are treated as fixed. The coefficients of `feature` are $\beta_2 + \zeta_{2j}$, where ζ_{2j} is a random deviation from the fixed effect β_2 for households j, and the coefficients of `pricec` become $\beta_3 + \zeta_{3j}$, where ζ_{3j} vary randomly over households. In marketing, this kind of heterogeneity in the effects of covariates is often called *response heterogeneity*.

The following model is specified for the conditional probability that brand s is chosen, given the covariates $\mathbf{x}_{ij}^{[s]}$ and the random coefficients ζ_{2j} and ζ_{3j}:

$$
\begin{aligned}
&\Pr(y_{ij} = s | \mathbf{x}_{ij}^{[1]}, \mathbf{x}_{ij}^{[2]}, \mathbf{x}_{ij}^{[3]}, \zeta_{2j}, \zeta_{3j}) \\
&= \frac{\exp\left\{\beta_1^{[1]} d_1^{[s]} + \beta_1^{[3]} d_3^{[s]} + (\beta_2 + \zeta_{2j}) x_{2ij}^{[s]} + (\beta_3 + \zeta_{3j}) x_{3ij}^{[s]}\right\}}{\sum_{c=1}^{3} \exp\left\{\beta_1^{[1]} d_1^{[c]} + \beta_1^{[3]} d_3^{[c]} + (\beta_2 + \zeta_{2j}) x_{2ij}^{[c]} + (\beta_3 + \zeta_{3j}) x_{3ij}^{[c]}\right\}}
\end{aligned}
\tag{12.10}
$$

Given the covariates and the random coefficients, this model is equivalent to a linear model for the utility $U_{ij}^{[s]}$ of alternative s for household j at the ith purchasing occasion,

$$
U_{ij}^{[s]} = \beta_1^{[1]} d_1^{[s]} + \beta_1^{[3]} d_3^{[s]} + (\beta_2 + \zeta_{2j}) x_{2ij}^{[s]} + (\beta_3 + \zeta_{3j}) x_{3ij}^{[s]} + \epsilon_{ij}^{[s]}
$$

where the $\epsilon_{ij}^{[s]}$ have Gumbel distributions that are independent over alternatives, occasions, and households. Because the random coefficients ζ_{2j} and ζ_{3j} are shared for different purchasing occasions of a household, the utilities are no longer independent, and the assumption of IIA is relaxed.

It is assumed that the random coefficients have a bivariate normal distribution,

$$
\begin{bmatrix} \zeta_{2j} \\ \zeta_{3j} \end{bmatrix} \sim N \left(\begin{bmatrix} 0 \\ 0 \end{bmatrix}, \begin{bmatrix} \psi_2 & \\ \psi_{32} & \psi_3 \end{bmatrix} \right)
$$

and that the random coefficients are independent of the observed covariates $\mathbf{x}_{ij}^{[1]}$, $\mathbf{x}_{ij}^{[2]}$, and $\mathbf{x}_{ij}^{[3]}$.

Let us now estimate (12.10) using `gllamm`. To specify random coefficients of the marketing variables, we first define equations by using the `eq` commands:

```
. eq pr: pricec
. eq fea: feature
```

The only changes that must be made to the previous `gllamm` command are to specify different equations in `eqs()` to let the two marketing variables have random coefficients.

```
. gllamm brand Yoplait WeightW feature pricec,
> i(house) link(mlogit) expanded(set choice o) noconstant nrf(2) eqs(pr fea) adapt
number of level 1 units = 7023
number of level 2 units = 100

Condition Number = 6.6305969

gllamm model

log likelihood = -1891.2796
```

brand	Coef.	Std. Err.	z	P>|z|	[95% Conf. Interval]	
Yoplait	.9850251	.0998629	9.86	0.000	.7892974	1.180753
WeightW	-.6631998	.0573897	-11.56	0.000	-.7756817	-.550718
feature	.792219	.3051963	2.60	0.009	.1940452	1.390393
pricec	-.5337412	.0841761	-6.34	0.000	-.6987233	-.3687592

```
Variances and covariances of random effects
------------------------------------------------------------------------------

***level 2 (house)

    var(1): .49884763 (.09903316)
    cov(2,1): .54118847 (.26753215) cor(2,1): .42572259

    var(2): 3.2394835 (1.1857948)
------------------------------------------------------------------------------
. estimates store rcoeff
```

We can again use a likelihood-ratio test to compare model (12.10) with the standard conditional logit model (12.8) by using the commands

```
. lrtest rcoeff condlogit
Likelihood-ratio test                              LR chi2(3)  =     930.80
(Assumption: condlogit nested in rcoeff)           Prob > chi2 =     0.0000
```

Again the asymptotic null distribution is not simply a chi-squared with 3 degrees of freedom (for two variances and one covariance) because the variances cannot be negative. The naïve p-value is less than 0.05 and we know that it is too large, so we can reject the standard conditional logit model in favor of the conditional logit model with random coefficients at the 5% significance level.

As before, we can obtain estimated odds ratios with confidence intervals without refitting the model by using

```
. gllamm, eform
number of level 1 units = 7023
number of level 2 units = 100

Condition Number = 6.6305969

gllamm model

log likelihood = -1891.2796
```

brand	exp(b)	Std. Err.	z	P>|z|	[95% Conf. Interval]	
Yoplait	2.677879	.2674208	9.86	0.000	2.201849	3.256825
WeightW	.5152001	.0295672	-11.56	0.000	.4603898	.5765357
feature	2.208291	.6739624	2.60	0.009	1.214151	4.016428
pricec	.586407	.0493614	-6.34	0.000	.4972197	.6915919

(Estimates for random part not shown)

The estimated coefficients and their standard errors as well as corresponding odds ratios with 95% confidence intervals were reported under "Random coefficients" in table 12.2 on page 658. According to the random-coefficient model, the use of newspaper feature advertising for a brand increases the odds of buying that brand by an estimated 121% $[121\% = 100\% \times (2.208291 - 1)]$ for a given household, keeping the prices of the brands and the advertising of the other brands constant. Increasing the price of a brand by 1 cent per oz reduces the odds of buying that brand by an estimated 41% $[-41\% = 100\% \times (0.586407 - 1)]$ for a given household, controlling for the prices of other brands and feature advertising. We see that the estimated household-specific effects of feature and pricec are quite similar to those reported for the model with alternative-specific random intercepts. Comparing the estimated conditional or household-specific effects of the marketing variables with their marginal or population-averaged counterparts, we again see that the latter are closer to 1, as would be expected.

Based on the estimated random-coefficient model, the 95% range of the household-specific odds ratios for feature can be obtained as (see section 4.7)

```
. display exp(.792219 - 1.96*sqrt(.49884762))
.55315764
. display exp(.792219 + 1.96*sqrt(.49884762))
8.8158413
```

We see that the 95% range from 0.55 to 8.82 includes 1, so the household-specific effect of feature advertising is not necessarily positive. Analogously, the 95% range of the household-specific odds ratios for `pricec` can be obtained as

```
. display exp(-.5337412 - 1.96*sqrt(1.1857948))
.06938618
. display exp(-.5337412 + 1.96*sqrt(1.1857948))
4.9559319
```

The 95% range from 0.07 to 4.96 is wide and also includes 1. The correlation between the two random coefficients is estimated as 0.43.

12.7.3 ❖ Preference and response heterogeneity

A final model to consider combines the models fit in the two previous sections by including both preference heterogeneity (brand-specific random intercepts $\zeta_{1j}^{[1]}$ and $\zeta_{1j}^{[3]}$) and response heterogeneity (random coefficients ζ_{2j} and ζ_{3j} for the marketing variables). Hence, the two random-effects models that we have previously discussed in this chapter are both nested in this model.

The conditional probability of choosing brand s ($s = 1, \ldots, 3$), given the covariates, the random alternative-specific intercepts, and the random coefficients, is specified as

$$\Pr(y_{ij} = s | \mathbf{x}_{ij}^{[1]}, \mathbf{x}_{ij}^{[2]}, \mathbf{x}_{ij}^{[3]}, \zeta_{1j}^{[1]}, \zeta_{1j}^{[3]}, \zeta_{2j}, \zeta_{3j})$$

$$= \frac{\exp\left\{(\beta_1^{[1]} + \zeta_{1j}^{[1]})d_1^{[s]} + (\beta_1^{[3]} + \zeta_{1j}^{[3]})d_3^{[s]} + (\beta_2 + \zeta_{2j})x_{2ij}^{[s]} + (\beta_3 + \zeta_{3j})x_{3ij}^{[s]}\right\}}{\sum_{c=1}^3 \exp\left\{(\beta_1^{[1]} + \zeta_{1j}^{[1]})d_1^{[c]} + (\beta_1^{[3]} + \zeta_{1j}^{[3]})d_3^{[c]} + (\beta_2 + \zeta_{2j})x_{2ij}^{[c]} + (\beta_3 + \zeta_{3j})x_{3ij}^{[c]}\right\}}$$

Given the covariates and the random coefficients, the corresponding model for the utility $U_{ij}^{[s]}$ is

$$U_{ij}^{[s]} = (\beta_1^{[1]} + \zeta_{1j}^{[1]})d_1^{[s]} + (\beta_1^{[3]} + \zeta_{1j}^{[3]})d_3^{[s]} + (\beta_2 + \zeta_{2j})x_{2ij}^{[s]} + (\beta_3 + \zeta_{3j})x_{3ij}^{[s]} + \epsilon_{ij}^{[s]}$$

where the $\epsilon_{ij}^{[s]}$ have independent Gumbel distributions.

As before, we assume that the random effects have a multivariate normal distribution:

$$\begin{bmatrix} \zeta_{1j}^{[1]} \\ \zeta_{1j}^{[3]} \\ \zeta_{2j} \\ \zeta_{3j} \end{bmatrix} \sim N\left(\begin{bmatrix} 0 \\ 0 \\ 0 \\ 0 \end{bmatrix}, \begin{bmatrix} \psi^{[1]} & & & \\ \psi^{[31]} & \psi^{[3]} & & \\ \psi_2^{[1]} & \psi_2^{[3]} & \psi_2 & \\ \psi_3^{[1]} & \psi_3^{[3]} & \psi_{32} & \psi_3 \end{bmatrix}\right)$$

The random effects are assumed to be independent from the observed covariates $\mathbf{x}_{ij}^{[1]}$, $\mathbf{x}_{ij}^{[2]}$, and $\mathbf{x}_{ij}^{[3]}$.

Estimation using gllamm

The model including both preference and response heterogeneity has a complicated random part with four random effects and altogether 10 variance and covariance parameters. Fitting this model by maximum likelihood is very computationally challenging using gllamm. Unless you have a powerful computer as well as a patient personality, our best advice is "don't try this at home!" (it can take a couple of days using a laptop).

For the four random effects of the model (two for the alternative-specific intercepts and two for the marketing variables), we define the following equations for gllamm:

```
. eq Yo: Yoplait
. eq We: WeightW
. eq pr: pricec
. eq fea: feature
```

We fit the model by maximum likelihood in gllamm using eight-point (the default) adaptive quadrature:

```
. gllamm brand Yoplait WeightW feature pricec, i(house) link(mlogit)
> expanded(set choice o) noconstant eqs(Yo We pr fea) nrf(4) adapt
number of level 1 units = 7023
number of level 2 units = 100

Condition Number = 11.388455

gllamm model

log likelihood = -986.19516
```

brand	Coef.	Std. Err.	z	P>|z|	[95% Conf. Interval]	
Yoplait	1.27877	.5334415	2.40	0.011	.2332441	2.324296
WeightW	-2.765252	.6674995	-4.14	0.000	-4.073527	-1.456977
feature	1.179442	.3337126	3.53	0.000	.5253771	1.833506
pricec	-.642399	.0943309	-6.81	0.000	-.8272841	-.4575139

```
Variances and covariances of random effects
------------------------------------------------------------------------------

***level 2 (house)

    var(1): 14.80929 (4.206074)
    cov(2,1): 9.9720435 (3.5426805) cor(2,1): .57027494

    var(2): 20.647407 (5.429534)
    cov(3,1): .1715578 (.47452189) cor(3,1): .09051308
    cov(3,2): .65328064 (.53710655) cor(3,2): .29190063

    var(3): .24258503 (.10729907)
    cov(4,1): 2.2522654 (1.8464909) cor(4,1): .83550404
    cov(4,2): .58244816 (1.5580014) cor(4,2): .18298716
    cov(4,3): -.14708314 (.16674517) cor(4,3): -.42631144

    var(4): .49069066 (.60326651)
------------------------------------------------------------------------------
```

Estimated odds ratios with confidence intervals for the model are obtained as

```
. gllamm, eform

number of level 1 units = 7023
number of level 2 units = 100

Condition Number = 11.388455

gllamm model

log likelihood = -986.19516
```

brand	exp(b)	Std. Err.	z	P>|z|	[95% Conf. Interval]	
Yoplait	3.592219	1.916239	2.40	0.017	1.26269	10.21949
WeightW	.0629603	.0420259	-4.14	0.000	.0170173	.2329394
feature	3.252558	1.085419	3.53	0.000	1.691096	6.255783
pricec	.526029	.0496208	-6.81	0.000	.4372352	.632855

(Estimates for random part not shown)

Regarding the estimates for the fixed part of the model, we see that the estimates for the alternative-specific intercepts of 1.28 and -2.77 are fairly similar to those reported for the model with just preference heterogeneity (see "Random alt.-specific intercepts" in table 12.2 on page 658). The fixed coefficients of feature and pricec are estimated as 1.18 and -0.64, respectively, corresponding to odds ratios of 3.25 and 0.53. Compared with the estimates for the special cases of this model shown under "Multilevel" in table 12.2, the household-specific odds ratio of feature advertising is considerably larger and the household-specific effect of price somewhat lower in this model.

We see that the estimated variances of the alternative-specific random intercepts of 14.81 and 20.65, and the estimated covariance between the intercepts of 9.97, are quite close to the corresponding estimates for the model with just preference heterogeneity (see "Random alt.-specific intercepts" in table 12.2). In contrast, the estimated variances of the random coefficients of feature and pricec of 0.24 and 0.49, and the estimated covariance between the coefficients of -0.15, are very different from those for the model with only response heterogeneity (see "Random coefficients" in table 12.2).

Estimation using mixlogit

Because maximum likelihood estimation using adaptive quadrature in `gllamm` takes a very long time for discrete-choice models with many random effects, it is worthwhile considering alternative approaches to estimation.

A popular approach for discrete choice models is *simulated maximum likelihood* (SML), where the required integration is performed by simulation instead of numerical integration. The simulations are performed by drawing quasi-random numbers in a clever way, using what are called Halton draws, to reduce the number of draws required to obtain a given Monte Carlo error. We refer to Train (2009) for the theory and practice of simulated maximum likelihood for discrete-choice models.

To fit the model with preference and response heterogeneity by simulated maximum likelihood in Stata, we will use the user-contributed program `mixlogit` by Hole (2007), which can be downloaded by using the command

```
. ssc install mixlogit, replace
```

(it is a good idea to obtain the latest version, which is faster than the original). The `mixlogit` command fits *mixed logit models* or, in other words, logit models with random effects such as those used in this chapter. `mixlogit` also has a `mixlpred` postestimation command for predicting choice probabilities.

The model can be fit by using the following `mixlogit` command:

```
. mixlogit choice, group(set) id(house) rand(Yoplait WeightW feature pricec) corr
Mixed logit model                               Number of obs   =      7023
                                                LR chi2(10)     =   2720.01
Log likelihood = -996.67586                     Prob > chi2     =    0.0000
```

choice	Coef.	Std. Err.	z	P>|z|	[95% Conf. Interval]	
Yoplait	1.619169	.2246412	7.21	0.000	1.17888	2.059457
WeightW	-2.050162	.3352067	-6.12	0.000	-2.707155	-1.393169
feature	1.336173	.2879728	4.64	0.000	.7717566	1.900589
pricec	-.6084425	.0655207	-9.29	0.000	-.7368607	-.4800242
/l11	2.956528	.3831274	7.72	0.000	2.205612	3.707444
/l21	2.261802	.3130134	7.23	0.000	1.648307	2.875297
/l31	.4597804	.3384704	1.36	0.174	-.2036094	1.12317
/l41	.1973935	.0852472	2.32	0.021	.030312	.364475
/l22	5.769872	.4939714	11.68	0.000	4.801706	6.738038
/l32	-.5196223	.3845178	-1.35	0.177	-1.273263	.2340187
/l42	.3530624	.0678431	5.20	0.000	.2200923	.4860326
/l33	.8147397	.3559046	2.29	0.022	.1171796	1.5123
/l43	-.4663116	.0562559	-8.29	0.000	-.5765711	-.356052
/l44	-.042378	.0441377	-0.96	0.337	-.1288862	.0441303

In `mixlogit`, `choice` is the response variable. The `group()` option is used to specify the purchasing occasions or choice sets, the `id()` option to specify the cluster variable, and the `rand()` option to specify the covariates that have random coefficients. A covariate designated to have a random coefficient is automatically given a fixed effect in `mixlogit` and is therefore not included in the specification of the fixed part of the model (this is the reason that no covariates are included after `choice` because all covariates have random coefficients in this case). The `corr` option specifies that we want to fit a model where the random effects are correlated and the `nrep()` option can be used to determine the number of Halton draws used (the default is 50). We refer to Hole (2007) for more information regarding the syntax of `mixlogit`.

`mixlogit` reports the estimated Cholesky decomposition of the covariance matrix of the random effects. To obtain the desired estimated variances and covariances of the random effects, we use the postestimation command `mixlcov`:

```
. mixlcov
```

choice	Coef.	Std. Err.	z	P>|z|	[95% Conf. Interval]	
v11	8.741058	2.265455	3.86	0.000	4.30085	13.18127
v21	6.687082	1.612455	4.15	0.000	3.526729	9.847435
v31	1.359354	1.047854	1.30	0.195	-.6944015	3.41309
v41	.5835993	.2153143	2.71	0.007	.1615911	1.005608
v22	38.40717	6.144645	6.25	0.000	26.36389	50.45046
v32	-1.958222	2.28916	-0.86	0.392	-6.444892	2.528449
v42	2.48359	.5644682	4.40	0.000	1.377253	3.589927
v33	1.145206	.7358702	1.56	0.120	-.2970729	2.587485
v43	-.472624	.2328278	-2.03	0.042	-.9289581	-.0162899
v44	.3828596	.0916587	4.18	0.000	.2032119	.5625074

Fitting the model is very fast in `mixlogit` with the default of 50 Halton draws, but we see that the estimates and the log likelihood are dramatically different from those produced by `gllamm`. The log likelihood is -986.20 in `gllamm` and -996.68 in `mixlogit`. The estimates for the fixed alternative-specific intercepts are 1.62 and -2.05 compared with 1.28 and -2.77, for `mixlogit` and `gllamm`, respectively, and the fixed coefficients of `feature` and `pricec` are estimated as 1.34 and -0.61 compared with 1.18 and -0.64. Regarding the random part of the model, we see that the variances of the alternative-specific random intercepts are estimated as 8.74 and 38.41 by `mixlogit` compared with 14.81 and 20.65 by `gllamm`, and the variance of the random coefficients of `feature` and `pricec` are estimated as 1.15 and 0.38 compared with 0.24 and 0.49.

As would be expected, the more Halton draws you use, the more reliable the estimates should become because Monte Carlo error is reduced at the cost of computation time. It is sometimes suggested that 50 Halton draws may suffice for model selection and 500 draws for the final model (Hole 2007). However, Hensher, Rose, and Greene (2005) point out that the number of draws required depends on the complexity of the model, such as the number of random effects and whether these are correlated. They also recommend refitting the models with an increasing number of Halton draws and monitoring the stability of the estimates to make sure that the results are reliable (analogous to the approach for assessing adequacy of the quadrature approximation). For the model considered in this section with four correlated random effects, we found that as many as 5,000 or 10,000 Halton draws may be required to give reliable results, which is very different from the default of just 50 Halton draws in `mixlogit`.

The following `mixlogit` command fits the model with preference and response heterogeneity using 10,000 Halton draws (this will take a long time):

```
. mixlogit choice,
> group(set) id(house) rand(Yoplait WeightW feature pricec) corr nrep(10000)
Mixed logit model                          Number of obs   =       7023
                                           LR chi2(10)     =    2742.32
Log likelihood = -985.51663                Prob > chi2     =     0.0000
```

choice	Coef.	Std. Err.	z	P>|z|	[95% Conf. Interval]	
Yoplait	1.056259	.4436564	2.38	0.017	.1867085	1.92581
WeightW	-2.730497	.5927365	-4.61	0.000	-3.89224	-1.568755
feature	1.152327	.3262482	3.53	0.000	.5128918	1.791761
pricec	-.6352615	.092048	-6.90	0.000	-.8156723	-.4548508
/l11	3.911001	.5862744	6.67	0.000	2.761924	5.060078
/l21	2.724117	.8150612	3.34	0.001	1.126627	4.321608
/l31	.5800756	.4833779	1.20	0.230	-.3673277	1.527479
/l41	.0642829	.1315759	0.49	0.625	-.1936012	.322167
/l22	3.961898	.5530667	7.16	0.000	2.877907	5.045889
/l32	-.2746034	.3283181	-0.84	0.403	-.918095	.3688882
/l42	.1107349	.093737	1.18	0.237	-.0729862	.2944561
/l33	.4247641	.3889848	1.09	0.275	-.3376321	1.18716
/l43	-.3219568	.0718756	-4.48	0.000	-.4628304	-.1810831
/l44	.3022503	.1211518	2.49	0.013	.0647971	.5397034

We use `mixlcov` to obtain the estimated variances and covariances of the random effects (`mixlcov` with the `sd` option reports the estimated standard deviations of the random effects instead of the covariance matrix):

```
. mixlcov
```

choice	Coef.	Std. Err.	z	P>|z|	[95% Conf. Interval]	
v11	15.29593	4.58584	3.34	0.001	6.307848	24.28401
v21	10.65403	4.201682	2.54	0.011	2.418879	18.88917
v31	2.268676	1.954882	1.16	0.246	-1.562822	6.100175
v41	2514104	.5072169	0.50	0.620	-.7427165	1.245537
v22	23.11745	5.38359	4.29	0.000	12.56581	33.66909
v32	.4922433	1.743574	0.28	0.778	-2.925099	3.909585
v42	.6138345	.4404169	1.39	0.163	-.2493667	1.477036
v33	.5923193	.6826357	0.87	0.386	-.745622	1.930261
v43	-.1298749	.1450726	-0.90	0.371	-.414212	.1544621
v44	.2114059	.0832908	2.54	0.011	.0481589	.3746529

The estimates from `mixlogit` based on 10,000 Halton draws are quite similar to those produced by `gllamm` with eight-point adaptive quadrature.

Simulated maximum likelihood is certainly an interesting alternative to maximum likelihood using adaptive quadrature when there are many random effects. However, we recommend that caution is exercised when using `mixlogit` to ensure that enough Halton draws are used to obtain reliable estimates. The number of draws should be increased until the difference in estimates is small. In summary, it is worth heeding the words of caution from Train (2009, 230):

> "The use of Halton draws and other quasi-random numbers in simulation-based estimation is fairly new and not completely understood."

12.8   Prediction of random effects and response probabilities

We illustrate prediction of random effects and response probabilities by returning to the model with response heterogeneity discussed in section 12.7.2. The model included fixed alternative-specific intercepts and random coefficients for `feature` and `pricec`.

We begin by restoring the `gllamm` estimates for this model:

```
. estimates restore rcoeff
```

For each household, we can obtain empirical Bayes predictions $\widetilde{\zeta}_{2j}$ and $\widetilde{\zeta}_{3j}$ of the random coefficients of `pricec` and `feature`, respectively, by using `gllapred` with the u option,

```
. gllapred eb, u
(means and standard deviations will be stored in ebm1 ebs1 ebm2 ebs2)
```

and empirical Bayes predictions of the household-specific coefficients $\widehat{\beta}_2 + \widetilde{\zeta}_{2j}$ and $\widehat{\beta}_3 + \widetilde{\zeta}_{3j}$ as

```
. generate eb_pricec = ebm1 + _b[pricec]
. generate eb_feature = ebm2 + _b[feature]
```

We can then plot the predicted household-specific coefficients,

```
. egen pickone = tag(house)
. twoway scatter eb_pricec eb_feature if pickone==1, mlabel(house)
> xtitle(EB prediction of feature coefficient)
> ytitle(EB prediction of price coefficient)
```

giving the graph in figure 12.3. The predicted coefficients of `feature` are mostly positive, whereas the predicted coefficients of `pricec` are mostly negative, as expected. However, some households appear to behave strangely. For instance, household 55 seems to like high prices and household 97 appears to be less likely to choose a brand when it has been advertised than when it has not been advertised. These predictions could of course also be due to chance or omitted covariates.

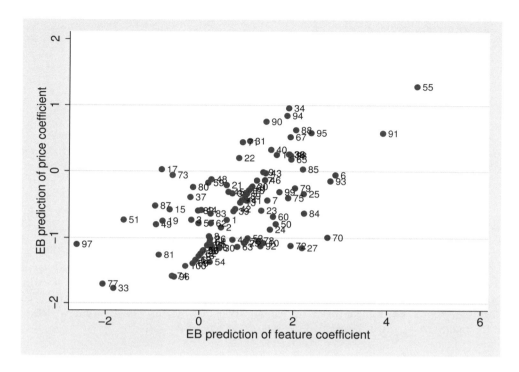

Figure 12.3:   Empirical Bayes (EB) predictions of household-specific coefficients of `pricec` and `feature`

Another way of aiding our understanding of the response heterogeneity is by plotting the brand choice probabilities for individual households against covariates. We consider the case where there is no feature advertising (`feature=0`) and where the price of Weight Watchers and Dannon is held constant at 8 cents/oz, whereas the price of Yoplait varies from 0.5 cent/oz to 20 cents/oz (approximately the range in the data). To obtain posterior mean probabilities for particular households, we need to create new rows of data for these households (prediction data), with covariates equal to the desired values. By setting the response variable to missing for invented data, we make sure that they do not affect the posterior distribution of the random effects (see section 10.13.2).

We first save the data and create a new dataset for the prediction data and then append the original data. For the predictions, we start by creating data for one household, initially 40 rows of data, with `pricec` increasing in steps of 0.5 cents/oz from 0.5 to 20:

```
. save temp, replace
(note: file temp.dta not found)
file temp.dta saved
. drop _all
. set obs 40
obs was 0, now 40
. generate pricec = _n*.5
```

We then expand the data to alternative sets comprising the three brands. The variable set should be unique for each alternative set for a given household, and we hence use values greater than 3,000 to be sure that these values do not already occur in the data:

```
. generate set = 3000 + _n
. expand 3
(80 observations created)
```

At this point, we create all the variables used in the original gllamm command so that gllapred can make the required predictions. In addition to the variable pricec, which should vary for Yoplait but be constant at 8 cents/oz for the other brands, we create a variable, yopprice, for the price of Yoplait (constant within the alternative sets) to use for making the graph later:

```
. by set, sort: generate brand=_n
. generate yopprice = pricec
. replace pricec = 8 if brand > 1
(78 real changes made)
. generate feature = 0
. tabulate brand, generate(br)
```

brand	Freq.	Percent	Cum.
1	40	33.33	33.33
2	40	33.33	66.67
3	40	33.33	100.00
Total	120	100.00	

```
. rename br1 Yoplait
. rename br2 Dannon
. rename br3 WeightW
. generate choice = Dannon
```

The choice variable is arbitrary but needs to be equal to 1 once per choice set in the prediction data. Here it was arbitrarily set equal to Dannon.

Predictions will be made for six households, namely, household 33, 15, 7, 48, 6, and 55. These households were deliberately selected to vary considerably in their household-specific coefficient of pricec by inspecting figure 12.3. We need to create identical prediction data for each of these households, which is easily accomplished using the expand command:

```
. expand 6
(600 observations created)
. by set brand, sort: generate house=_n
. recode house 1=33 2=15 3=7 4=48 5=6 6=55
(house: 720 changes made)
```

The response variable is set to missing as explained above, and the original data (`temp.dta`) are appended to the prediction data:

```
. replace brand = .
(720 real changes made, 720 to missing)
. append using temp
(label b already defined)
```

To keep track of which subset of the data is prediction data, we generate a dummy variable, `preddata`:

```
. generate preddata = brand ==.
```

We are now ready to obtain posterior mean purchasing probabilities by using the `gllapred` command with the `mu` option to specify that we want predicted probabilities (because the `marginal` option is not used, we obtain the default posterior mean probabilities instead of marginal probabilities) and the `fsample` option to obtain predictions for observations that are not in the estimation sample:

```
. gllapred probs, mu fsample
(mu will be stored in probs)
```

The posterior mean probabilities for the six selected households can then be plotted using

```
. egen select = anymatch(house), values(33 15 7 48 6 55)
. twoway (line probs yopprice if Yoplait==1, sort)
> (line probs yopprice if WeightW==1, sort lpatt(longdash))
> (line probs yopprice if Dannon==1, sort lpatt(shortdash))
> if feature==0&select==1&preddata==1,
> by(house) legend(order(1 "Yoplait" 2 "WeightW" 3 "Dannon") rows(1))
> xtitle(Price of Yoplait in cents/oz) ytitle(Probability of purchase)
```

and the resulting graph is shown in figure 12.4.

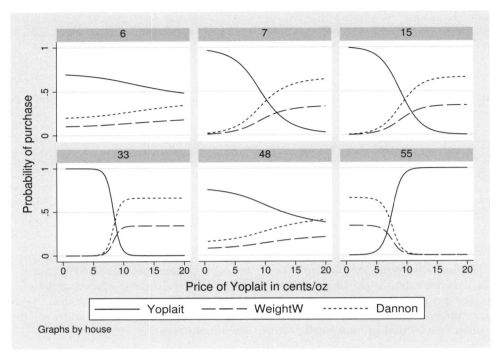

Figure 12.4: Posterior mean choice probabilities versus price in cents/oz of Yoplait for six households. Based on conditional logit model with response heterogeneity when there is no feature advertising and when the price of Weight Watchers and Dannon is held constant at 8c/oz.

Households 7, 15, and 33 had large negative empirical Bayes predictions of the household-specific coefficient of `pricec`, and this strong price sensitivity is apparent in the figure where the solid curve for the probability of purchasing Yoplait decreases rapidly. Households 6 and 48 are less price sensitive. Household 55, which had an extremely large empirical Bayes prediction of the household-specific coefficient of `pricec`, appears to find Yoplait more attractive (compared with the other brands) when it is more expensive (keeping the price of the other brands constant).

12.9 Summary and further reading

We have introduced the most commonly used models for nominal responses or discrete choices: the multinomial logit model and the conditional logit model. The models can be derived from a utility-maximization perspective that makes them attractive for applications involving choices made by individuals. The multinomial logit and conditional logit models possess a property called independence of irrelevant alternatives (IIA) which is often deemed to be unrealistic in practice. IIA is relaxed in the multilevel versions of the models that we have described in this chapter. We will return to the multinomial logit model in section 14.2.5, where it is used for discrete-time survival modeling of multiple competing events.

We have not discussed multinomial probit models in this chapter. In these models, the error terms $\epsilon_{ij}^{[s]}$ in the utility-maximization formulation have normal distributions instead of the Gumbel distributions assumed for logit models. Stata's `mprobit` program fits multinomial probit models with independent $\epsilon_{ij}^{[s]}$ by maximum likelihood using ordinary quadrature. The `asmprobit` (for "alternative-specific multinomial probit") program also fits multinomial probit models where the $\epsilon_{ij}^{[s]}$ are correlated or heteroskedastic (different covariance structures can be specified) and alternative-specific covariates can be included. Estimation proceeds by maximum simulated likelihood using the Geweke–Hajivassiliou–Keane (GHK) algorithm. We have also not considered models for rankings in this chapter. The Stata program `rologit` can be used to fit such models without random effects, and `gllamm` can be used for models including random effects (and other kinds of latent variables); see chapter 9 of the `gllamm` manual (Rabe-Hesketh et al. 2004).

For a readable introduction to models for nominal responses (without random effects), we refer the reader to Long (1997, chap. 6). The book by Train (2009) is an excellent introduction to modeling nominal responses or discrete choices, both with and without random effects, with an emphasis on simulation-based inference. We also recommend the papers by Skrondal and Rabe-Hesketh (2003a) and Skrondal and Rabe-Hesketh (2003c), the latter paper being quite demanding, and the book chapter by Hedeker (2008).

The exercises involve choices of cars (exercise 12.2) and electricity suppliers (exercise 12.6) in conjoint experiments, purchases of saltine crackers (exercise 12.1), type of housing found for homeless people (exercise 12.3), type of contraception used by Bangladeshi women (exercise 12.5), and parties voted for in British national elections (exercise 12.4).

12.10 Exercises

12.1 Buying crackers data

Jain, Vilcassim, and Chintagunta (1994) analyzed data on 2,509 purchases of saltine crackers by 100 households. The method of data collection was the same as for the yogurt data described in section 12.5, via optical scanners and identification cards. There were three major national brands: Sunshine, Keebler, and Nabisco. Local brands were grouped together as "Private" labels.

The variables in the dataset `crackers.dta` are

- `id`: identifier for household
- `occ`: purchasing occasion

- **brand**: brands
  (1: Sunshine; 2: Keebler; 3: Nabisco; 4: Private)
- **choice**: dummy variable for the chosen brand
- **feature**: dummy variable for newspaper feature advertisement
- **display**: dummy variable for brand being on special end-of-aisle or middle-of-aisle display
- **price**: price in U.S.\$/oz at occasion. For the brand purchased, the price is the actual price (shelf price net of value of coupons redeemed) and for all other brands it is the shelf price.

1. Create a unique identifier for each combination of **id** and **occ**. Also create dummy variables for the brands, treating Private as the reference category.

2. Use **clogit** to fit a conditional logit model with the covariates **feature**, **display**, and **price** and dummies for the brands Sunshine, Keebler, and Nabisco.

3. Interpret the estimates.

4. Fit the same model using **gllamm**.

5. Include a random coefficient for **feature** only, and compare this model with the model from step 4 by using a likelihood-ratio test.

6. Include a random coefficient for **display** only, and compare this model with the model from step 4 by using a likelihood-ratio test.

7. Include a random coefficient for **price** only, and compare this model with the model from step 4 by using a likelihood-ratio test.

8. Comment on the findings.

12.2 Hybrid car data

We consider a subset of the hybrid car data described by Train and Sonnier (2005). The data are provided with Kenneth Train's matlab program for mixed logit estimation by maximum simulated likelihood.

Each survey respondent was presented with 15 choice situations in each of which they were asked to choose one among three hypothetical cars. The cars were described in terms of several attributes including price and fuel type: gasoline (petrol), electric, and hybrid (gasoline–electric) cars. Such data are referred to as stated preference data in contrast to revealed preference data where the purchasing behavior is observed. An advantage of stated preference data is that the attributes can be varied in so-called conjoint designs to obtain the required information on the importance and trade-offs of attributes in making decisions. It is also possible to include alternatives that do not exist yet, such as a new subway. However, a problem with stated preference data is that the choices have no real implications for the respondent and are therefore fundamentally different from real purchasing decisions.

Random-digit dialing was used to contact potential respondents throughout the state of California, U.S.A. Those intending to buy a new vehicle within the next three years were asked to participate in the study. If they were willing to participate, they were sent a packet of materials, including information sheets that described the new vehicles and the choice experiments. The respondents were later called to go over the information and obtain their choices and demographic details.

The variables in the dataset `hybrid.dta` are

- `id`: identifier for respondent
- `situation`: choice situation (unique identifier for each respondent–situation combination)
- `chosen`: dummy variable for choosing the alternative
- `price`: price in U.S.\$10,000
- `electric`: dummy variable for electric car
- `hybrid`: dummy variable for hybrid car
- `cost`: running cost in U.S.\$ per month
- `range`: number of miles between refueling/recharging
- `highperf`: dummy variable for high-performance car (top speed 120 miles per hour, 8 seconds to reach 60 miles per hour)
- `medperf`: dummy variable for medium-performance car (top speed 100 miles per hour, 12 seconds to reach 60 miles per hour)
  (The low category had a top speed of 80 miles per hour and 16 seconds to reach 60 miles per hour.)

1. Use the `clogit` command to fit a model with `chosen` as the response variable and `price`, `cost`, `range`, `electric`, `hybrid`, `highperf`, and `medperf` as explanatory variables (one coefficient per covariate).

2. Interpret the estimates.

3. Fit the same model in `gllamm`. Note that `gllamm` requires an identifier for the alternatives as response variable, but here each choice set has a different set of hypothetical alternatives characterized by the attributes. You can arbitrarily label the four alternatives for each situation from 1 to 4.

4. Extend the model to include correlated random coefficients for `hybrid` and `electric`, and fit the model using `gllamm`. Use `ip(m)` and `nip(11)` for degree-11 spherical quadrature to speed up estimation (this will take some time to run).

12.3   Housing the homeless data

The McKinney Homeless Research Project study (Hough et al. 1997; Hurlburt, Wood, and Hough 1996) was conducted in San Diego, California, to evaluate an intervention that aimed to provide independent housing for severely mentally ill homeless people. The intervention was a program of federal assistance to provide subsidized housing for low-income families and individuals, known as Section 8. Eligibility for the study was restricted to individuals diagnosed with severe and persistent mental illness who were either homeless or at high risk of becoming homeless at the start of the study. Study participants were randomly assigned to one of two levels of access to independent housing using Section 8 certificates. They were followed up at 6, 12, and 24 months to determine their housing status (streets or shelters, community housing, or independent housing). The data are provided with the book by Hedeker and Gibbons (2006) and with the `supermix` software.

The variables in the dataset `homeless.dta` are

- `id`: person identifier
- `time`: occasion (0, 1, 2, 3), corresponding to 0, 6, 12, and 24 months, respectively
- `housing`: housing status (0: Streets or shelters; 1: Community housing; 2: Independent housing)
- `section8`: dummy variable for Section 8 group (1: Yes; 0: No)

1. Explore the missing-data patterns.
2. Create dummy variables for the three postrandomization time points and interactions between the time dummies and the treatment (Section 8) dummy.
3. Use `mlogit` to fit a multinomial logit model for `housing`, treating streets or shelters as the base outcome, with the treatment, the three time dummies, and the three time by treatment interactions as covariates.
4. Expand the data and fit the model from step 3 in `gllamm`.
5. Fit the model with two correlated alternative-specific random intercepts (use five-point adaptive quadrature—this will take a while).
6. Interpret the estimates.

12.4 British election data $\boxed{\text{Solutions}}$

Skrondal and Rabe-Hesketh (2003c, 2004) modeled data from the 1987–1992 panel of the British Election Study (Heath et al. 1993) and were given permission by Anthony Heath to make the data available. In the 1987 and 1992 panel waves, respondents were asked to state which party they had voted for in the British general elections earlier the same year. The three major political parties were Conservative, Labour, and Liberal (Alliance), and votes for other parties were treated as missing.

The variables in the dataset `elections.dta` are

- `serialno`: identifier for respondent
- `occ`: unique identifier for each combination of `serialno` and panel wave
- `party`: political party (1: Conservative; 2: Labour; 3: Liberal)
- `rank`: rank ordering of parties, where first ranking is party actually voted for

- lrdist: the distance between a voter's position on the left–right political dimension and the mean position of the party voted for. The mean positions of the parties over voters were used to avoid rationalization problems (Brody and Page 1972). The placements were constructed from four scales, where respondents located themselves and each of the parties on an 11-point scale anchored by two contrasting statements (priority should be unemployment versus inflation, increase government services versus cut taxation, nationalization versus privatization, more effort to redistribute wealth versus less effort).

- yr87: dummy variable for the 1987 national elections
- yr92: dummy variable for the 1992 national elections
- male: dummy for the voter being male
- age: age of the voter in 10-year units
- manual: dummy variable for father of voter being a manual worker
- inflation: rating of perceived inflation since the last election on a five-point scale

1. Create a variable, chosen, equal to 1 for the party voted for (rank equal to 1) and 0 for the other parties.
2. Standardize lrdist and inflation to have mean 0 and variance 1. Produce all the dummy variables and interactions necessary to fit a conditional logistic regression model (using clogit) for chosen, with the following covariates: the standardized versions of lrdist and inflation, and the dummy variables yr87, yr92, male, and manual. All variables except the standardized version of lrdist should have party-specific coefficients.
3. Fit the model using clogit and gllamm, using Conservatives as the base outcome.
4. Extend the model to include a person-level random slope for lrdist, and fit the extended model in gllamm.
5. Write down the model and interpret the estimates.
6. Instead of including a random slope for lrdist, include correlated person-level random intercepts for Labour and Liberal. Use the options ip(m) and nip(15) to use degree-15 spherical quadrature. This problem will take quite a long time to run.

See also exercise 16.8 for three-level modeling of these data.

12.5 Contraceptive method data

Here we analyze a subsample of the Bangladesh Fertility Survey (Huq and Cleland 1990) that was previously analyzed by Amin, Diamond, and Steele (1998) and made available with the MLwiN software (Rasbash et al. 2009).

The variables in the dataset contraceptive.dta that will be considered here are

- woman: woman identifier

- **district**: district identifier
- **method**: contraceptive method (1: sterilization; 2: modern reversible method; 3: traditional method; 4: not using contraception)
- **children**: number of living children at the time of the survey
- **urban**: dummy variable for living in an urban area (1: urban; 0: rural)
- **educ**: level of education (1: none; 2: lower primary; 3: upper primary; 4: secondary or higher)
- **hindu**: dummy variable for being Hindu (1: Hindu; 0: Muslim)
- **d_lit**: proportion of women in district who are literate
- **d_pray**: proportion of Muslim women in district who pray every day

1. Produce dummy variables, **one** and **twoplus**, for having one child and two or more children, respectively. Produce a dummy variable, **educat**, for having any education. Drop women who use the traditional method (to keep the model simpler and make it faster to estimate), and recode **method** so that it takes the values 1, 2, and 3 for sterilization, modern reversible method, and no contraception, respectively.

2. Use **mlogit** to fit a multinomial logit model with **method** as response variable and **one**, **twoplus**, **urban**, **educat**, **hindu**, **d_lit**, and **d_pray** as covariates.

3. Expand the data and fit the same model as in step 2 using **gllamm**.

4. Fit a two-level model that includes the same covariates as in steps 2 and 3. Include correlated district-level random intercepts for sterilization and the modern method. To speed up estimation in **gllamm**, use the options **ip(m)** and **nip(7)** for degree-7 spherical quadrature.

5. Interpret the estimates.

12.6 Electricity supplier data

Here we will analyze stated preference data on choice of electricity supplier originally used by Revelt and Train (2000). The data were made available by Kenneth Train and also accompanies Hole (2007). A sample of residential electricity customers were presented with up to 12 experiments. Each experiment consisted of choosing among four hypothetical electricity suppliers that differed in terms of the characteristics described in the variable list below. (See exercise 12.2 for a brief discussion of stated preference data.)

The variables in the dataset **electricity.dta** are

- **price**: price in cents per kilowatt-hour (kWh), 0 if time-of-day or seasonal rates
- **contract**: contract length in years (prices are guaranteed for duration of contract and customer pays penalty for switching suppliers during contract period; 0 means either side can stop the agreement at any time)
- **local**: dummy variable for company being local
- **known**: dummy variable for company being well known

- `tod`: dummy variable for time-of-day rates (11 cents per kWh between 8 a.m. and 8 p.m., and 5 cents per kWh from 8 p.m. to 8 a.m.)
- `seasonal`: dummy variable for seasonal rates (10 cents per kWh in the summer, 8 cents per kWh in the winter, and 6 cents per kWh in the spring and fall)

1. Fit a conditional logit model using `clogit` with `price`, `contract`, `local`, `known`, `tod`, and `seasonal` as explanatory variables that have constant coefficients.

2. Fit the same model using `gllamm`. This requires an identifier of the alternatives as response variable. In this example, a variable that arbitrarily labels the four alternatives for each experiment from 1 to 4 can be used.

3. Extend the model by including a random coefficient of `local`, and fit the model using `gllamm`.

4. Use a likelihood-ratio test at the 5% level of significance to compare the models in steps 2 and 3.

5. Extend the model further by including another random coefficient for `known`. Use `ip(m)` and `nip(11)` to speed up estimation.

6. Compare the models in steps 3 and 5 using a likelihood-ratio test at the 5% level of significance.

7. Interpret the estimates for the chosen model.

Part VI

Models for counts

13 Counts

13.1 Introduction

An important outcome in many investigations is a count of how many times some event has occurred. For instance, we could count the number of epileptic seizures in a week for a patient, the number of patents awarded to a company in a 5-year period, or the number of murders in a year in a city. Counts are nonnegative, integer-valued responses, taking on values 0, 1, 2, ...

In this chapter, we will discuss Poisson models for counts. After introducing single-level Poisson regression models, we describe multilevel Poisson regression models, which include random effects to model dependence and unobserved heterogeneity. Some new issues that arise in modeling counts, such as using offsets and dealing with overdispersion, will be considered. We also briefly describe negative binomial models for count data, as well as fixed-effects models and generalized estimating equations.

Two applications are considered: longitudinal modeling of number of doctor visits and small area estimation or disease mapping of lip cancer incidence. On first reading, it might be a good idea to concentrate on the first application.

13.2 What are counts?

13.2.1 Counts versus proportions

In this chapter, we consider counts of events that could in principle occur any time during a time interval (or anywhere within a spatial region). For instance, von Bortkiewicz (1898) observed 14 corps of the Prussian army from 1875 to 1894 and counted the number of deaths from horse kicks (see the logo on the spine of this book). Other examples of such counts are the number of times a person visits a doctor during a year, the number of times a person blinks in an hour, and the number of violent fights occurring in a school during a week. The term *count* without further qualification usually implies this type of count. As we discuss in section 13.3, a Poisson distribution is often appropriate in this case.

Another type of count is a count of events or *successes* that can only occur at a predetermined number of *trials*. The count y is then less than or equal to the number of trials n, and the response can be expressed as a sample proportion $p = y/n$. For instance, when counting the number of retired people in a city block at a given point in time, each of n inhabitants is either retired (success) or not (failure), and the count cannot exceed n. A meaningful summary then is the proportion of the city block's population that is retired. An appropriate distribution for the corresponding count is often the binomial distribution for n trials. If there are covariates, the same models as those for dichotomous responses are used, such as the logit or probit model. The only difference is that a binomial denominator n_i must be specified for each unit i. In `xtmelogit`, the `binomial()` option can be used to specify a variable containing the values n_i; in `gllamm`, the `denom()` option does this. For low probabilities of success and large n_i, the binomial distribution is well approximated by the Poisson distribution. We do not consider proportions further in this chapter.

13.2.2 Counts as aggregated event-history data

Counts can be thought of as aggregated versions or summaries of more detailed data on the occurrences of some kind of event. For instance, when considering the number of crimes, we usually aggregate over spatial regions and time intervals instead of retaining the original information on the locations and timings of the individual crimes. The size of spatial regions and time intervals determines the resolution at which we can investigate spatial and temporal variation. When counts are broken down by time intervals, the analysis can be referred to as survival analysis, a common model being the piecewise exponential model we will discuss in chapter 15.

Covariate information can be considered by aggregating within categories of categorical covariates, for example, producing separate counts of crimes for each type of victim classified by race and gender. If spatial and temporal variation are also of interest, we could obtain counts by region, time interval, race, and gender so that the observations are just cells of a contingency table. In this case, the observations for analysis are not units in the usual sense. As we will see in the next section, in Poisson regression it does not matter how aggregated the data are as long as we do not aggregate over covariate values. For a given set of covariate values, there can be several counts or one count equal to the sum of the counts. The parameter estimates will be the same in either case.

However, some of the variables defining the cells may be viewed as defining units or clusters that could be characterized by unobserved covariates and hence modeled by including random effects. In the crime example, regions or neighborhoods might be considered as such clusters. As we will see in section 13.10, it is possible to include random intercepts, say, for regions, even if the data have been aggregated to one count per region. Although such aggregated data provide no information on the dependence among counts within regions (because there is only one count per region), they do provide information on the random-intercept variance via the phenomenon known as overdispersion, to be discussed later.

13.3   Single-level Poisson models for counts

It is often assumed that events occur independently of each other and at a constant *incidence rate* λ (pronounced "lambda"), defined as the instantaneous probability of a new event per time interval. It follows that the number of events y occurring in a time interval of length t has a Poisson distribution

$$\Pr(y;\mu) \;=\; \frac{\exp(-\mu)\mu^y}{y!}$$

where μ is the expectation of y and is given by

$$\mu \;=\; \lambda t$$

The incidence rate λ is also often called the *intensity*, and the time interval t is sometimes referred to as the *exposure*.

If we observe two independent counts y_1 and y_2 during two successive time intervals of length t, and if the incidence rate λ is the same for both intervals, we can either add the counts together, thus obtaining an interval of length $2t$, or consider the counts separately. A convenient property of the Poisson distribution is that both approaches yield the same likelihood, up to a multiplicative constant.

$$
\begin{aligned}
\Pr(y_1;\mu{=}\lambda t)\Pr(y_2;\mu{=}\lambda t) \;&=\; \frac{\exp(-\lambda t)(\lambda t)^{y_1}}{y_1!}\frac{\exp(-\lambda t)(\lambda t)^{y_2}}{y_2!} \\[4pt]
&=\; \frac{\exp(-\lambda 2t)(\lambda t)^{y_1+y_2}}{y_1!\,y_2!} \\[4pt]
&=\; \frac{\exp(-\lambda 2t)(\lambda 2t)^{y_1+y_2}}{(y_1+y_2)!}\times\frac{(y_1+y_2)!}{2^{y_1+y_2}y_1!\,y_2!} \\[4pt]
&\propto\; \Pr(y_1+y_2;\mu{=}\lambda 2t) \qquad\qquad\qquad\quad (13.1)
\end{aligned}
$$

The multiplicative constant does not affect parameter estimation because it does not depend on the parameters. Therefore, no information regarding λ is lost by aggregating the data. We must, however, keep track of the exposure by using offsets, as discussed below.

When counts are observed for different units or subjects i characterized by covariates, the expectation μ_i is usually modeled using a log-linear model. For one covariate x_i, a multiplicative regression model for the expected counts is specified as

$$\mu_i \equiv E(y_i|x_i) \;=\; \exp(\beta_1 + \beta_2 x_i) \;=\; \exp(\beta_1)\times\exp(\beta_2 x_i) \qquad (13.2)$$

where we note that the exponential function precludes negative expected counts. The model can alternatively be written as an additive log-linear model:

$$\ln(\mu_i) \;=\; \beta_1 + \beta_2 x_i$$

If we think of this as a generalized linear model, the link function $g(\cdot)$ is just the natural logarithm, and the inverse link function $h(\cdot)\equiv g^{-1}(\cdot)$ is the exponential.

A nice feature of the log link is that if the exposure time t is the same for all persons, then the exponentiated coefficient $\exp(\beta_2)$ can be interpreted as the *incidence-rate ratio* (IRR) for a unit increase in x_i. This can be seen by substituting $\mu_i = \lambda_i t$ in (13.2):

$$\lambda_i t = \exp(\beta_1 + \beta_2 x_i) = \exp(\beta_1)\exp(\beta_2 x_i)$$

Then the ratio for two persons i and i' with covariate values x_i and $x_{i'}$ becomes

$$\frac{\lambda_i}{\lambda_{i'}} = \frac{\lambda_i t}{\lambda_{i'} t} = \frac{\exp(\beta_1)\exp(\beta_2 x_i)}{\exp(\beta_1)\exp(\beta_2 x_{i'})} = \exp\{\beta_2(x_i - x_{i'})\} = \exp(\beta_2)^{(x_i - x_{i'})}$$

When the covariate increases by 1 unit, $x_i - x_{i'} = 1$, the ratio of the incidence rates, $\lambda_i/\lambda_{i'}$, or incidence-rate ratio is hence $\exp(\beta_2)$.

The log-linear model for the expectation is combined with the assumption that conditional on the covariate x_i, the count y_i has a Poisson distribution with mean μ_i:

$$\Pr(y_i|x_i) = \frac{\exp(-\mu_i)\mu_i^{y_i}}{y_i!}$$

If different persons have different exposures t_i, the natural logarithm of the exposure must be included as an *offset*, a covariate with regression coefficient set to 1:

$$\lambda_i t_i = \exp\{\beta_1 + \beta_2 x_i + \underbrace{\ln(t_i)}_{\text{offset}}\} = \exp(\beta_1)\exp(\beta_2 x_i)t_i$$

Then the incidence rate becomes

$$\lambda_i = \exp(\beta_1)\exp(\beta_2 x_i)$$

so $\exp(\beta_2)$ still represents the incidence-rate ratio, $\lambda_i/\lambda_{i'}$, when $x_i - x_{i'} = 1$.

Following the logic of (13.1), it is clear that we can sum the counts for all persons sharing the same covariate value and let i denote the corresponding group of persons. If we also sum the corresponding exposures, we obtain the same maximum likelihood estimates of the parameters β_1 and β_2 using the aggregated data as using the unit-level data.

For the Poisson distribution, the conditional variance of the counts, given the covariate, equals the conditional expectation

$$\text{Var}(y_i|x_i) = \mu_i \tag{13.3}$$

As mentioned in section 10.2.1, such a relationship between the variance and the mean is called a variance function. In practice, the conditional sample variance is often larger or smaller than the mean; phenomena known as *overdispersion* or *underdispersion*, respectively. Overdispersion could be due to variability in the incidence rates λ_i that is not fully accounted for by the included covariates and is more common than underdispersion. Underdispersion could be due to a regularity in the events being counted, such as buses arriving in approximately 20-minute intervals. In this case, the number of buses counted in, say, 2-hour intervals would vary less than that expected for the Poisson distribution. We will return to overdispersion in section 13.9.

13.4 Did the German health-care reform reduce the number of doctor visits?

Government expenditures on health care surged in Germany in the 80s and 90s. To reduce the expenditure, a major health-care reform took place in 1997. The reform raised the copayments for prescription drugs by up to 200% and imposed upper limits on reimbursement of physicians by the state insurance. Given the large of gross domestic product (GDP) spent on health, it is of interest to investigate whether the reform was a success in the sense that the number of doctor visits decreased after the reform.

To address this research question, Winkelmann (2004) analyzed data from the German Socio-Economic Panel (SOEP Group 2001) that can be downloaded from the *Journal of Applied Econometrics Data Archive*. We will consider a subset of his data, comprised of women working full time in the 1996 panel wave preceding the reform and the 1998 panel wave following the reform.

The dataset `drvisits.dta` has the following variables:

- `id`: person identifier (j)
- `numvisit`: self-reported number of visits to a doctor during the 3 months before the interview (y_{ij})
- `reform`: dummy variable for interview being during the year after the reform versus the year before the reform (x_{2i})
- `age`: age in years (x_{3ij})
- `educ`: education in years (x_{4ij})
- `married`: dummy variable for being married (x_{5ij})
- `badh`: dummy variable for self-reported current health being classified as "very poor" or "poor" (versus "very good", "good", or "fair") (x_{6ij})
- `loginc`: logarithm of household income (in 1995 German Marks, based on OECD weights for household members) (x_{7ij})

We read in the German health-care data by typing

```
. use http://www.stata-press.com/data/mlmus3/drvisits
```

13.5 Longitudinal data structure

As pointed out in *Introduction to models for longitudinal and panel data (part III)*, it is useful to describe longitudinal data before statistical modeling. We start by exploring the participation patterns for the two panel waves by using the `xtdescribe` command:

```
. quietly xtset id reform
. xtdescribe if numvisit<.
        id:  3, 4, ..., 9189                              n =       1518
    reform:  0, 1, ..., 1                                 T =          2
             Delta(reform) = 1 unit
             Span(reform)  = 2 periods
             (id*reform uniquely identifies each observation)
Distribution of T_i:    min      5%    25%    50%    75%    95%    max
                          1       1      1      1      2      2      2

      Freq.   Percent   Cum.  |  Pattern
     ─────────────────────────┼──────────
        709     46.71  46.71  |  11
        418     27.54  74.24  |  .1
        391     25.76 100.00  |  1.
     ─────────────────────────┼──────────
       1518    100.00         |  XX
```

Fewer than half the persons provide responses for both occasions (having the pattern "11"). Some of the missing data is due to attrition (dropout) and nonresponse, and some is due to younger people entering and older people leaving the sample because the sample is restricted to those aged between 20 and 60 years at any point in time (a *rotating panel*). We could also summarize the variables using xtsum.

13.6 Single-level Poisson regression

Before including random effects to model longitudinal dependence, we consider ordinary Poisson regression for the number of doctor visits.

13.6.1 Model specification

The expected number of visits μ_{ij} at occasion i for person j is specified as a log-linear model,

$$\ln(\mu_{ij}) = \nu_{ij} = \beta_1 + \beta_2 x_{2i} + \cdots + \beta_7 x_{7ij}$$

or equivalently as an exponential model for the expected number of visits:

$$\mu_{ij} = \exp(\nu_{ij})$$

The number of doctor visits y_{ij} is assumed to have a Poisson distribution with expectation μ_{ij}, given the covariates. We do not have to include an offset because doctor visits were counted for the same interval, namely, 3 months, for all persons at both occasions. The exponentiated regression coefficients can therefore be interpreted as rate ratios or as ratios of expected counts for any length of interval we like to think about.

13.6.2   Estimation using Stata

We first fit the Poisson regression model with the `poisson` command, using the `irr` option (for "incidence-rate ratio") to obtain exponentiated estimates. We specify the `vce(cluster id)` option to produce standard errors based on the sandwich estimator taking clustering into account, and we store the estimates for later use:

```
. poisson numvisit reform age educ married badh loginc summer, irr vce(cluster id)
Poisson regression                              Number of obs   =       2227
                                                Wald chi2(7)    =     248.40
                                                Prob > chi2     =     0.0000
Log pseudolikelihood = -5942.6924               Pseudo R2       =     0.1073

                                  (Std. Err. adjusted for 1518 clusters in id)
```

| | | Robust | | | | |
numvisit	IRR	Std. Err.	z	P>|z|	[95% Conf. Interval]	
reform	.8689523	.048284	-2.53	0.011	.7792885	.9689328
age	1.004371	.0033158	1.32	0.186	.9978928	1.01089
educ	.9894036	.0115247	-0.91	0.360	.9670715	1.012251
married	1.042542	.0722475	0.60	0.548	.9101354	1.194212
badh	3.105111	.2642862	13.31	0.000	2.628019	3.668814
loginc	1.160559	.091528	1.89	0.059	.9943449	1.354558
summer	1.010269	.0939635	0.11	0.913	.8419145	1.212288
_cons	.6617582	.3796245	-0.72	0.472	.2149803	2.037042

```
. estimates store ordinary
```

We can alternatively view the Poisson regression model as a generalized linear model and use the `glm` command with the `family(poisson)`, `link(log)`, and `vce(cluster id)` options. (The log link is the default link for the Poisson distribution, so the `link()` option could be omitted.) To facilitate interpretation, we use the `eform` option to get exponentiated regression coefficients that represent incidence-rate ratios:

```
. glm numvisit reform age educ married badh loginc summer,
> family(poisson) link(log) vce(cluster id) eform
Generalized linear models                    No. of obs       =        2227
Optimization     : ML                        Residual df      =        2219
                                             Scale parameter  =           1
Deviance         =    7419.853221            (1/df) Deviance  =    3.343782
Pearson          =    9688.740471            (1/df) Pearson   =    4.366264
Variance function: V(u) = u                  [Poisson]
Link function    : g(u) = ln(u)              [Log]
                                             AIC              =    5.344133
Log pseudolikelihood =  -5942.69244          BIC              =    -9685.11
```

                      (Std. Err. adjusted for 1518 clusters in id)

numvisit	IRR	Robust Std. Err.	z	P>|z|	[95% Conf. Interval]	
reform	.8689523	.048284	-2.53	0.011	.7792885	.9689328
age	1.004371	.0033158	1.32	0.186	.9978928	1.01089
educ	.9894036	.0115247	-0.91	0.360	.9670715	1.012251
married	1.042542	.0722475	0.60	0.548	.9101354	1.194212
badh	3.105111	.2642862	13.31	0.000	2.628019	3.668814
loginc	1.160559	.091528	1.89	0.059	.9943449	1.354558
summer	1.010269	.0939635	0.11	0.913	.8419145	1.212288
_cons	.6617582	.3796245	-0.72	0.472	.2149803	2.037042

The estimates from `poisson` and `glm` are identical, as would be expected, and are displayed under the heading "Poisson" in table 13.1. The confidence intervals are based on the sandwich estimator taking the dependence of the repeated counts (given the covariates) into account. The estimated incidence-rate ratio for `reform` is 0.87, implying a 13% reduction in the number of doctor visits per time unit (such as month or year) between 1996 and 1998 for given covariate values.

The variable `reform` changes from 0 in 1996 to 1 in 1998 for everyone, so it is completely confounded with any other changes that may have occurred in Germany in that time period (secular trends). A superior design for causal inference would be to have a control group that does not experience the treatment at the second occasion, in which case a difference-in-difference approach such as that described in exercise 5.3 could be used (including treatment group, time, and time × treatment group in the model, where treatment group is a time-constant covariate for ever receiving the treatment). It should therefore be kept in mind that when we talk about the effect of reform, we mean this casually and not necessarily causally.

Table 13.1: Estimates for different kinds of Poisson regression: Ordinary, GEE, random-intercept (RI), and fixed-intercept (FI)

| | Marginal effects | | | | Conditional effects | | | |
| | Poisson | | GEE* Poisson | | RI Poisson | | FI§ Poisson | |
	Est	(95% CI)†	Est	(95% CI)†	Est	(95% CI)	Est	(95% CI)
Fixed part								
Incidence-rate ratios								
$\exp(\beta_2)$ [reform]	0.87	(0.78, 0.97)	0.88	(0.80, 0.98)	0.95	(0.90, 1.02)	1.02	(0.95, 1.09)
$\exp(\beta_3)$ [age]	1.00	(1.00, 1.01)	1.01	(1.00, 1.01)	1.01	(1.00, 1.01)		
$\exp(\beta_4)$ [educ]	0.99	(0.97, 1.01)	0.99	(0.97, 1.01)	1.01	(0.98, 1.03)	0.98	(0.71, 1.36)
$\exp(\beta_5)$ [married]	1.04	(0.91, 1.19)	1.04	(0.90, 1.19)	1.08	(0.97, 1.20)	1.05	(0.84, 1.30)
$\exp(\beta_6)$ [badh]	3.11	(2.63, 3.67)	3.02	(2.54, 3.58)	2.47	(2.19, 2.78)	1.77	(1.51, 2.08)
$\exp(\beta_7)$ [loginc]	1.16	(0.99, 1.35)	1.15	(0.98, 1.34)	1.10	(0.96, 1.25)	0.97	(0.77, 1.21)
$\exp(\beta_8)$ [summer]	1.01	(0.84, 1.21)	0.97	(0.82, 1.16)	0.87	(0.76, 0.98)	0.81	(0.69, 0.95)
Random part								
$\sqrt{\psi_{11}}$					0.90			
Log likelihood	−5,942.69				−4,643.36		−2,903.51•	

* Using exchangeable working correlation
† Based on the sandwich estimator
§ age omitted from model because within-effects of reform and age are perfectly confounded
• Log conditional likelihood

13.7   Random-intercept Poisson regression

We now explicitly model the dependence between the number of doctor visits before the reform y_{1j} and the number of visits after the reform y_{2j} for a person j, given included covariates.

13.7.1   Model specification

One way to model the dependence within persons is to include a person-specific random intercept ζ_{1j} in the Poisson regression model:

$$
\begin{aligned}
\mu_{ij} \equiv E(y_{ij}|\mathbf{x}_{ij}, \zeta_{1j}) &= \exp(\beta_1 + \beta_2 x_{2i} + \cdots + \beta_7 x_{7ij} + \zeta_{1j}) \\
&= \exp\{(\beta_1 + \zeta_{1j}) + \beta_2 x_{2i} + \cdots + \beta_7 x_{7ij}\} \\
&= \exp(\zeta_{1j})\exp(\beta_1 + \beta_2 x_{2i} + \cdots + \beta_7 x_{7ij})
\end{aligned}
$$

It is assumed that $\zeta_{1j}|\mathbf{x}_{ij} \sim N(0, \psi_{11})$, which implies that ζ_{1j} and $\mathbf{x}_{ij}$ are independent, and it is assumed that ζ_{1j} are independent across persons j. The exponential of the random intercept, $\exp(\zeta_{1j})$, is sometimes called a *frailty*. The number of visits y_{1j} and y_{2j} for a person j at the two occasions are specified as conditionally independent given the random intercept ζ_{1j} (and the covariates $\mathbf{x}_{1j}$ and $\mathbf{x}_{2j}$).

As always in random-effects models, the regression coefficients have conditional or cluster-specific interpretations. In the present application, the clusters are persons, so the coefficients represent person-specific effects. Interestingly, we can also interpret the coefficients as marginal or population-averaged effects because the relationship between the marginal expectation of the count (given $\mathbf{x}_{ij}$ but averaged over ζ_{1j}) and the covariates is

$$
\mu_{ij}^M = \exp\{(\beta_1 + \psi_{11}/2) + \beta_2 x_{2i} + \cdots + \beta_7 x_{7ij}\}
$$

The intercept is the only parameter that is not the same in the marginal and conditional models [in general the marginal intercept is $E\{\exp(\beta_1 + \zeta_{1j})\}$, which becomes $\exp(\beta_1 + \psi_{11}/2)$ for a normally distributed random intercept]. For random-coefficient models, the fixed coefficients of the covariates that have random coefficients are also not the same in conditional and marginal models (see section 13.8).

Because marginal and conditional effects coincide for random-intercept Poisson regression models, in contrast to random-intercept logistic regression, consistent estimation of regression parameters (apart from the intercept) does not hinge on the correct choice of random-intercept distribution. What is required is correct specification of the mean structure and lack of correlation between the random intercept and the covariates.

The marginal or population-averaged variance (given $\mathbf{x}_{ij}$ but averaged over ζ_{1j}) is

$$
\mathrm{Var}(y_{ij}|\mathbf{x}_{ij}) = \mu_{ij}^M + (\mu_{ij}^M)^2\{\exp(\psi_{11}) - 1\} \tag{13.4}
$$

If $\psi_{11} > 0$, this variance is greater than the marginal mean. Therefore, the variance–mean relationship for Poisson models in (13.3) is relaxed, and the variance is greater than that implied by a Poisson model, producing overdispersion.

13.7.2 Measures of dependence and heterogeneity

The intraclass correlation between the counts y_{1j} and y_{2j} at the two occasions depends on the values of the observed covariates (and the exposure if this is included as an offset in the linear predictor) and therefore cannot be used as a simple measure of dependence. In contrast to logit, probit, or complementary log-log models for dichotomous and ordinal responses, Poisson models cannot be formulated in terms of latent responses underlying the observed responses. Thus we cannot define an intraclass correlation in terms of latent responses as shown in section 10.9.1 for dichotomous responses.

Stryhn et al. (2006) show that the conditional (given the covariates) intraclass correlation of the counts y_{ij} and $y_{i'j}$ for two units within the same cluster that have a common fixed part of the linear predictor $\mathbf{x}'\boldsymbol{\beta} \equiv \beta_1 + \beta_2 x_2 + \cdots + \beta_7 x_7$, is given by

$$\rho = \frac{\exp(2\mathbf{x}'\boldsymbol{\beta} + 2\psi) - \exp(2\mathbf{x}'\boldsymbol{\beta} + \psi)}{\exp(2\mathbf{x}'\boldsymbol{\beta} + 2\psi) - \exp(2\mathbf{x}'\boldsymbol{\beta} + \psi) + \exp(\mathbf{x}'\boldsymbol{\beta} + \psi/2)}$$

and hence depends on the covariates in a quite complex manner.

As a measure of heterogeneity, we can consider randomly drawing pairs of persons j and j' with the same covariate values and forming the incidence-rate ratio, or ratio of expected counts for the same exposure, comparing the person who has the larger random intercept with the person who has the smaller random intercept, given by $\exp(|\zeta_j - \zeta_{j'}|)$. Following the derivation in section 10.9.2, the median incidence-rate ratio is given by

$$\mathrm{IRR}_{\mathrm{median}} = \exp\left\{\sqrt{2\psi_{11}}\Phi^{-1}(3/4)\right\}$$

13.7.3 Estimation using Stata

Random-intercept Poisson models can be fit using `xtpoisson`, `xtmepoisson`, and `gllamm`. All three programs use numerical integration, as discussed in section 10.11.1. The details of implementation and speed considerations (see section 10.11.2) are the same as those for logistic models, with `xtpoisson` corresponding to `xtlogit` and `xtmepoisson` to `xtmelogit`. The predictions available for each of the commands also correspond to their logistic counterparts described in section 10.12.

Using `xtpoisson`

The syntax for `xtpoisson` is similar to that for `xtlogit`, but we must specify the `normal` option because `xtpoisson` otherwise assumes a gamma distribution for the frailty $\exp(\zeta_{1j})$, as briefly described in section 13.9.2. The command for fitting the random-intercept model therefore is

```
. quietly xtset id
. xtpoisson numvisit reform age educ married badh loginc summer, normal irr
```

Random-effects Poisson regression   Number of obs   =   2227
Group variable: id   Number of groups   =   1518

Random effects u_i ~ Gaussian   Obs per group: min =   1
                                     avg =   1.5
                                     max =   2

                               Wald chi2(7)   =   253.16
Log likelihood = -4643.3608   Prob > chi2   =   0.0000

numvisit	IRR	Std. Err.	z	P>|z|	[95% Conf. Interval]	
reform	.9547597	.0310874	-1.42	0.155	.895733	1.017676
age	1.006003	.0028278	2.13	0.033	1.000475	1.01156
educ	1.008656	.0127767	0.68	0.496	.9839221	1.034011
married	1.077904	.0595749	1.36	0.175	.9672414	1.201228
badh	2.466573	.1519255	14.66	0.000	2.186076	2.783061
loginc	1.097455	.0747067	1.37	0.172	.9603805	1.254095
summer	.8672783	.0562749	-2.19	0.028	.763707	.9848956
_cons	.5159176	.2645129	-1.29	0.197	.1888714	1.409271
/lnsig2u	-.2016963	.0611823	-3.30	0.001	-.3216115	-.0817812
sigma_u	.9040703	.0276566			.8514575	.9599341

```
Likelihood-ratio test of sigma_u=0: chibar2(01) =  2598.66 Pr>=chibar2 = 0.000
```

According to this model, the estimated effect of `reform` is to reduce the incidence rate, and therefore the expected number of visits in a given period, by about 5% (controlling for the other variables), but this is not significant at the 5% level. Each extra year of age is associated with an estimated 0.6% increase in the incidence rate; the incidence-rate ratio for a 10-year increase in age is estimated as $1.006^{10} = 1.06$, corresponding to a 6% increase, for given values of the other covariates. The estimates also suggest that being in bad health more than doubles the incidence rate and that the incidence rate decreases by 13% in the summer, controlling for the other variables. The other coefficients are not significant at the 5% level. The estimates were also reported under the heading "RI Poisson" in table 13.1.

From the last line of output, we see that we can reject the null hypothesis that the random-intercept variance is zero, $\psi_{11} = 0$, with a likelihood-ratio statistic of 2,598.66 and $p < 0.001$. (As discussed in section 2.6.2, the `chibar2(01)` notation indicates that the p-value reported in the output takes into account that the null hypothesis is on the border of parameter space.)

The median incidence-rate ratio is estimated as

```
. display exp(sqrt(2)*.90407*invnormal(3/4))
2.3687622
```

Half the time, the ratio of the expected number of visits (in a given period) for two randomly chosen persons with the same covariate values, comparing the one who has the larger expected value with the other one, will exceed 2.37. We could avoid having to say "comparing the one who has the larger expected value with the other one" by saying the following: Half the time, the ratio of the expected number of visits will lie in the range from 0.42 (= 1/2.37) to 2.37, and the other half the time, the ratio will lie outside that range.

Using xtmepoisson

The syntax for `xtmepoisson` is exactly as that for `xtmelogit` except that we use the `irr` option to get exponentiated regression coefficients instead of the `or` option. We fit the model and store the estimates using the following commands:

```
. xtmepoisson numvisit reform age educ married badh loginc summer || id:, irr
Mixed-effects Poisson regression          Number of obs      =      2227
Group variable: id                        Number of groups   =      1518

                                          Obs per group: min =         1
                                                         avg =       1.5
                                                         max =         2

Integration points =    7                 Wald chi2(7)       =    253.06
Log likelihood = -4643.2823               Prob > chi2        =    0.0000
```

numvisit	IRR	Std. Err.	z	P>|z|	[95% Conf. Interval]	
reform	.9547758	.0310889	-1.42	0.155	.8957463	1.017695
age	1.006003	.0028283	2.13	0.033	1.000475	1.011562
educ	1.008659	.0127792	0.68	0.496	.9839206	1.034019
married	1.077902	.059584	1.36	0.175	.9672229	1.201245
badh	2.466399	.1519349	14.65	0.000	2.185887	2.782909
loginc	1.09743	.0747155	1.37	0.172	.96034	1.254089
summer	.8672584	.0562778	-2.19	0.028	.7636823	.9848822
_cons	.5159046	.264545	-1.29	0.197	.1888388	1.409443

Random-effects Parameters	Estimate	Std. Err.	[95% Conf. Interval]	
id: Identity				
sd(_cons)	.9043832	.0276826	.8517217	.9603006

```
LR test vs. Poisson regression:  chibar2(01) =  2598.82 Prob>=chibar2 = 0.0000
. estimates store xtri
```

The estimates are nearly identical to those produced by `xtpoisson`.

We could also fit the model more quickly by using one quadrature point, giving the Laplace method. This method works well in this example, but as discussed in section 10.11.2 we do not recommend it in general because the estimates can be poor.

Using gllamm

The random-intercept Poisson model can be fit as follows using `gllamm`:

```
. generate cons = 1
. eq ri: cons
. gllamm numvisit reform age educ married badh loginc summer,
> family(poisson) link(log) i(id) eqs(ri) eform adapt
number of level 1 units = 2227
number of level 2 units = 1518

Condition Number = 723.77605

gllamm model

log likelihood = -4643.3427
```

numvisit	exp(b)	Std. Err.	z	P>|z|	[95% Conf. Interval]	
reform	.9547481	.0310831	-1.42	0.155	.8957293	1.017656
age	1.006002	.0028266	2.13	0.033	1.000477	1.011557
educ	1.008646	.0127702	0.68	0.497	.9839247	1.033988
married	1.077896	.059554	1.36	0.175	.9672696	1.201174
badh	2.466857	.15192	14.66	0.000	2.186367	2.78333
loginc	1.097486	.0746823	1.37	0.172	.9604527	1.25407
summer	.8673159	.0562616	-2.19	0.028	.7637672	.9849033
_cons	.5157228	.2643128	-1.29	0.196	.188872	1.408203

```
Variances and covariances of random effects
------------------------------------------------------------------------------

***level 2 (id)

    var(1): .81691979 (.04972777)
------------------------------------------------------------------------------
. estimates store glri
```

Here we specified an equation, `ri`, for the random intercept by using the `eqs()` option. This is not necessary because `gllamm` would include a random intercept by default. However, defining the intercept explicitly from the start makes it easier to use estimates from this model as starting values for random-coefficient models because the parameter matrix will have the correct column labels.

The estimates are close to those by using `xtlogit` and `xtmepoisson`, including the random-intercept standard deviation, estimated by `gllamm` as

```
. display sqrt(.81691979)
.90383615
```

We could also obtain robust standard errors using the command

```
gllamm, robust eform
```

There is no likelihood-ratio test for the random-intercept variance in the `gllamm` output, but we can perform this test ourselves by comparing the random-intercept model (with estimates stored under `glri`) with the ordinary Poisson model (with estimates stored under `ordinary`):

```
. lrtest ordinary glri, force
Likelihood-ratio test                             LR chi2(1)  =    2598.70
(Assumption: ordinary nested in glri)             Prob > chi2 =     0.0000
```

We used the `force` option to force Stata to perform the test, which is necessary because the two models being compared were fit using different commands. We must divide the *p*-value by 2, as discussed in section 2.6.2, which makes no difference to the conclusion here that the random-intercept variance is highly significant.

13.8   Random-coefficient Poisson regression

13.8.1   Model specification

The random-intercept Poisson regression model accommodates dependence among the repeated counts. However, it assumes that the effect of the health-care reform is the same for all persons. In this section, we relax this assumption by introducing an additional person-level random coefficient ζ_{2j} for `reform` (x_{2i}):

$$
\begin{aligned}
\ln(\mu_{ij}) &= \beta_1 + \beta_2 x_{2i} + \cdots + \beta_7 x_{7ij} + \zeta_{1j} + \zeta_{2j} x_{2ij} \\
&= (\beta_1 + \zeta_{1j}) + (\beta_2 + \zeta_{2j}) x_{2i} + \cdots + \beta_7 x_{7ij}
\end{aligned}
$$

The above specification allows the effect of the health-care reform $\beta_2 + \zeta_{2j}$ to vary over persons j. We assume that, given the covariates $\mathbf{x}_{ij}$, the random intercept and random coefficient have a bivariate normal distribution with zero means and covariance matrix

$$
\mathbf{\Psi} = \begin{bmatrix} \psi_{11} & \psi_{12} \\ \psi_{21} & \psi_{22} \end{bmatrix}, \qquad \psi_{21} = \psi_{12}
$$

The correlation between the random intercept and random coefficient becomes

$$
\rho_{21} = \frac{\psi_{21}}{\sqrt{\psi_{11}\psi_{22}}}
$$

It follows from the normality assumption for the random effects, given the covariates, and the homoskedasticity assumptions above, that the random effects are independent of the covariates.

The marginal expected count is given by

$$
\begin{aligned}
E(y_{ij}|\mathbf{x}_{ij}) &= \exp\{\beta_1 + \beta_2 x_{2i} + \cdots + \beta_7 x_{7ij} + (\psi_{11} + 2\psi_{21} x_{2i} + \psi_{22} x_{2i}^2)/2\} \\
&= \exp\{\beta_1 + \psi_{11}/2 + \beta_2 x_{2i} + \psi_{21} x_{2i} + (\psi_{22}/2) x_{2i}^2 + \cdots + \beta_7 x_{7ij}\} \\
&= \exp\{\beta_1 + \psi_{11}/2 + (\beta_2 + \psi_{21} + \psi_{22}/2) x_{2i} + \cdots + \beta_7 x_{7ij}\}
\end{aligned}
$$

where the last equality holds only if x_{2i} is a dummy variable, because $x_{2i} = x_{2i}^2$ in this case.

Hence, the multiplicative effect of the reform on the *marginal* expected count now no longer equals $\exp(\beta_2)$ as in the random-intercept model but $\exp(\beta_2 + \psi_{21} + \psi_{22}/2)$. In general, the marginal expected count is given by the exponential of the sum of the linear predictor and half the variance of the random part of the model. Unfortunately, this variance is usually not a linear function of the variable x_{2ij} having a random coefficient (except when x_{2i} is a dummy variable), so the variable no longer has a simple multiplicative effect on the marginal expectation.

13.8.2   Estimation using Stata

Using xtmepoisson

xtpoisson cannot be used for random-coefficient models. The syntax for random-coefficient models in xtmepoisson is the same as that for xtmixed or xtmelogit, apart from the irr option that is used to obtain exponentiated regression coefficients:

```
. xtmepoisson numvisit reform age educ married badh loginc summer ||
> id: reform, covariance(unstructured) irr

Mixed-effects Poisson regression          Number of obs      =      2227
Group variable: id                        Number of groups   =      1518

                                          Obs per group: min =         1
                                                         avg =       1.5
                                                         max =         2

Integration points =    7                 Wald chi2(7)       =    241.11
Log likelihood = -4513.7299               Prob > chi2        =    0.0000
```

numvisit	IRR	Std. Err.	z	P>|z|	[95% Conf. Interval]	
reform	.9022792	.0483866	-1.92	0.055	.812257	1.002278
age	1.003456	.0028317	1.22	0.222	.997921	1.009021
educ	1.008894	.0128121	0.70	0.486	.9840931	1.034321
married	1.086874	.064118	1.41	0.158	.9681982	1.220097
badh	3.028323	.2322948	14.44	0.000	2.605606	3.519619
loginc	1.135636	.0866487	1.67	0.096	.9778964	1.31882
summer	.9140246	.0741941	-1.11	0.268	.7795846	1.071649
_cons	.4060731	.2305667	-1.59	0.112	.1334429	1.2357

Random-effects Parameters	Estimate	Std. Err.	[95% Conf. Interval]	
id: Unstructured				
sd(reform)	.9303105	.0561786	.8264687	1.047199
sd(_cons)	.9541108	.0357016	.8866412	1.026714
corr(reform,_cons)	-.4908242	.0506047	-.583535	-.3854839

```
LR test vs. Poisson regression:     chi2(3) =   2857.93   Prob > chi2 = 0.0000
Note: LR test is conservative and provided only for reference.
```

We store the estimates,

```
. estimates store xtrc
```

for use in the likelihood-ratio test below.

The random-intercept Poisson regression model is nested in the random-coefficient model (with $\psi_{22} = \psi_{21} = 0$). We can therefore perform a likelihood-ratio test using

```
. lrtest xtri xtrc
Likelihood-ratio test                          LR chi2(2)  =    259.10
(Assumption: xtri nested in xtrc)              Prob > chi2 =    0.0000
Note: The reported degrees of freedom assumes the null hypothesis is not on the
      boundary of the parameter space.  If this is not true, then the reported
      test is conservative.
```

As discussed in section 4.6, the asymptotic null distribution for this test is $1/2\chi_1^2 + 1/2\chi_2^2$, and we conclude that the random-intercept Poisson model is rejected in favor of the random-coefficient Poisson model.

Maximum likelihood estimates for the random-coefficient Poisson model are reported under "RC Poisson" in table 13.2, where we for comparison also report estimates for the random-intercept Poisson model.

Table 13.2: Estimates for different kinds of random-effects Poisson regression: random-intercept (RI) and random-coefficient (RC) models

	RI Poisson		RC Poisson	
	Est	(95% CI)	Est	(95% CI)
Fixed part				
Incidence-rate ratios				
$\exp(\beta_2)$ [reform]	0.95	(0.90, 1.02)	0.90	(0.81, 1.00)
$\exp(\beta_3)$ [age]	1.01	(1.00, 1.01)	1.00	(1.00, 1.01)
$\exp(\beta_4)$ [educ]	1.01	(0.98, 1.03)	1.01	(0.98, 1.03)
$\exp(\beta_5)$ [married]	1.08	(0.97, 1.20)	1.09	(0.97, 1.22)
$\exp(\beta_6)$ [badh]	2.47	(2.19, 2.78)	3.03	(2.61, 3.52)
$\exp(\beta_7)$ [loginc]	1.10	(0.96, 1.25)	1.14	(0.98, 1.32)
$\exp(\beta_8)$ [summer]	0.87	(0.76, 0.98)	0.91	(0.78, 1.07)
Random part				
$\sqrt{\psi_{11}}$	0.90		0.95	
$\sqrt{\psi_{22}}$			0.93	
$\rho_{21} = \psi_{21}/(\sqrt{\psi_{11}}\sqrt{\psi_{22}})$			-0.49	
Log likelihood	-4,643.36		-4,513.73	

Using gllamm

We now fit the random-coefficient Poisson regression model in `gllamm`, using the estimates from the random-intercept model (stored under `glri`) as starting values:

```
. estimates restore glri
. matrix a = e(b)
. eq rc: reform
. gllamm numvisit reform age educ married badh loginc summer,
> family(poisson) link(log) i(id) nrf(2) eqs(ri rc) from(a) eform adapt

number of level 1 units = 2227
number of level 2 units = 1518

Condition Number = 812.85865

gllamm model

log likelihood = -4513.8005
```

numvisit	exp(b)	Std. Err.	z	P>|z|	[95% Conf. Interval]	
reform	.9023139	.048376	-1.92	0.055	.8123103	1.00229
age	1.003457	.0028304	1.22	0.221	.9979246	1.00902
educ	1.008889	.0128058	0.70	0.486	.9841002	1.034303
married	1.086858	.0640872	1.41	0.158	.9682361	1.220013
badh	3.02813	.2322062	14.45	0.000	2.605564	3.519226
loginc	1.13564	.0866072	1.67	0.095	.9779701	1.318731
summer	.9140484	.0741615	-1.11	0.268	.7796627	1.071597
_cons	.406047	.2304432	-1.59	0.112	.1335043	1.234973

```
Variances and covariances of random effects
--------------------------------------------------------------------------------

***level 2 (id)

    var(1): .90914636 (.06767417)
    cov(2,1): -.43462184 (.07121035) cor(2,1): -.49034787

    var(2): .8641332 (.10415944)
--------------------------------------------------------------------------------
```

We store the estimates for later use:

```
. estimates store glrc
```

It was not necessary to add two new elements to the matrix of starting values for the two new parameters being estimated (as we did in section 11.7.2). This is because the column labels of the matrix `a` let `gllamm` know which parameters are supplied, and `gllamm` sets the starting values of the other parameters to zero. By specifying the same equation `ri` for the random intercept in both models, we ensure that `gllamm` recognizes the previous estimate of the (square root of the) random-intercept variance as the starting value for that parameter in the random-coefficient model.

Robust standard errors and confidence intervals can be obtained using

```
gllamm, robust eform
```

13.8.3   Interpretation of estimates

The estimated incidence-rate ratios have changed somewhat compared with the random-intercept Poisson model. Note in particular that the estimated incidence-rate ratio for `reform` now implies a 10% reduction in the expected number of visits per year *for a given person* and is nearly significant at the 5% level.

Instead of thinking of this model as a random-coefficient model, we could view it as a model with a random intercept ζ_{1j} for 1996 and a random intercept $\zeta_{1j} + \zeta_{2j}$ for 1998, because x_{2i} is 0 in 1996 and 1 in 1998. In 1996, the random-intercept variance is

$$\text{Var}(\zeta_{1j}|\mathbf{x}_{1j}) \;=\; \psi_{11}$$

which is estimated as 0.91. In 1998, the variance is

$$\text{Var}(\zeta_{1j} + \zeta_{2j}|\mathbf{x}_{2j}) \;=\; \psi_{11} + \psi_{22} + 2\psi_{21}$$

and is estimated as 0.90. The covariance between the intercepts is

$$\text{Cov}(\zeta_{1j}, \zeta_{1j} + \zeta_{2j}|\mathbf{x}_{1j}, \mathbf{x}_{2j}) \;=\; \psi_{11} + \psi_{21}$$

which is estimated as 0.48.

In contrast, the conventional random-intercept model (13.4) with the same random intercept ζ_{1j} in 1996 and 1998 had a single parameter ψ_{11} representing both the random-intercept variance at the two occasions as well as the covariance across time. Assuming that this parameter "tries" to produce variances and covariance close to the estimated values using the more flexible model, it is not surprising that the estimate $\widehat{\psi}_{11} = 0.82$ is between 0.91 and 0.48.

The reason we can identify the three parameters of the random part is that the variance parameters for 1996 and 1998 determine the relationship between marginal mean and marginal variance (given the covariates) at these two time points, as shown in (13.4), and the covariance parameter affects the correlation between the counts. In the random-intercept model, all three properties were determined by one parameter, whereas the random-coefficient model uses separate parameters for overdispersion at the two time points and for dependence across time.

As discussed in section 4.10, an analogous *linear* random-coefficient model with only two time points is not identified because it includes a fourth parameter, the level-1 variance θ, in the random part. The analogous random-coefficient logistic or probit model for dichotomous responses is also not identified because random intercepts have no effect on the marginal variance–mean relationship in these models (see section 10.8).

13.9   Overdispersion in single-level models

We now temporarily return to single-level models to discuss models for overdispersion in the absence of clustering. Subsequently, we allow for overdispersion at level 1 in two-level models in section 13.10.

The assumption of the Poisson model that the variance of the count is equal to the expectation (given the covariates) is often violated. The most common violation is overdispersion or extra-Poisson variability, meaning that the variance is larger than the expectation. Overdispersion could be due to unobserved covariates that vary between the units of observation. For this to be meaningful, the "cases" or records of data must correspond to units, such as persons, occasions, houses, or schools, that could potentially be characterized by omitted covariates. If the data have been aggregated over units or disaggregated within units, the methods described here are not appropriate.

In this section, we consider methods for dealing with overdispersion in single-level models as a preparation for the next section, where we discuss overdispersion in two-level models. We therefore ignore clustering for simplicity, but note that this is inappropriate for the health-care reform data. We will use model-based standard errors so that the impact of modeling overdispersion on standard errors can be seen, but these standard errors are not correct because clustering is ignored.

13.9.1   Normally distributed random intercept

As discussed in section 13.7.1, random intercepts induce overdispersion. Even when we do not have clustered data, random intercepts can therefore be included to model overdispersion. The model and its implied marginal mean and variance are exactly as those for two-level models except that the random intercept varies between the level-1 units and hence does not produce any dependence among groups of observations.

The model can be written as

$$\ln(\mu_{ij}) \; = \; \beta_1 + \beta_2 x_{2i} + \cdots + \beta_7 x_{7ij} + \zeta_{ij}^{(1)} \tag{13.5}$$

where $\zeta_{ij}^{(1)}|\mathbf{x}_{ij} \sim N(0, \psi^{(1)})$ and we have included a (1) superscript to denote that the random intercept varies at level 1 (and not at level 2 or higher as elsewhere in this book).

The marginal expectation becomes

$$\mu_{ij}^M \equiv E(y_{ij}|\mathbf{x}_{ij}) \; = \; \exp(\underbrace{\beta_1 + \psi^{(1)}/2}_{\beta_1^M} + \beta_2 x_{2i} + \cdots + \beta_7 x_{7ij}) \tag{13.6}$$

and the marginal variance is

$$\mathrm{Var}(y_{ij}|\mathbf{x}_{ij}) \; = \; \mu_{ij}^M + (\mu_{ij}^M)^2 \{\exp(\psi^{(1)}) - 1\} \tag{13.7}$$

which is larger than the marginal expectation μ_{ij}^M if $\psi^{(1)} > 0$.

To fit the model with a normally distributed random intercept at level 1, we first generate an identifier, obs, for the level-1 observations,

```
. generate obs = _n
```

and then specify obs as the "clustering" variable in the xtpoisson command:

```
. quietly xtset obs
. xtpoisson numvisit reform age educ married badh loginc summer, normal irr
Random-effects Poisson regression          Number of obs      =      2227
Group variable: obs                        Number of groups   =      2227

Random effects u_i ~ Gaussian              Obs per group: min =         1
                                                          avg =       1.0
                                                          max =         1

                                           Wald chi2(7)       =    272.60
Log likelihood  = -4546.8881               Prob > chi2        =    0.0000
```

numvisit	IRR	Std. Err.	z	P>|z|	[95% Conf. Interval]	
reform	.881623	.0466248	-2.38	0.017	.7948166	.97791
age	1.002419	.0026053	0.93	0.353	.9973256	1.007538
educ	1.005101	.0117969	0.43	0.665	.9822433	1.02849
married	1.084023	.0602281	1.45	0.146	.9721784	1.208735
badh	3.203538	.2429967	15.35	0.000	2.760985	3.717027
loginc	1.151836	.0840599	1.94	0.053	.9983224	1.328956
summer	.9576537	.0786351	-0.53	0.598	.8152943	1.124871
_cons	.3949809	.2143587	-1.71	0.087	.13634	1.144271
/lnsig2u	-.1143904	.0548408	-2.09	0.037	-.2218765	-.0069043
sigma_u	.9444097	.0258961			.894994	.9965538

```
Likelihood-ratio test of sigma_u=0: chibar2(01) =  2791.61 Pr>=chibar2 = 0.000
```

We see from the estimated standard deviation of the level-1 random intercept of 0.94 and the highly significant likelihood-ratio test that there is evidence for overdispersion. The factor $\exp(\psi^{(1)}) - 1$, that multiplies the squared marginal expectation to obtain the additive overdispersion component [see expression (13.7)], is estimated as about 1.44.

Another consequence of including a random intercept, in addition to overdispersion, is that it produces a larger marginal probability of zeros than the ordinary Poisson model. Thus the problem of excess zeros often observed in count data is addressed to some extent. So-called zero-inflated Poisson (ZIP) models are tailor-made to address this problem, but these models will not be discussed here.

13.9.2 Negative binomial models

As previously mentioned, the random-intercept model with a normally distributed random intercept does not have a closed-form likelihood and is hence fit using numerical integration. An appealing and computationally more efficient approach is to alter the model specification so that a closed-form likelihood is achieved. This can be accomplished by specifying a gamma distribution either for the exponentiated random intercept $\exp(\zeta_j)$ (NB2) or for the cluster-specific mean μ_{ij} (NB1), yielding two different kinds of negative binomial models.

Mean dispersion or NB2

The *mean dispersion* or NB2 version of the negative binomial model can be written as a random-intercept model as in (13.5), but instead of assuming a normal distribution for the level-1 random intercept $\zeta_{ij}^{(1)}$, we assume a gamma distribution for the frailty $\exp(\zeta_{ij}^{(1)})$ with mean 1 and variance α (pronounced "alpha"). The gamma distribution is then said to have scale parameter α and shape parameter $1/\alpha$.

It follows from this specification that the marginal mean is

$$\mu_{ij}^M \equiv E(y_{ij}|\mathbf{x}_{ij}) = \exp(\beta_1 + \beta_2 x_{2i} + \cdots + \beta_7 x_{7ij}) \tag{13.8}$$

and that the marginal variance has the same quadratic form as (13.7), with

$$\text{Var}(y_{ij}|\mathbf{x}_{ij}) = \mu_{ij}^M + (\mu_{ij}^M)^2 \alpha$$

so that α corresponds to $\exp(\psi^{(1)}) - 1$ in (13.7).

This model can be fit using the `nbreg` command with the `dispersion(mean)` option:

```
. nbreg numvisit reform age educ married badh loginc summer,
> irr dispersion(mean)
Negative binomial regression                    Number of obs   =       2227
                                                LR chi2(7)      =     303.15
Dispersion       = mean                         Prob > chi2     =     0.0000
Log likelihood = -4562.0459                     Pseudo R2       =     0.0322
```

numvisit	IRR	Std. Err.	z	P>|z|	[95% Conf. Interval]	
reform	.8734045	.0447241	-2.64	0.008	.7900022	.9656119
age	1.004806	.0024754	1.95	0.052	.9999656	1.009669
educ	.9971352	.0115579	-0.25	0.805	.9747375	1.020047
married	1.081049	.057606	1.46	0.144	.9738393	1.200062
badh	3.118932	.2335543	15.19	0.000	2.693181	3.611987
loginc	1.142179	.081637	1.86	0.063	.9928749	1.313934
summer	.9424437	.074725	-0.75	0.455	.8067981	1.100895
_cons	.6648993	.3553603	-0.76	0.445	.2332518	1.895339
/lnalpha	.0007291	.0475403			-.0924481	.0939064
alpha	1.000729	.047575			.9116965	1.098457

```
Likelihood-ratio test of alpha=0:  chibar2(01) = 2761.29 Prob>=chibar2 = 0.000
```

The same estimates can be obtained using `xtpoisson`, as shown in section 13.9.1 but without the `normal` option. We see that α is estimated as practically 1, which is considerably lower than the corresponding factor $\exp(\psi^{(1)}) - 1$ from Poisson regression with a normally distributed random intercept at level 1, which is estimated as about 1.44.

Constant dispersion or NB1

The *constant dispersion* or NB1 version of the negative binomial model cannot be derived from a random-intercept model but is instead obtained by assuming that the person-specific expected count μ_{ij} has a gamma distribution with expectation $\exp(\beta_1 + \beta_2 x_{2i} + \cdots + \beta_7 x_{7ij})$ and variance $\exp(\beta_1 + \beta_2 x_{2i} + \cdots + \beta_7 x_{7ij})\delta$.

From this specification, it follows that the count y_{ij} has the same expectation as the NB2 version of the model in (13.8) but the variance function now is

$$\mathrm{Var}(y_{ij}|\mathbf{x}_{ij}) \;=\; \mu_{ij}^M(1+\delta)$$

Unlike the random-intercept models, the variance for the constant-dispersion negative binomial model is a constant multiple of the expectation. This model can be fit using the `nbreg` command with the `dispersion(constant)` option:

```
. nbreg numvisit reform age educ married badh loginc summer,
> irr dispersion(constant)

Negative binomial regression                    Number of obs    =        2227
                                                LR chi2(7)       =      226.58
Dispersion      = constant                      Prob > chi2      =      0.0000
Log likelihood = -4600.3276                     Pseudo R2        =      0.0240
```

numvisit	IRR	Std. Err.	z	P>|z|	[95% Conf. Interval]	
reform	.9002629	.0401958	-2.35	0.019	.8248293	.9825952
age	1.001107	.0022372	0.50	0.621	.9967317	1.005501
educ	1.007226	.0097883	0.74	0.459	.9882232	1.026595
married	1.038237	.0490102	0.79	0.427	.9464885	1.138879
badh	2.596117	.148113	16.72	0.000	2.321462	2.903265
loginc	1.131328	.0685746	2.04	0.042	1.004601	1.274041
summer	1.00864	.0693101	0.13	0.900	.8815455	1.154058
_cons	.7705889	.3470568	-0.58	0.563	.318757	1.862884
/lndelta	.9885984	.0532775			.8841765	1.09302
delta	2.687465	.1431814			2.42099	2.983271

```
Likelihood-ratio test of delta=0:   chibar2(01) = 2684.73 Prob>=chibar2 = 0.000
```

We see that δ is estimated as 2.69, and the multiplicative overdispersion term $1+\delta$ is hence estimated as 3.69.

13.9.3 Quasilikelihood

Generalized linear models are often fit using an algorithm called iteratively reweighted least squares that depends only on the expectation and variance of the response as a function of the covariates (although the log likelihood is used to monitor convergence). In the quasilikelihood approach, this same algorithm (or set of estimating equations) is used with a variance-expectation relationship (or variance function) of our choice, even when there is no statistical model that implies such a relationship. For the expectation, we retain the expression from the conventional Poisson regression model,

$$\ln(\mu_{ij}) \;=\; \beta_1 + \beta_2 x_{2i} + \cdots + \beta_7 x_{7ij}$$

However, for the variance we relax the assumption inherent in the Poisson distribution that it is equal to the expectation by introducing the proportionality parameter ϕ (pronounced "phi")

$$\mathrm{Var}(y_{ij}|\mathbf{x}_{ij}) \;=\; \phi\mu_{ij}$$

This variance function has the same form as that for the negative binomial model type NB1, whereas random-intercept models have variances that are quadratic functions of the mean.

The estimates of the regression parameters are the same as the maximum likelihood estimates for an ordinary Poisson model, and the proportionality parameter ϕ is estimated by a simple moment estimator, most commonly by the Pearson chi-squared statistic divided by the residual degrees of freedom. The scale parameter affects the standard errors only; they are multiplied by $\sqrt{\widehat{\phi}}$.

We can use the `glm` command with the `scale()` option to obtain maximum quasi-likelihood estimates. Here we use the `scale(x2)` option to estimate the proportionality parameter by the Pearson chi-squared statistic divided by the residual degrees of freedom:

```
. glm numvisit reform age educ married badh loginc summer,
> family(poisson) link(log) eform scale(x2)

Generalized linear models                        No. of obs      =        2227
Optimization     : ML                            Residual df     =        2219
                                                 Scale parameter =           1
Deviance         =    7419.853221                (1/df) Deviance =    3.343782
Pearson          =    9688.740471                (1/df) Pearson  =    4.366264

Variance function: V(u) = u                      [Poisson]
Link function    : g(u) = ln(u)                  [Log]

                                                 AIC             =    5.344133
Log likelihood   =    -5942.69244                BIC             =    -9685.11
```

		OIM				
numvisit	IRR	Std. Err.	z	P>|z|	[95% Conf.	Interval]
reform	.8689523	.0482622	-2.53	0.011	.7793268	.9688851
age	1.004371	.0027347	1.60	0.109	.9990249	1.009745
educ	.9894036	.0124256	-0.85	0.396	.9653472	1.014059
married	1.042542	.0607123	0.72	0.474	.9300882	1.168593
badh	3.105111	.1966385	17.89	0.000	2.742665	3.515454
loginc	1.160559	.0874758	1.98	0.048	1.001173	1.34532
summer	1.010269	.0853037	0.12	0.904	.8561784	1.192091
_cons	.6617582	.3721437	-0.73	0.463	.2197967	1.992405

(Standard errors scaled using square root of Pearson X2-based dispersion.)

We see from the output next to `(1/df)` `Pearson` that the proportionality parameter is estimated as $\widehat{\phi} = 4.37$, which can be compared with the corresponding parameter for the negative binomial model of type NB1, $1 + \widehat{\delta} = 3.69$. It follows that the estimated model-based standard errors for the naïve Poisson model that does not accommodate overdispersion would be $1/\sqrt{\hat{\phi}} = 0.48$ times the estimated standard errors from the quasilikelihood approach. This can also be seen by fitting the model with model-based standard errors by using the `glm` command above (without the `scale(x2)` option) or the `poisson` command (output not shown).

Because the quasilikelihood method changes only the standard errors, an alternative but similar approach is to use the sandwich estimator for the standard errors by specifying the `vce(robust)` option in the `glm` command.

13.10   Level-1 overdispersion in two-level models

We now return to two-level Poisson regression models with random effects. As seen in (13.4), we would expect that including a random intercept ζ_{1j} at level 2 has, at least to some degree, addressed the problem of overdispersion. However, as discussed in section 13.8.3, the model uses a single parameter to induce both overdispersion for the level-1 units and dependence among level-1 units in the same cluster.

Sometimes there may be additional overdispersion at level 1 not accounted for by the random effect at level 2. For instance, in the health-care reform data, there may be unobserved heterogeneity between occasions within persons, such as medical problems (representing unobserved time-varying covariates) that can lead to several extra doctor visits within the same 3-month period. After conditioning on the person-level random effect, the counts at the occasions are then overdispersed.

The most natural approach to handling level-1 overdispersion in a two-level model is by including a level-1 random intercept in addition to the level-2 random effect(s). The model then becomes a three-level model, with random effects at two nested levels, as discussed in chapter 8 (see exercise 16.6).

Hausman, Hall, and Griliches (1984) suggested another approach. Their model, which can be fit using Stata's `xtnbreg` command, has a closed-form likelihood and is specified as follows. The conditional expectation μ_{ij} is assumed to have a gamma distribution with mean $\exp(\beta_1 + \beta_2 x_{2i} + \cdots + \beta_7 x_{7ij})\delta_j$ and variance $\exp(\beta_1 + \beta_2 x_{2i} + \cdots + \beta_7 x_{7ij})\delta_j^2$. For a cluster j with a given δ_j, the conditional expectation of the count becomes

$$
\begin{aligned}
E(y_{ij}|\mathbf{x}_{ij},\delta_j) &= \exp(\beta_1 + \beta_2 x_{2i} + \cdots + \beta_7 x_{7ij})\delta_j \\
&= \exp\left\{(\beta_1 + \ln\delta_j) + \beta_2 x_{2i} + \cdots + \beta_7 x_{7ij}\right\}
\end{aligned}
\tag{13.9}
$$

and the conditional variance becomes

$$
\text{Var}(y_{ij}|\mathbf{x}_{ij},\delta_j) = E(y_{ij}|\mathbf{x}_{ij},\delta_j)(1 + \delta_j)
\tag{13.10}
$$

The counts for the level-1 units in cluster j therefore have a cluster-specific intercept $\beta_1 + \ln\delta_j$ and are subject to multiplicative overdispersion (of the NB1 form) with a cluster-specific overdispersion factor $(1 + \delta_j)$. It is then assumed that $1/(1 + \delta_j)$ has a beta distribution. The syntax for fitting this model is

```
xtset id
xtnbreg numvisit reform age educ married badh loginc summer, irr
```

A weakness of this model is that the person-specific intercept and level-1 overdispersion factor are both determined by the same parameter δ_j. It is therefore not possible to have heterogeneity at level 2 without having overdispersion at level 1 or vice versa. A consequence is that although the single-level Poisson model is a special case of the single-level negative binomial model (with $\delta = 0$ for NB1 and $\alpha = 0$ for NB2), the two-level Poisson model is not a special case of the two-level negative binomial model. Another problem with the model is that the parameters of the beta distribution are difficult to interpret. We therefore do not recommend using this model.

Perhaps the simplest approach to handling overdispersion at level 1 in a two-level random-intercept Poisson model is to use the sandwich estimator for the standard errors. At the time of writing this book, the only command that can provide these standard errors is `gllamm` with the `robust` option. After restoring the random-intercept model estimates,

```
. estimates restore glri
```

we can simply issue the command

```
. gllamm, robust eform

number of level 1 units = 2227
number of level 2 units = 1518

Condition Number = 723.77605

gllamm model

log likelihood = -4643.3427

Robust standard errors
```

numvisit	exp(b)	Std. Err.	z	P>|z|	[95% Conf. Interval]	
reform	.9547481	.0503036	-0.88	0.379	.8610748	1.058612
age	1.006002	.0031322	1.92	0.055	.9998817	1.01216
educ	1.008646	.0127823	0.68	0.497	.9839016	1.034012
married	1.077896	.0708484	1.14	0.254	.9476075	1.226097
badh	2.466857	.2880487	7.73	0.000	1.962236	3.101249
loginc	1.097486	.095603	1.07	0.286	.9252306	1.301811
summer	.8673159	.0722128	-1.71	0.087	.7367263	1.021053
_cons	.5157228	.3259741	-1.05	0.295	.1494154	1.780071

```
Variances and covariances of random effects
------------------------------------------------------------------------------

***level 2 (id)

    var(1): .81691979 (.0523264)
------------------------------------------------------------------------------
```

We see that the robust confidence intervals are somewhat wider than those based on model-based standard errors.

13.11   Other approaches to two-level count data

13.11.1   Conditional Poisson regression

Instead of using random intercepts ζ_{1j} for persons, we could instead treat the intercepts as fixed parameters α_j. An advantage of such a fixed-effects approach is that it does not make any assumptions regarding the person-specific intercepts. The estimated regression coefficients represent within effects, where persons serve as their own controls. A disadvantage is that effects of variables that do not vary within persons, such as gender, cannot be estimated.

As discussed for fixed-intercept logistic regression in section 10.14.1, we can eliminate the person-specific intercepts by constructing a likelihood that is conditional on the sum of the counts for each person, a sufficient statistic for the person-specific intercept. Only persons having observations in both years contribute to the conditional likelihood analysis; in the current application, this means that we lose the information from as many as 809 persons. We also lose the information from 95 persons who had zero doctor visits at both occasions (conditional on the sum being 0, the probability that both counts are 0 is 1, regardless of the parameter values. Hence, these observations do not provide information on the parameters).

Recall that within-person effects are estimated in the fixed-intercept approach. Attempting to fit the fixed-intercept Poisson regression model by conditional maximum likelihood using `xtpoisson` with the `fe` option,

```
xtset id
xtpoisson numvisit reform age educ married badh loginc summer, fe irr
```

fails. The reason is that the within-person variation in the covariates `reform` and `age` is perfectly confounded, in the sense that `reform` changes from 0 to 1 for all persons from 1996 to 1998 and `age` increases by 2 years for all persons from 1996 to 1998. It follows that the effects of `reform` and `age` are not separately identified in the fixed-intercept approach. In contrast, these effects are separately identified in ordinary Poisson regression and random-effects Poisson regression because `age` also varies *between* patients. This issue is similar to that discussed for different time scales in the fixed-effects approach to continuous longitudinal data in section 5.4.

Because of the confounding of the within effects of `reform` and `age`, we need to omit one of these covariates from the model; we omit `age`. This means that the estimated within effect of `reform` represents the combined within effects of `reform` and `age`. After omitting `age`, we can fit the fixed-intercept Poisson regression model by using the `fe` option:

```
. quietly xtset id

. xtpoisson numvisit reform educ married badh loginc summer, fe irr
note: 809 groups (809 obs) dropped because of only one obs per group
note: 95 groups (190 obs) dropped because of all zero outcomes
Conditional fixed-effects Poisson regression    Number of obs      =       1228
Group variable: id                              Number of groups   =        614

                                                Obs per group: min =          2
                                                               avg =        2.0
                                                               max =          2

                                                Wald chi2(6)       =      57.97
Log likelihood  =  -1080.974                    Prob > chi2        =     0.0000
```

numvisit	IRR	Std. Err.	z	P>|z|	[95% Conf. Interval]	
reform	1.015137	.0359576	0.42	0.671	.9470517	1.088116
educ	.9839241	.1610757	-0.10	0.921	.7138608	1.356156
married	1.045286	.1181487	0.39	0.695	.8375752	1.304507
badh	1.772883	.1450189	7.00	0.000	1.510265	2.081167
loginc	.9680603	.1119556	-0.28	0.779	.7717229	1.214349
summer	.8079478	.0676067	-2.55	0.011	.6857365	.9519396

The estimates were reported together with estimates from other models in table 13.1 on page 695. The nonsignificant point estimate of 1.02 for the IRR suggests that the German health-care reform did not affect the number of doctor visits. However, the estimated effect of `reform` from the fixed-intercept Poisson regression model should be interpreted with extra caution, as discussed above.

We can alternatively fit fixed-intercept Poisson models by maximum likelihood (instead of conditional maximum likelihood) if we include dummy variables for persons j. This method would be analogous to the fixed-effects estimator of within-person effects discussed for linear models in section 3.7.2. In contrast to logistic regression, maximum likelihood estimation of Poisson regression models with log links that include person dummies yields consistent estimates of the within-person effects (identical to the conditional maximum likelihood estimates), even when the cluster sizes are small as in the German health-care data. To replicate the estimates above from `fe`, we use the `poisson` command for maximum likelihood estimation and use `i.id` to include dummy variables for all persons (apart from one, who is represented by the intercept). Because a very large number of parameters is estimated, we need to increase the matrix size allowed in Stata by using the `set matsize` command before fitting the model. For Stata/MP and Stata/SE, the command to type is

```
. set matsize 2000
```

For Stata/IC and Small Stata, this example cannot be run because of size limits.

```
. poisson numvisit reform educ married badh loginc summer i.id, irr
Poisson regression                                  Number of obs   =       2227
                                                    LR chi2(1523)   =    7507.36
                                                    Prob > chi2     =     0.0000
Log likelihood = -2903.5113                         Pseudo R2       =     0.5639
```

numvisit	IRR	Std. Err.	z	P>|z|	[95% Conf. Interval]	
reform	1.015137	.0359576	0.42	0.671	.9470518	1.088116
educ	.9839241	.1610757	-0.10	0.921	.7138608	1.356156
married	1.045286	.1181487	0.39	0.695	.8375752	1.304507
badh	1.772883	.1450189	7.00	0.000	1.510265	2.081167
loginc	.9680603	.1119556	-0.28	0.779	.7717229	1.214349
summer	.8079478	.0676067	-2.55	0.011	.6857365	.9519396

(Estimates for the very large number of person dummies not shown)

We see that the estimates are identical to those produced by xtpoisson. However, this approach is somewhat impractical when there are many clusters as here because a large number of parameters must be estimated.

13.11.2   Conditional negative binomial regression

Hausman, Hall, and Griliches (1984) proposed a conditional negative binomial model where the parameter δ_j in (13.9) and (13.10) is treated as fixed. However, while the fixed-effects Poisson model does not allow for overdispersion, the fixed-effects negative binomial model does not allow for lack of overdispersion. Another oddity of the fixed-effects negative binomial model is that effects of cluster-specific covariates can be estimated (in contrast to other fixed-effects models) because their inclusion affects the variance function. See also Allison and Waterman (2002).

The syntax for fitting this model in Stata is

```
xtnbreg numvisit reform age educ married badh loginc summer, fe irr
```

13.11.3   Generalized estimating equations

We now consider a multivariate extension of the quasilikelihood approach discussed in section 13.9.3 called generalized estimating equations (GEE). GEE can be used to estimate *marginal effects*, just as in ordinary Poisson regression but taking the dependence among units nested in clusters into account. The ideas are the same as discussed for dichotomous responses in section 10.14.2.

GEE estimates using the default exchangeable correlation structure but with robust sandwich-based standard errors are obtained using

```
. quietly xtset id
. xtgee numvisit reform age educ married badh loginc summer,
> family(poisson) link(log) vce(robust) eform

GEE population-averaged model          Number of obs      =      2227
Group variable:                    id  Number of groups   =      1518
Link:                             log  Obs per group: min =         1
Family:                       Poisson                 avg =       1.5
Correlation:              exchangeable                 max =         2
                                       Wald chi2(7)       =    235.37
Scale parameter:                    1  Prob > chi2        =    0.0000

                               (Std. Err. adjusted for clustering on id)
```

		Robust			
numvisit	IRR	Std. Err.	z	P>|z|	[95% Conf. Interval]
reform	.8849813	.0467574	-2.31	0.021	.7979238 .9815371
age	1.005253	.0033549	1.57	0.116	.9986986 1.01185
educ	.9906936	.0117057	-0.79	0.429	.9680144 1.013904
married	1.038283	.072958	0.53	0.593	.904698 1.191593
badh	3.020459	.264062	12.64	0.000	2.544821 3.584997
loginc	1.150186	.0914521	1.76	0.078	.9842114 1.34415
summer	.9741462	.0862448	-0.30	0.767	.8189626 1.158735
_cons	.6818911	.3926379	-0.66	0.506	.2205906 2.107867

These estimates were reported under the heading "GEE Poisson" in table 13.1.

We display the fitted working correlation matrix using `estat wcorrelation`,

```
. estat wcorrelation
Estimated within-id correlation matrix R:
               c1          c2
      r1        1
      r2  .2135204           1
```

and see that the correlation between the Pearson residuals at the two occasions is estimated as 0.21.

13.12 Estimating marginal and conditional effects when responses are missing at random

From the discussion in section 13.7.1, we would expect the estimated conditional effects using random-intercept Poisson models to be similar to the estimated marginal effects using ordinary Poisson regression or GEE (apart from the intercept). However, it is evident from table 13.1 that the estimates are quite different.

This discrepancy is probably due to the extremely unbalanced nature of the data. As discussed for linear models in section 5.8.1, if the responses are missing at random (missingness does not depend on unobserved responses, given the observed responses and covariates) and if the probability of a response being missing at an occasion depends on the observed response at the other occasion, then maximum likelihood estimation of the correct model gives consistent parameter estimates. However, maximum likelihood estimation of an incorrect model, such as Poisson regression that ignores within-person dependence, gives inconsistent estimates. GEE is also not consistent if the probability of a response being missing at an occasion depends on the observed response at another occasion after controlling for covariates. It is often claimed that GEE requires data to be missing completely at random (MCAR) for consistency, but missingness can actually depend on the covariates.

❖ Simulation

We now simulate complete data from a random-intercept Poisson model (with parameter values equal to those we have estimated for the data) and produce data that are missing at random (MAR) by letting the probability of being missing at the second occasion depend on the observed response at the first occasion. We then compare estimates from GEE using an exchangeable correlation structure with maximum likelihood estimates for ordinary and random-intercept Poisson models. This simulation is similar to the one described in section 5.8.1 (see also exercise 6.6).

We can use the postestimation command `gllasim` for `gllamm` to simulate responses from the model just fit in `gllamm`. To create missing data according to a process of our choice, we must first simulate complete data. To accomplish this, we must first ensure that each person has two rows of data, one for 1996 and one for 1998. (see also section 10.13.2 where we used `fillin` to produce balanced data). We create a variable, `num`, containing the number of rows of data per person by typing

```
. egen num = count(numvisit), by(id)
```

For those with one row only, we want to expand the data by 2, and for those with two rows, we do not want to expand the data, which is equivalent to expanding them by a factor of 1.

```
. generate mult = 3 - num
. expand mult
(809 observations created)
```

We now generate an occasion identifier, `occ`, equal to `reform` for those who already have complete data and equal to 0 and 1 for arbitrarily chosen rows otherwise.

```
. by id (reform), sort: generate occ = _n - 1
```

To keep track of which responses were actually missing in the original data, we create the dummy variable `missing`:

```
. generate missing = (occ!=reform & num==1)
```

People with `num` equal to 1 originally had only one observation. If the observation was from 1996, `reform` takes on the value 0 twice in the expanded data (because the `expand` command just clones all variables) whereas `occ` is 0 and 1, so the 1998 observation is flagged as missing. For someone with missing data for 1996, reform is 1 twice, and the

1996 response is flagged as missing. Now we replace responses where `missing` equals 1 with missing values:

```
. replace numvisit = . if missing==1
(809 real changes made, 809 to missing)
```

For persons with missing data, the variable `reform` takes on the same value at both occasions. We rectify this by replacing the value of `reform` at the occasion when the response was missing with `occ`:

```
. replace reform = occ if missing==1
(809 real changes made)
```

We are now ready to simulate new responses y. To ensure that the results can be replicated, we first sort the data and then set the random-number seed to an arbitrary number:

```
. sort id reform
. set seed 1211
```

We then retrieve the `gllamm` estimates for the random-intercept Poisson model and use the `gllasim` command with the `fsample` option to simulate responses for the full sample, not just the estimation sample

```
. estimates restore glri
. gllasim y, fsample
(simulated responses will be stored in y)
```

For those who visited the doctor at least twice in 1996, we replace the response for 1998 with a missing value with a probability of 0.9:

```
. by id (reform), sort: generate drop = (y[1]>2 & runiform()<.9)
. replace y = . if drop==1 & reform==1
(466 real changes made, 466 to missing)
```

Because the responses are missing at random (MAR), maximum likelihood estimation of the model that generated the data should yield consistent estimates.

```
. quietly xtset id
. xtpoisson y reform age educ married badh loginc summer, normal irr
```

```
Random-effects Poisson regression          Number of obs     =        2570
Group variable: id                         Number of groups  =        1518

Random effects u_i ~ Gaussian              Obs per group: min =           1
                                                          avg =         1.7
                                                          max =           2

                                           Wald chi2(7)      =      213.48
Log likelihood   = -4840.7573              Prob > chi2       =      0.0000
```

y	IRR	Std. Err.	z	P>|z|	[95% Conf. Interval]	
reform	1.029429	.036059	0.83	0.408	.9611257	1.102586
age	1.006054	.0028652	2.12	0.034	1.000453	1.011685
educ	1.020096	.0126798	1.60	0.109	.9955447	1.045254
married	1.151819	.0669211	2.43	0.015	1.027849	1.290742
badh	2.660478	.2005024	12.98	0.000	2.295146	3.083962
loginc	1.04616	.0756244	0.62	0.532	.9079598	1.205395
summer	.8274459	.0645118	-2.43	0.015	.7101919	.9640588
_cons	.6261601	.3363712	-0.87	0.383	.2184856	1.794519
/lnsig2u	-.1261919	.0571691	-2.21	0.027	-.2382412	-.0141425
sigma_u	.9388534	.0268367			.8877007	.9929537

```
Likelihood-ratio test of sigma_u=0: chibar2(01) =  2871.89 Pr>=chibar2 = 0.000
```

We see that the estimates above are close to the "true" parameter values, which are the estimates for the original data.

However, ordinary Poisson regression produces quite different estimates:

```
. poisson y reform age educ married badh loginc summer, irr
```

```
Poisson regression                         Number of obs     =        2570
                                           LR chi2(7)        =     1426.49
                                           Prob > chi2       =      0.0000
Log likelihood = -6276.7036                Pseudo R2         =      0.1020
```

y	IRR	Std. Err.	z	P>|z|	[95% Conf. Interval]	
reform	.6131784	.0177075	-16.94	0.000	.5794363	.6488855
age	1.004305	.0012989	3.32	0.001	1.001763	1.006854
educ	1.01852	.0057876	3.23	0.001	1.007239	1.029926
married	1.128861	.031214	4.38	0.000	1.069311	1.191727
badh	2.719263	.0854277	31.84	0.000	2.556879	2.891961
loginc	1.108184	.0392507	2.90	0.004	1.033863	1.187847
summer	.8433938	.0343386	-4.18	0.000	.7787066	.9134545
_cons	.6891216	.1810127	-1.42	0.156	.4118218	1.153141

Specifically, the intervention now looks effective because those who visited the doctor frequently in 1998 were more likely to drop out, and because of the within-person dependence between responses, it is these same people who tended to visit the doctor frequently in 1996. Consequently, the mean number of doctor visits in 1998 is underestimated by ordinary Poisson regression. In contrast, the random-intercept model "knows" about the within-person dependence and hence makes the correct adjustment.

Interestingly, generalized estimating equations (GEE) does not work well in this case although we allow for within-person dependence by specifying an exchangeable correlation structure:

```
. quietly xtset id
. xtgee y reform age educ married badh loginc summer, family(poisson)
> link(log) vce(robust) eform
GEE population-averaged model          Number of obs      =      2570
Group variable:                    id  Number of groups   =      1518
Link:                             log  Obs per group: min =         1
Family:                       Poisson                 avg =       1.7
Correlation:              exchangeable                 max =         2
                                       Wald chi2(7)       =    242.22
Scale parameter:                    1  Prob > chi2        =    0.0000

                               (Std. Err. adjusted for clustering on id)
```

y	IRR	Robust Std. Err.	z	P>|z|	[95% Conf. Interval]
reform	.7609233	.0251568	-8.26	0.000	.7131804	.8118624
age	1.004157	.0030258	1.38	0.169	.9982446	1.010105
educ	1.017779	.0134357	1.33	0.182	.9917829	1.044456
married	1.131598	.0699859	2.00	0.046	1.002416	1.277428
badh	2.718051	.2167291	12.54	0.000	2.324799	3.177824
loginc	1.119422	.0891531	1.42	0.157	.9576402	1.308535
summer	.8264793	.0788856	-2.00	0.046	.6854672	.9964999
_cons	.646465	.3697563	-0.76	0.446	.2107105	1.98337

13.13 Which Scottish counties have a high risk of lip cancer?

We now consider models for disease mapping or small-area estimation. Clayton and Kaldor (1987) presented and analyzed data on lip cancer for each of 56 Scottish counties over the period 1975–1980. These data have also been analyzed by Breslow and Clayton (1993) and Leyland (2001) among many others.

The dataset lips.dta has the following variables:

- county: county identifier (1 to 56)
- o: observed number of lip cancer cases
- e: expected number of lip cancer cases
- x: percentage of population working in agriculture, fishing, or forestry

The expected number of lip cancer cases is based on the age-specific lip cancer rates for the whole of Scotland and the age distribution of the counties. We read in the lip cancer data by typing

```
. use http://www.stata-press.com/data/mlmus3/lips, clear
```

13.14 Standardized mortality ratios

The standardized mortality ratio (SMR) for a county is defined as the ratio of the incidence rate to that expected if the age-specific incidence rates were equal to those of a reference population (for example, Breslow and Day [1987]), here the whole of Scotland. The crude estimate of the SMR for county j is obtained using

$$\widehat{\text{SMR}}_j = \frac{o_j}{e_j}$$

where o_j is the observed number of cases and e_j is the expected number of cases.

We first calculate the crude SMRs as percentages for the lip cancer data:

```
. generate smr = 100*o/e
```

The number of observed and expected lip cancer cases, and crude SMRs for the 56 counties are presented in table 13.3. The crude SMRs are also shown in the map in figure 13.1. When the SMR exceeds 100, the county has more cases than would be expected given the age distribution.

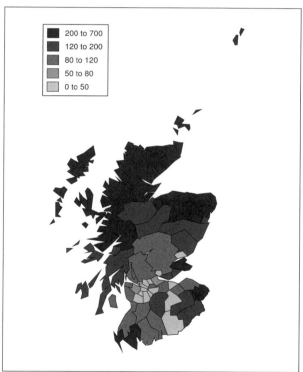

Figure 13.1: Map of crude SMR as percentage (*Source:* Skrondal and Rabe-Hesketh 2004)

Table 13.3: Observed and expected numbers of lip cancer cases and various SMR estimates (in percentages) for Scottish counties

County	#	Obs o_j	Exp e_j	Crude SMRs	Predicted SMRs Norm.	NPMLE
Skye. Lochalsh	1	9	1.4	652.2	470.7	342.6
Banf. Buchan	2	39	8.7	450.3	421.8	362.4
Caithness	3	11	3.0	361.8	309.4	327.1
Berwickshire	4	9	2.5	355.7	295.2	321.6
Ross. Cromarty	5	15	4.3	352.1	308.5	327.6
Orkney	6	8	2.4	333.3	272.0	311.1
Moray	7	26	8.1	320.6	299.9	322.2
Shetland	8	7	2.3	304.3	247.8	292.5
Lochaber	9	6	2.0	303.0	239.0	280.1
Gordon	10	20	6.6	301.7	279.1	319.9
W. Isles	11	13	4.4	295.5	262.5	315.5
Sutherland	12	5	1.8	279.3	219.2	254.3
Nairn	13	3	1.1	277.8	198.4	222.7
Wigtown	14	8	3.3	241.7	210.9	249.6
NE. Fife	15	17	7.8	216.8	204.6	245.3
Kincardine	16	9	4.6	197.8	178.9	171.4
Badenoch	17	2	1.1	186.9	151.9	163.2
Ettrick	18	7	4.2	167.5	154.7	136.7
Inverness	19	9	5.5	162.7	154.2	128.4
Roxburgh	20	7	4.4	157.7	149.1	130.3
Angus	21	16	10.5	153.0	147.8	117.1
Aberdeen	22	31	22.7	136.7	135.0	116.4
Argyll. Bute	23	11	8.8	125.4	123.3	116.4
Clydesdale	24	7	5.6	124.6	122.9	116.7
Kirkcaldy	25	19	15.5	122.8	121.6	116.4
Dunfermline	26	15	12.5	120.1	119.1	116.3
Nithsdale	27	7	6.0	115.9	116.3	115.5
E. Lothian	28	10	9.0	111.6	111.4	115.9
Perth. Kinross	29	16	14.4	111.3	111.1	116.3
W. Lothian	30	11	10.2	107.8	108.5	115.9
Cumnock-Doon	31	5	4.8	105.3	107.2	111.5
Stewartry	32	3	2.9	104.2	109.1	109.4
Midlothian	33	7	7.0	99.6	102.7	113.1
Stirling	34	8	8.5	93.8	97.3	112.9
Kyle. Carrick	35	11	12.3	89.3	92.2	114.1
Inverclyde	36	9	10.1	89.1	92.6	112.4
Cunninghame	37	11	12.7	86.8	89.7	113.3
Monklands	38	8	9.4	85.6	89.4	109.4
Dumbarton	39	6	7.2	83.3	89.4	104.9
Clydebank	40	4	5.3	75.9	85.7	94.5

Table 13.3: Observed and expected numbers of lip cancer cases and various SMR estimates (in percentages) for Scottish counties (cont.)

County	#	Obs o_j	Exp e_j	Crude SMRs	Predicted SMRs Norm.	NPMLE
Renfrew	41	10	18.8	53.3	59.1	40.6
Falkirk	42	8	15.8	50.7	57.9	40.9
Clackmannan	43	2	4.3	46.3	68.8	65.5
Motherwell	44	6	14.6	41.0	50.7	37.5
Edinburgh	45	19	50.7	37.5	40.8	36.2
Kilmarnock	46	3	8.2	36.6	53.2	42.2
E. Kilbride	47	2	5.6	35.8	57.9	49.7
Hamilton	48	3	9.3	32.1	48.5	38.8
Glasgow	49	28	88.7	31.6	33.8	36.2
Dundee	50	6	19.6	30.6	39.8	36.2
Cumbernauld	51	1	3.4	29.1	63.1	57.8
Bearsden	52	1	3.6	27.6	61.1	55.4
Eastwood	53	1	5.7	17.4	46.4	40.6
Strathkelvin	54	1	7.0	14.2	40.8	37.8
Tweeddale	55	0	4.2	0.0	43.2	40.8
Annandale	56	0	1.8	0.0	64.9	60.6

Source: Clayton and Kaldor (1987)

13.15 Random-intercept Poisson regression

An important limitation of crude SMRs is that estimates for counties with small populations are very imprecise. This problem can be addressed by using random-intercept Poisson models in conjunction with empirical Bayes prediction. The resulting SMRs are shrunken toward the overall SMR, thereby borrowing strength from other counties (see section 2.11.2).

Although we only have one observation per county, we can introduce a county-level random intercept to model overdispersion, as discussed in section 13.9.

13.15.1 Model specification

We consider a random-intercept Poisson regression model where the observed number of lip cancer cases in county j is assumed to have a Poisson distribution with mean μ_j,

$$\ln(\mu_j) = \ln(e_j) + \beta_1 + \zeta_j$$

Here $\zeta_j \sim N(0, \psi)$ is a random intercept representing unobserved heterogeneity between counties and $\ln(e_j)$ is an offset, a covariate with regression coefficient set to 1. The purpose of the offset is to ensure that $\beta_1 + \zeta_j$ can be interpreted as a model-based county-specific log SMR. This interpretation becomes clear by subtracting the offset from both sides of the equation:

$$\ln(\mu_j) - \ln(e_j) = \underbrace{\ln(\mu_j/e_j)}_{\text{SMR}_j} = \beta_1 + \zeta_j$$

13.15.2   Estimation using gllamm

We will use `gllamm` for estimation because this is the only command that can provide the required posterior expectations of the standardized mortality ratios at the time of writing this book.[1] We first generate a variable for the offset,

```
. generate lne = ln(e)
```

and then we pass this variable to `gllamm` using the `offset()` option. We also specify a Poisson distribution and log link using the `family()` and `link()` options, respectively:

```
. gllamm o, i(county) offset(lne) family(poisson) link(log) adapt
number of level 1 units = 56
number of level 2 units = 56

gllamm model

log likelihood = -181.32392
```

o	Coef.	Std. Err.	z	P>|z|	[95% Conf. Interval]
_cons	.080211	.1167416	0.69	0.492	-.1485983 .3090204
lne	1	(offset)			

```
Variances and covariances of random effects
------------------------------------------------------------------------------

***level 2 (county)

    var(1): .58432672 (.14721754)
------------------------------------------------------------------------------
```

The maximum likelihood estimates for the random-intercept model are given in table 13.4 under "Normal ML". We could also include the percentage of the population in county j working in agriculture, fishing, or forestry as a county-level covariate; see exercise 13.6.

---

1. `xtmepoisson` could be used to obtain posterior modes of ζ_j and hence posterior modes of SMRs.

Table 13.4: Estimates for random-intercept models for Scottish lip cancer data

	Normal ML		NPMLE (C=4)	
	Est	(SE)	Est	(SE)
β_1 [_cons]	0.08	(0.12)	0.08	(0.12)
ψ	0.58		$0.63^\dagger$	
Log likelihood	-181.32		-174.39	

†Derived from discrete distribution

13.15.3   Prediction of standardized mortality ratios

We now consider predictions of the SMRs using empirical Bayes(see also section 10.13.2). Specifically, we would like to estimate the posterior expectation of the SMR,

$$\widetilde{\text{SMR}}_j = \int \exp(\widehat{\beta}_1 + \zeta_j)\, \text{Posterior}(\zeta_j | o_j, e_j)\, d\zeta_j \tag{13.11}$$

It may be tempting to obtain the predicted SMRs by simply plugging in the estimate $\widehat{\beta}_1$ and the predicted random intercept $\widetilde{\zeta}_j$ in the exponential function. However, this would be incorrect because the expectation of a nonlinear function of a random variable is not equal to the nonlinear function of the expectation of the random variable.

Empirical Bayes predictions of the SMRs can be obtained using `gllapred` with the `mu` option to get posterior means on the scale of the response and the `nooffset` option to omit the offset $\ln(e_j)$ from the linear predictor,

```
. gllapred mu, mu nooffset
. generate thet = 100*mu
```

where `thet` are the predicted SMRs in percent. We then list these predicted SMRs for the first 10 counties:

```
. sort county
. list county thet in 1/10, clean noobs
    county       thet
         1   470.7203
         2   421.7975
         3   309.4238
         4   295.1885
         5   308.5293
         6   272.0485
         7   299.8875
         8   247.8383
         9   238.9533
        10   279.1406
```

These empirical Bayes predictions were given under "Norm." in table 13.3 and are displayed as a map in figure 13.2. The maps in figures 13.1 and 13.2 were drawn using Stata, and an annotated do-file to create the maps can be obtained by typing

```
copy http://www.stata-press.com/data/mlmus3/scotmaps.do scotmaps.do
```

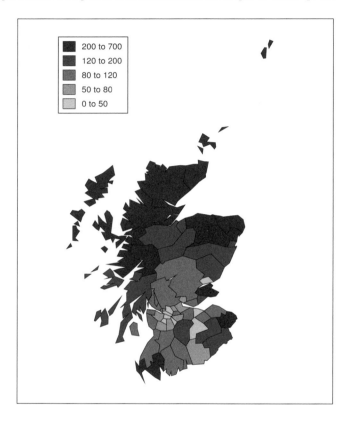

Figure 13.2: Map of SMRs assuming normally distributed random intercept (no covariate) (*Source:* Skrondal and Rabe-Hesketh 2004)

We can plot the empirical Bayes predictions against the crude SMRs by typing

```
. twoway (scatter thet smr, msymbol(none) mlabpos(0) mlabel(county))
> (function y=x, range(0 600)), xline(108) yline(108)
> xtitle(Crude SMR) ytitle(Empirical Bayes SMR) legend(off)
```

The $y = x$ line has been superimposed, as well as lines $x = 108$ and $y = 108$, representing the SMR in percent, $100 \times \exp(\widehat{\beta}_1)$, when ζ_j equals its mean 0 (these are median SMRs). The resulting graph is given in figure 13.3. Shrinkage is apparent because counties with particularly high crude SMRs lie below the $y = x$ line (have predictions lower than the crude SMR) and counties with particularly low crude SMRs lie above the $y = x$ line.

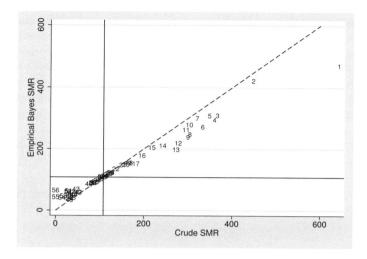

Figure 13.3: Empirical Bayes SMRs versus crude SMRs

13.16 ❖ Nonparametric maximum likelihood estimation

13.16.1   Specification

Instead of assuming that the random intercept ζ_j is normally distributed, we can relax this assumption using nonparametric maximum likelihood estimation (NPMLE).

It can been shown that the nonparametric estimator of the random-intercept distribution is discrete with locations $\zeta_j = e_c$ and probabilities π_c $(c = 1, \ldots, C)$, where the number of locations C with nonzero probabilities is determined to make the likelihood as large as possible.

Obviously, the probabilities must add to one:

$$\pi_1 + \cdots + \pi_C \;=\; 1 \tag{13.12}$$

The discrete distribution of ζ_j can be parameterized so that it has zero mean,

$$\pi_1 e_1 + \cdots + \pi_C e_C \;=\; 0 \tag{13.13}$$

which has the advantage that we need not omit the intercept β_1 from the model:

$$\ln(\mu_j) \;=\; \ln(e_j) + \beta_1 + \zeta_j$$

13.16.2   Estimation using gllamm

To obtain the nonparametric maximum likelihood estimates, we must use the maximum possible number of locations so that it is not possible to achieve a larger likelihood by introducing more locations.

This maximum number can be determined using the following iterative approach. First, we fit the model with a given number of locations C, for instance, $C=2$. Keeping all the parameters equal to the resulting maximum likelihood estimates, we introduce another point with a small probability π_{C+1} and evaluate the likelihood for a large number of different locations for this point, for example, by setting $e_{C+1} = -5 + l/100$, $(l = 1, \ldots, 1000)$. If the likelihood increases for any of these locations, we have evidence that another point is required. The model is then refit with $C+1$ points, and we repeat the search for a location at which introducing a new point increases the likelihood. These cycles are repeated until the search does not identify any more locations.

For the Scottish lip cancer data, we start with $C=2$ locations and maximize the likelihood with respect to β_1, e_1, and π_1, with e_2 and π_2 determined by the constraints in (13.12) and (13.13). This is accomplished in `gllamm` using the `ip(f)` and `nip(2)` options.

```
. gllamm o, i(county) offset(lne) family(poisson) link(log) ip(f) nip(2)
number of level 1 units = 56
number of level 2 units = 56

Condition Number = 9.5651502

gllamm model

log likelihood = -205.40139
```

o	Coef.	Std. Err.	z	P>|z|	[95% Conf. Interval]
_cons	.0314632	.1037764	0.30	0.762	-.1719348 .2348612
lne	1	(offset)			

```
Probabilities and locations of random effects
-------------------------------------------------------------------------------

***level 2 (county)

    loc1: -.85205, .54307
 var(1): .46272368
   prob: 0.3893, 0.6107
-------------------------------------------------------------------------------
```

We see that the locations are estimated as $\widehat{e}_1 = -0.85$ and $\widehat{e}_2 = 0.54$ with estimated probabilities $\widehat{\pi}_1 = 0.39$ and $\widehat{\pi}_2 = 0.61$. The variance of this discrete distribution,

$$\widehat{\text{Var}}(\zeta_j) \;=\; \widehat{\pi}_1\widehat{e}_1^2 + \widehat{\pi}_2\widehat{e}_2^2$$

is given as 0.46 next to `var(1)`.

We now gradually increase the number of locations until the likelihood cannot be increased any further. For $C = 3$ locations, we use the `nip(3)` option and the previous estimates for $C = 2$ as starting values. We also use the `gateaux()` option (for the *Gâteaux derivative* or directional derivative) to specify the smallest and largest locations to be considered in the search, as well as the number of equally spaced steps to take within these limits. Finally, we must also supply the previous log likelihood using the `lf0()` option, which requires as its first argument the number of parameters in the previous model. It is convenient to obtain the parameter estimates, log likelihood, and number of parameters in the previous model as follows:

```
. matrix a = e(b)
. local ll = e(ll)
. local k = e(k)
. gllamm o, i(county) offset(lne) family(poisson) link(log) ip(f) nip(3)
> gateaux(-5 5 100) from(a) lf0('k' 'll')
.....................................................................................
> ....................
maximum gateaux derivative is 6.872835

number of level 1 units = 56
number of level 2 units = 56

Condition Number = 9.2086581


gllamm model                              Number of obs     =        56
                                          LR chi2(2)        =     61.92
Log likelihood = -174.44216               Prob > chi2       =    0.0000

---------------------------------------------------------------------------
           o |    Coef.    Std. Err.      z    P>|z|    [95% Conf. Interval]
-------------+-------------------------------------------------------------
       _cons |  .0860003    .1189174    0.72   0.470    -.1470736    .3190742
         lne |         1    (offset)
---------------------------------------------------------------------------


Probabilities and locations of random effects
---------------------------------------------------------------------------

***level 2 (county)

   loc1: -1.1012, 1.1149, .07126
 var(1): .63591774
   prob: 0.2753, 0.241, 0.4836
---------------------------------------------------------------------------
```

To increase the number of locations to $C = 4$, we use the same commands and options as before except that we change the `nip()` option from `nip(3)` to `nip(4)`:

```
. matrix a = e(b)
. local ll = e(ll)
. local k = e(k)
```

```
. gllamm o, i(county) offset(lne) family(poisson) link(log) ip(f) nip(4)
> gateaux(-5 5 100) from(a) lf0('k' 'll')
.......................................................................
> ....................
maximum gateaux derivative is .00020442

number of level 1 units = 56
number of level 2 units = 56

Condition Number = 72.542683
```

```
gllamm model                           Number of obs    =        56
                                       LR chi2(2)       =      0.11
Log likelihood = -174.38535            Prob > chi2      =    0.9448
```

o	Coef.	Std. Err.	z	P>|z|	[95% Conf. Interval]
_cons	.0847742	.1188573	0.71	0.476	-.1481819 .3177303
lne	1	(offset)			

```
Probabilities and locations of random effects
--------------------------------------------------------------------------------

***level 2 (county)

    loc1: -1.1002, 1.0426, 1.2628, .06676
  var(1): .6339741
    prob: 0.2749, 0.1847, 0.0617, 0.4788
--------------------------------------------------------------------------------
```

For $C=5$, we obtain

```
. matrix a = e(b)
. local ll = e(ll)
. local k = e(k)
. gllamm o, i(county) offset(lne) family(poisson) link(log) nip(5) ip(f)
> gateaux(-5 5 100) from(a) lf0('k' 'll')
.......................................................................
> ....................
maximum gateaux derivative is -.00005128
maximum gateaux derivative less than 0.00001
```

The output suggests that the likelihood cannot be increased by including a fifth point. However, we have only tried 100 locations. To make sure that there is no location at which the likelihood can be increased, we also used the option gateaux(-5 5 1000), but the maximum Gâteaux derivative was still negative.

We therefore conclude that nonparametric maximum likelihood estimation (NPMLE) has $C=4$ locations. The nonparametric maximum likelihood estimates were shown in table 13.4 under "NPMLE". The estimates from NPMLE are similar to those obtained by assuming normality.

To visualize the discrete distribution, we can access the matrices of location and log-probability estimates stored by `gllamm` by using

```
. matrix locs = e(zlc2)'
. matrix lp = e(zps2)'
```

Here the apostrophe in `e(zlc2)'` transposes the row matrix to a column matrix that can be stored as a variable in the dataset by using `svmat`:

```
. svmat locs
. svmat lp
```

We then transform the estimated log probabilities to probabilities and the log-incidence rates e_c to SMRs by using

```
. generate p = exp(lp1)
. generate smrloc = 100*exp(_b[_cons] + locs1)
```

Finally, we produce a graph of the discrete distribution:

```
. twoway (dropline p smrloc), xtitle(Location) ytitle(Probability)
```

The graph is shown in figure 13.4.

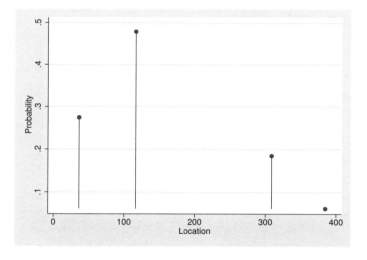

Figure 13.4: Lip cancer in Scotland: SMRs (locations) and probabilities for nonparametric maximum likelihood estimate of random-intercept distribution

13.16.3   Prediction

Empirical Bayes predictions of the SMRs based on nonparametric maximum likelihood estimation can be obtained as

```
. gllapred mu2, mu nooffset
. generate thet2 = 100*mu2
```

where `thet2` are the predicted SMRs in percentages. We list these predicted SMRs for the first 10 counties together with predictions based on the normality assumption for the random intercept `thet`:

```
. sort county
. list county thet thet2 in 1/10, clean noobs
    county       thet       thet2
         1   470.7203    342.6493
         2   421.7975    362.4183
         3   309.4238    327.1122
         4   295.1885    321.6138
         5   308.5293    327.5669
         6   272.0485     311.061
         7   299.8875    322.2489
         8   247.8383    292.5375
         9   238.9533    280.1219
        10   279.1406    319.8951
```

Predicted SMRs based on nonparametric maximum likelihood estimation (NPMLE) for all Scottish counties were reported under "NPMLE" in table 13.3.

13.17   Summary and further reading

In this chapter, we have discussed Poisson regression modeling of count data. Multilevel Poisson regression models with random effects were fit to longitudinal data on number of doctor visits in Germany and small-area estimation or disease mapping of standardized mortality ratios (SMRs) for lip cancer in Scotland.

We have outlined the motivation and some of the basic properties of the Poisson distribution, such as the equality of the expectation and variance. Different approaches for handling overdispersion in Poisson regression were considered such as quasilikelihood, robust standard errors, and the introduction of a level-1 random intercept. We have introduced random effects at the cluster level to accommodate within-cluster dependence, which must be tackled, for instance, when modeling longitudinal or panel data of counts. Both random-intercept and random-coefficient models were considered. Other approaches to modeling clustered count data were briefly considered, such as conditional maximum likelihood, GEE, and NPMLE where the distributional assumption for the random intercept is relaxed (see also Rabe-Hesketh, Pickles, and Skrondal [2003]).

Most of the issues discussed for prediction of random effects for dichotomous responses persist for counts, such as the choice between using the mean or the mode of the posterior distribution. We also experience shrinkage in empirical Bayes prediction as before. In small-area estimation or disease mapping, inclusion of random effects is beneficial for "borrowing strength" from other counties in predicting SMRs. Empirical Bayes predictions for disease mapping can be improved by modeling spatial dependence (see Skrondal and Rabe-Hesketh, [2004, sec 11.4]), which results in an even smoother disease map.

Poisson regression is often used to model survival or durations in continuous time. If events occur according to a Poisson process, the durations between events have an exponential distribution, and piecewise exponential survival models can be fit via Poisson regression, as described in section 15.4.1. Chapter 15 uses Poisson and random-effects Poisson regression throughout.

Extensive treatments of methods for count data are given by Cameron and Trivedi (1998) and Winkelmann (2008), both of which also discuss some models for clustered data. A brief introduction to single-level regression models for counts is provided by Long (1997, chap. 8), and Hilbe (2011) is devoted to the negative binomial model. Use of Poisson regression for disease mapping is described in Lawson et al. (1999) and Lawson, Browne, and Vidal Rodeiro (2003).

The exercises cover a wide range of applications in various disciplines. Poisson regression models with random intercepts are used in exercise 13.1 on a longitudinal randomized controlled clinical trial of treatments for epilepsy, exercise 13.2 on a randomized crossover trial investigating the effects of an artificial sweetener on the number of headaches, exercise 13.3 on the number of police stops and ethnicity in New York precincts, exercise 13.4 on panel data to investigate the effects of research and development expenditures on the number of patents awarded to firms, exercise 13.5 on days absent from school among aboriginal and white Australian children, and exercise 13.7 on skin cancer mortality in the European Economic Community. Exercises 13.1, 13.2, and 13.3 also include random-coefficient models. Many of the exercises consider empirical Bayes prediction of random effects and the use of offsets to take varying exposure into account. Exercise 13.5 is on modeling overdispersion in single-level data and exercise 13.5 is on including cluster-specific covariates in small-area estimation and disease mapping. Exercise 13.2 includes a comparison of random-effects and fixed-effects approaches.

13.18 Exercises

13.1 Epileptic-fit data $\boxed{\text{Solutions}}$

We now consider the famous longitudinal epilepsy data used by Thall and Vail (1990), also analyzed in the seminal paper by Breslow and Clayton (1993). The data come from a randomized controlled trial comparing the drug progabide for the treatment of epilepsy with a placebo. The outcomes are counts of epileptic seizures during the 2 weeks before each of four clinic visits.

The dataset `epilep.dta` has the following variables:

- `subj`: subject identifier (j)
- `y`: count of epileptic seizures over 2-week period (y_{ij})
- `visit`: visit time (z_{ij}) coded -0.3, -0.1, 0.1, 0.3
- `treat`: treatment group (x_{2j}) (1: progabide; 0: placebo)
- `lbas`: logarithm of quarter of number of seizures in 8 weeks preceding entry to trial (x_{3j})
- `lbas_trt`: interaction between `lbas` and `treat` (x_{4j})
- `lage`: logarithm of age (x_{5j})
- `v4`: dummy variable for fourth visit (x_{6ij})

The covariates `lbas`, `lbas_trt`, `lage`, and `v4` have all been mean centered, which will affect only the intercept β_1 in the models below.

1. Model II in Breslow and Clayton is a log-linear (Poisson regression) model with covariates `lbas`, `treat`, `lbas_trt`, `lage`, and `v4` and a normally distributed random intercept for subjects. Fit this model using `gllamm`.

2. Breslow and Clayton also considered a random-coefficient model (Model IV) using the variable `visit` instead of `v4`. The effect of `visit` z_{ij} varies randomly between subjects. The model can be written as

$$\log(\mu_{ij}) \;=\; \beta_1 + \beta_2 x_{2j} + \cdots + \beta_5 x_{5j} + \beta_6 z_{ij} + \zeta_{1j} + \zeta_{2j} z_{ij}$$

where the subject-specific random intercept ζ_{1j} and slope ζ_{2j} have a bivariate normal distribution, given the covariates. Fit this model using `gllamm`.

3. Plot the posterior mean counts versus time for 12 patients in each treatment group.

13.2 Headache data

Here we consider data that were originally presented, described, and analyzed by McKnight and van den Eeden (1993), a subset of which were also analyzed by Hedeker (1999). A multiple-period, two-treatment, double-blind crossover trial was conducted to investigate if aspartame, an artificial sweetener, causes headaches.

The basic idea of a crossover trial is to administer both the active treatment and the placebo to each subject, hence letting the subjects serve as their own controls. Different sequences of the treatments are assigned so that period effects can be controlled for. Typically, treatment periods are separated by a washout period in an attempt to preclude carryover effects from treatment assignment in one period to the response in a following period. We refer to Jones and Kenward (2003) and Senn (2002) for detailed treatments of the design and analysis of crossover trials.

Twenty-seven patients were randomized to different sequences of aspartame (A) and placebo (P). Each sequence was preceded by a 7-day placebo run-in period followed by four treatment periods of 7 days each. Each treatment period was separated by a washout day. Both aspartame, given at a dose of 30mg/kg/day, and placebo were administered in capsules in three doses per day. The four possible orderings of treatments after the run-in period were APAP, APPA, PAPA, and PAAP. Before beginning treatment, each subject was asked how certain he or she was of experiencing headaches after ingesting aspartame.

The dataset `headache.dta` has the following variables:

- `id`: subject identifier
- `y`: number of headaches in week i for subject j, for some subjects counted over fewer than 7 days
- `days`: number of days for which headaches were counted (t_{ij})
- `aspartame`: dummy variable for aspartame versus placebo (x_{1ij})
- `belief`: dummy variable taking the value 1 for patients who were very sure that aspartame caused him or her headaches (before treatment) and 0 otherwise (x_{2j})
- `sequence`: the sequence of treatments administered over the four treatment periods
- `period`: taking the value 1 for the run-in period and 2–5 for the four treatment periods

1. Fit the random-intercept model given below using `xtmepoisson`,

$$\log(\mu_{ij}) = \underbrace{\ln(t_{ij})}_{\text{Offset}} + \tau_i + \beta_2 x_{1ij} + \zeta_{1j}$$

$$\mu_{ij} = t_{ij} \underbrace{\exp(\tau_i + \beta_2 x_{1ij} + \zeta_{1j})}_{\text{Rate}}$$

where $\ln(t_{ij})$, the logarithm of `days`, is an offset; τ_i is a period-specific parameter; and $\zeta_{1j}|x_{ij} \sim N(0, \psi_{11})$. Interpret the estimated coefficients.

2. Extend the model to investigate whether the effect of `aspartame` depends on the prior beliefs regarding its effect (at the 5% level). Interpret your results.

3. Extend the retained model from step 2 to investigate whether there is a carryover effect of `aspartame` (again at the 5% level). Make sure that the lagged treatment takes the value 0 in period 1 for all subjects (assuming that none of the subjects used the drug in the period preceding the run-in period).

4. In the discussion section of McKnight and van den Eeden (1993), the authors highlight that a limitation of their approach is that the treatment effect is not allowed to be subject specific. Extending the model retained in step 3, investigate whether the treatment effect varies randomly between subjects.

5. Fit the selected model using `gllamm` and produce a graph showing the empirical Bayes predictions of the headache rate per day for each patient (except patient 26, who never took aspartame) versus treatment. To plot the graph, it is helpful to pick out one observation for each combination of `id` and `aspartame`

(using the egen function tag()) and to use the connect(ascending) option in the twoway command.

6. Fit the fixed-effects version of the model from step 2. Comment on any differences in the parameter estimates.

13.3 Police stops data

Gelman and Hill (2007) analyzed data on the number of times the police stopped individuals for questioning or searching on the streets of New York City. In particular, Gelman and Hill wanted to investigate whether ethnic minorities were stopped disproportionately often compared with whites. They used data collected by the police over a 15-month period in 1998–1999 in which the stops were classified by ethnicity, type of suspected crime, and the precinct (area) in which it occurred. Only the three largest ethnic groups (black, Hispanic, and white) are included here.

It could be argued that it would be reasonable if the police stopped individuals from the different ethnic groups in proportion to their population size or in proportion to the number of crimes they have committed in the past. As a proxy for the latter, Gelman and Hill (2007) use the number of arrests in 1997.

The rows in the data in police.dta correspond to each possible combination of ethnicity, type of crime, and precinct. The dataset contains the following variables:

- eth: ethnicity (1: black; 2: Hispanic; 3: white)
- crime: type of crime (1: violent crimes; 2: weapons offenses; 3: property crimes; 4: drug crimes)
- precinct: New York City precinct (1–75)
- stops: number of police stops over 15-month period in 1998–1999
- arrests: number of arrests in New York City in 1997
- pop: population size of each ethnic group in the precinct
- prblack: proportion of precinct population that is black
- prhisp: proportion of precinct population that is Hispanic

These data, provided with Gelman and Hill (2007), have had some noise added to protect confidentiality. In this exercise, we will analyze the data on violent crimes only, so you can delete the data on the other types of crimes.

1. For stops because of suspected violent crimes, fit a Poisson model with the number of police stops as the response variable, dummies for the ethnic minority groups (blacks and Hispanics) as explanatory variables, and with

   a. pop as an exposure
   b. arrests as an exposure

   Use robust standard errors that take the clustering within precincts into account. Interpret the estimated incidence-rate ratios for the two types of exposure, and comment on the difference.

2. Fit the model from step 1 with `arrests` as the exposure but also including `prblack` and `prhisp` as further covariates.

3. Compare the estimated incidence-rate ratios for the minority groups between the model considered in step 2 and a model that uses fixed effects for precincts instead of the covariates `prblack` and `prhisp`. Does there appear to be much precinct-level confounding in step 2?

4. Fit the model from step 2 but also include a normally distributed random intercept for precincts. Use `xtmepoisson`.

5. For the model in step 4, calculate the estimated median incidence-rate ratio, comparing the precinct that has the larger random intercept with the precinct that has the smaller random intercept for two randomly chosen precincts having the same covariate values. Interpret this estimate.

6. ❖ Extend the model from step 4 further by including random coefficients for the ethnic minority dummy variables, specifying a trivariate normal distribution for the random intercept and slopes with a freely estimated covariance matrix.

   a. Write down the model using a two-stage formulation as discussed in section 4.9.
   b. Fit the model using `xtmepoisson` (this will take a long time).
   c. Obtain empirical Bayes modal predictions of the intercepts and slopes, and plot them using a scatterplot matrix.

13.4 Patent data

Hall, Griliches, and Hausman (1986) analyzed panel data on the number of patents awarded to 346 firms between 1975 and 1979. In particular, they considered the effect of research and development (R&D) expenditures in the current and previous years. The dataset has been analyzed many times. Good discussions can be found in Cameron and Trivedi (1998; 2005), who made the data available on the web page for their 1998 book.

The variables in the dataset `patents.dta` that we will use here are

- `cusip`: Compustat's (Committee on Uniform Security Identification Procedures) identifying number for firm (j)
- `year`: year (i) (1–5)
- `pat`: number of patents applied for during the year that were eventually granted (y_{ij})
- `scisect`: dummy variable for firm being in the scientific sector
- `logk`: logarithm of the book value of capital in 1972
- `logr`: logarithm of R&D spending during the year (in 1972 dollars) [$\ln(r_{ij})$]
- `logr1`, `logr2`, ..., `logr5`: lagged `logr` variable 1 year, 2 years, ..., 5 years ago [$\ln(r_{i-1,j})$, $\ln(r_{i-2,j})$, ..., $\ln(r_{i-5,j})$]

Consider the following model, which allows the expected number of patents successfully applied for by firm j in year i to depend on the logarithms of R&D expenditure for the current and previous 5 years:

$$\ln(\mu_{ij}) \;=\; \alpha + \beta_0 \ln(r_{ij}) + \beta_1 \ln(r_{i-1,j}) + \cdots + \beta_5 \ln(r_{i-5,j}) \qquad (13.14)$$

We can consider the effect of multiplying the expenditures for all years for a given firm by a constant a:

$$\ln(\mu_{ij}^a) \;=\; \alpha + \beta_0 \ln(ar_{ij}) + \beta_1 \ln(ar_{i-1,j}) + \cdots + \beta_5 \ln(ar_{i-5,j})$$

Taking derivatives with respect to a (using the chain rule), we find that

$$\frac{\partial \ln(\mu_{ij}^a)}{\partial a} = \frac{1}{\mu_{ij}^a}\frac{\partial \mu_{ij}^a}{\partial a} \;=\; \beta_0 \frac{r_{ij}}{ar_{ij}} + \beta_1 \frac{r_{i-1,j}}{ar_{i-1,j}} + \cdots + \beta_5 \frac{r_{i-5,j}}{ar_{i-5,j}}$$

$$= \frac{1}{a}(\beta_0 + \beta_1 + \cdots + \beta_5)$$

so that the *elasticity* is (see also display 6.2)

$$\left(\frac{\partial \mu_{ij}^a}{\partial a}\right)\left(\frac{a}{\mu_{ij}^a}\right) \;=\; \frac{\partial \mu_{ij}^a}{\mu_{ij}^a}\bigg/ \frac{\partial a}{a} \;=\; \beta_0 + \beta_1 + \cdots + \beta_5$$

The sum of the coefficients therefore represents the relative or percentage change in the expected number of patents per percentage change in total expenditures over the last 5 years (a given percentage change in a amounts to the same percentage change in ax).

1. Fit a Poisson model with the mean modeled as shown in (13.14) but also include a normally distributed random intercept for firm j and four dummy variables for years 2 to 5. Use `gllamm` with the `robust` option to obtain reasonable inferences even if there is overdispersion at level 1.

2. Use `lincom` to obtain the sum of the estimated coefficients of the contemporaneous and lagged log expenditures (the elasticity) and its 95% confidence interval. Also use the `test` or `lincom` command to test the null hypothesis that the elasticity is 1 at the 5% level of significance. Comment on your finding.

3. Repeat the above analysis (steps 1 and 2) also controlling for `logk`, a measure of the size of the firm, and `scisect`, a dummy variable for the firm being in the scientific sector.

13.5 School-absenteeism data

These data come from a sociological study by Quine (1973) and have previously been analyzed by Aitkin (1978). The sample included Australian aboriginal and white children from four age groups (final year in primary school and first 3 years in secondary school) who were classified as slow or average learners. The number of days absent from school during the school year was recorded for each child (children who had suffered a serious illness during the year were excluded).

The dataset `absenteeism.dta` has the following variables:

- `id`: child identifier
- `days`: number of days absent from school in one year
- `aborig`: dummy variable for child being aboriginal (versus white)
- `girl`: dummy variable for child being a girl
- `age`: age group (1: last year in primary school; 2: first year in secondary school; 3: second year in secondary school; 4: third year in secondary school)
- `slow`: dummy variable for child being a slow learner (1: slow; 0: average)

1. Use the `glm` command to fit an ordinary Poisson model with `aborig`, `girl`, dummy variables for age groups 2 to 4, and `slow` as covariates.

2. Repeat the analysis but this time use quasilikelihood estimation to include an overdispersion parameter ϕ. How have the estimates and standard errors changed?

3. Fit a negative binomial model of type NB1 and compare the estimated overdispersion factor with that estimated in step 2.

4. Use `xtpoisson` to include a normally distributed random intercept for children.

5. Write down the model of step 4, and interpret the estimates.

13.6 Lip-cancer data

1. Extend the model considered in section 13.15 by including x, the percentage of the population working in agriculture, fishing, or forestry, as a covariate.

    a. Write down the model and state all assumptions.
    b. Fit the model using `gllamm`.
    c. Interpret the exponentiated estimated coefficient of x.

2. Obtain the predicted standardized mortality ratios (SMRs) for this model.

13.7 Skin-cancer data

Langford and Lewis (1998) analyzed data from the Atlas of Cancer Mortality in the European Economic Community (Smans, Muir, and Boyle 1993). Malignant-melanoma (skin cancer) mortalities were defined as deaths recorded and certified by a medical practitioner as ICD-8 172 (ICD is an abbreviation of International Classification of Diseases). The data were collected between 1971 and 1980, although for the United Kingdom, Ireland, Germany, Italy, and The Netherlands, data were only available from 1975–1976 onward and are aggregated over the period of data collection. Because incidence was generally rising during this period, it is important to include nation as a variable in the analysis.

The geographical resolution of the data, referred to here as counties, was European Economic Community (EEC) levels II or III (identified by EEC statistical services).

For example, these units are counties in England and Wales, départements in France and Regierungsbezirke in (West) Germany. Regions are the EEC level-I areas.

The main research question is how malignant melanoma (skin cancer) is associated with ultraviolet (UV) radiation exposure.

Seven variables are included in the dataset `skincancer.dta`:

- `nation`: nation identifier (labeled)

    1. Belgium
    2. W. Germany
    3. Denmark
    4. France
    5. UK
    6. Italy
    7. Ireland
    8. Luxembourg
    9. Netherlands

- `region`: region identifier (EEC level-I areas)
- `county`: county identifier (EEC level-II and level-III areas)
- `deaths`: number of male deaths due to malignant melanoma (skin cancer) during 1971–1980 (1976–1980 for some countries)
- `expected`: expected number of male deaths due to malignant melanoma during 1971–1980 (1976–1980 for some countries), calculated from crude rates for all countries combined
- `uv`: epidemiological index of UV dose reaching the earth's surface in each county (see Langford and Lewis [1998]), mean centered

1. Fit a Poisson model for the number of male deaths from skin cancer using the log expected number of male deaths as an offset, dummy variables for the nations as covariates, and a normally distributed random intercept for counties. Use `gllamm`.

2. Write down the model, and interpret the estimates.

3. Refit the model including `uv` and squared `uv` as further covariates.

4. Obtain predictions of the fixed part of the linear predictor without the offset (using the `xb` and `nooffset` options), and plot these against `uv` by nation. Interpret the graph.

5. For the model in step 3, obtain empirical Bayes predictions of the SMR for each county (as a percentage). Summarize the SMRs by nation.

See also exercise 16.7 for a four-level model for these data.

Part VII

Models for survival or duration data

Introduction to models for survival or duration data (part VII)

In this part, we consider survival data, also referred to as failure-time, time-to-event, event-history, or duration data. Examples include time between surgery and death, duration in first employment, time to failure of light bulbs, and time to dropout from school.

The variable of main interest in survival analysis is the time from some event, such as the beginning of employment, to another event, such as unemployment. A unit or subject is said to become *at risk* of (or be eligible for) the event of interest after the initial event has occurred. Alternatively, we can view the response variable as the duration spent in origin state (employment) until transition to the destination state (unemployment). A state is called *absorbing* if it is impossible to leave the state, a canonical example being death.

Special methods are needed for survival data mainly because of right-censoring, where the time of the event is not known exactly for some subjects because the event has not happened by the end of the observation period. Left-truncation can also occur, where subjects are not included in the study if the event happened before the beginning of the study. A nice feature of survival analysis is that time-varying covariates can be included.

Defining analysis time: The origin

When analyzing the time to an event, perhaps the most important consideration is the origin of the time scale, the point in time when the clock starts ticking and the subject becomes at risk of experiencing the event. For instance, in studies of breast cancer mortality, time from birth until death from breast cancer may be less meaningful than time from menarche (onset of menstruation) since the risk of breast cancer before menarche is negligible. For divorce, the relevant duration may be time since marriage or perhaps the time since cohabitation, because the process leading to break-up may well begin when the couple starts living together. In general mortality studies, the origin is usually not conception, but birth, or in perinatal mortality, the origin is foetal viability (22 weeks of gestation) and death up to 7 days after birth is considered. Perinatal mortality can obviously have very different causes than death after infancy.

The origin may differ from the start of observation, depending on the research design. We use the term *analysis time* for the time scale that takes the value 0 at the origin, which is consistent with Stata's terminology. In clinical trials, the origin is typically the time at which treatment and placebo are administered, and the time at which patients become at risk coincides with the start of the observation period. In observational studies, such as epidemiological cohort studies, analysis time will often be age. For this reason, there is often delayed entry, where subjects have already been at risk before they enter the study. In Stata's `stset` command for defining survival data, the `origin()` option is used to specify the origin (when subjects become at risk), whereas the `enter()` option is used to specify the start of the observation period. When the entry time is later than the origin, we have delayed entry.

In some studies, several time scales may be relevant, for instance, both the age (time since birth) and the time from onset of exposure to a risk factor. In this case, one time scale is chosen as analysis time and the other time scales can be used as time-varying covariates.

Defining the event: Competing events and censoring

Another important consideration is the definition of the event itself. For instance, when studying teacher turnover, the event could be leaving the school, and right-censoring occurs if the teacher is still employed by the same school at the end of the observation period or is lost to follow-up. An important assumption in standard survival analysis is *independent censoring*, which means that the censoring time is independent of the survival time given the covariates in the model.

Unfortunately, it is not always obvious how to define the event. Teachers could leave the school because of dismissal, retirement, change of career, or change of school, among other reasons. We could define the event as exiting employment at the school regardless of the reason. However, by doing so, we implicitly assume that effects of covariates do not depend on the reason for exiting, and this may be unrealistic. Alternatively, we could define the event as change of school or change of career. The other reasons (for example, dismissal and retirement) can be thought of as competing events because their occurrence precludes occurrence of the event of interest. Analysis of the times to several events is called competing-risk analysis, and the models are called competing-risk or multistate models. If the survival times of the different events are not conditionally independent given the covariates, joint modeling is complex and requires unverifiable assumptions. Under conditional independence and in the case of continuous time, each event can be analyzed separately, treating all other events as censoring. In the discrete-time case, it is not possible to analyze the events separately, but joint modeling is straightforward.

Different kinds of censoring and truncation

The left panel of figure VII.1 shows the event histories of several hypothetical units or subjects in terms of calendar time. The lines start when the subjects become at risk of experiencing the event of interest (the origin). For most subjects, the line ends when the event occurs, as indicated by a •. The vertical lines represent the beginning and end of the observation period, from the recruitment of subjects into the study until the end of the study. The right panel of the figure shows the same event histories as the left panel but now the time axis represents *analysis time* with origin defined as the onset of risk. On the analysis time scale, the time when the event occurs is the survival time or duration. Here B and E represent the beginning and end of the time spent in the study, in terms of analysis time.

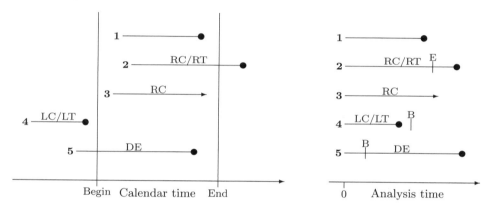

Figure VII.1: Illustration of different types of censoring and truncation in calendar time (left panel) and analysis time (right panel). Dots represent events and arrowheads represent censoring.

Subject 1, represented by the top line in the figure, is ideal in the sense that he becomes at risk within the observation period and experiences the event within the observation period, so the survival time is known.

When *censoring* occurs, the survival time is not exactly known. In one form of right-censoring (RC), often called generalized type I censoring, the event does not occur before the end of the observation period, and all we know is that the survival time exceeds the time between becoming at risk and the end of the observation period (situation 2 in the figure). Another form of right-censoring occurs when the subject stops being at risk of the event under investigation before the end of the observation period; for instance, he may experience a competing event such as dying from a disease other than the one under investigation or he may drop out of the study. For this type of censoring, the time the subject ceases to be at risk is indicated by an arrowhead (situation 3). It is usually assumed that censoring is noninformative in the sense that the survival times for the competing events are conditionally independent of the survival time of interest, given the covariates (also known as *independent censoring*).

When all that is known is that the event occurred before observation began (situation 4), the survival time is called left-censored (LC). For example, the event of interest may be onset of drug use, but all we may know at the start of observation is that a person has used drugs but not when he or she started this practice. Left-censoring is less common than right-censoring because in most studies observation begins when subjects become at risk for the event or those who have already experienced the event are not eligible for inclusion in the study.

In *current status data*, each subject is either left- or right-censored. For instance, surveys sometimes ask individuals whether they have had their sexual debut, producing a left-censored response at their current age if the answer is yes and a right-censored response if the answer is no.

When we only know that the event occurred within a time interval, but not precisely when it occurred, we say that the time is *interval-censored*. For instance, time of HIV infection may be known only to lie between the dates of a negative and positive HIV test, employment status may be known only at each wave of a panel survey, and age at first marriage may be rounded to years. Interval-censoring is also sometimes referred to as grouping and the resulting data as grouped-time survival data if the censoring limits are the same for all subjects.

Truncation occurs when a subject is not included in the study because of the timing of his or her event. *Left-truncation* (LT) occurs when subjects are excluded from the study because the event of interest occurred *before* observation began (situation 4). This can happen only if subjects become at risk before observation begins (situations 4 and 5 in the figure). Situation 5, where the subject enters the study after having already been at risk for a period, is called *delayed entry* (DE). Delayed entry and left-truncation are common in epidemiological cohort studies where only disease-free individuals are typically followed up, thus including subjects who have already been at risk and excluding subjects who developed the disease before observation began. We will discuss delayed entry and left-truncation in more detail in section 14.2.6.

Right-truncation (RT) occurs when a subject is excluded from the study because his event happened *after* the end of the observation period (situation 2). This can happen in retrospective studies, for instance, when investigating the incubation period of AIDS in patients who have developed the disease. Any person who has yet to experience the event is then not included in the study because he is not known to be at risk.

In chapters 14 and 15, we discuss methods that are suitable when left-truncation and right-censoring occur, but we assume that there is no left-censoring or right-truncation. As discussed in the *Discrete- versus continuous-time survival data* section of this introduction, interval-censoring can also be handled and is indeed one of the reasons why discrete-time methods may be appropriate, rather than continuous-time methods.

Time-varying covariates and different time scales

A special feature of survival data is *time-varying covariates*—covariates that can change between the time a person becomes at risk and experiences the event. For instance, if the event is reoffending after release from prison, employment status could be a time-varying covariate. A person's risk of reoffending could be lower during periods of employment than during periods of unemployment. Any aspect of the covariate history might be of interest, such as ever being employed after release from prison, the number of times employed, or the proportion of time or length of time employed. All of these aspects are time varying as the employment history unfolds.

In epidemiology, the risk of lung cancer among underground miners may be affected by exposure to radioactive radon gas, and therefore, the cumulative radon exposure (if regular measurements have been taken) or the length of employment as a miner, are potential time-varying covariates.

As discussed in *Introduction to models for longitudinal and panel data (part III)*, several different time scales can be of interest. The survival time itself is an age-like time scale because it is the time since the subject-specific event of becoming at risk (which is birth in studies of longevity). Period (current calendar time) and cohort (calendar time when becoming at risk) may also be of interest. Unless all subjects became at risk at the same calendar time, one of these time scales can be included, and period would be a time-varying covariate. The subject-specific timing of other events may also be of interest. For instance, if analysis time is age, time since start of employment as a miner (or exposure to other risk factors) could be included. Similarly, if the origin is time of surgery, age could be included as a time-varying covariate, or alternatively, age at the time of surgery could be included as a time-constant covariate.

Discrete- versus continuous-time survival data

It is useful to distinguish between discrete-time survival data and continuous-time survival data.

In *continuous-time survival data*, the exact survival and censoring times are recorded in relatively fine time units. Methods for continuous time can then be used, some of which assume that all survival times are unique and that there are no pairs of individuals with identical or *tied* survival times. We will discuss modeling of continuous-time survival data in chapter 15.

In chapter 14, we will consider *discrete-time survival data* characterized by relatively few possible survival (or censoring) times with many subjects sharing the same survival time. Discrete-time data can result from interval-censoring, where the event occurs in continuous time but we only know the time interval within which it occurred. In this case, we assume that the beginning and end of each interval is the same for all subjects in analysis time. Alternatively, the time scale is sometimes inherently or intrinsically discrete, examples being the number of menstrual cycles to conception and the number of elections to a change of government.

Multilevel and recurrent-event survival data

In one type of multilevel survival data, the subjects experience an event (for instance, death) that is *absorbing* (can only occur once), and subjects are nested in clusters, such as hospitals. An alternative type of multilevel survival data, sometimes called multivariate survival data, multiple spell data, or multiepisode data, arises if subjects can experience multiple events, either events of different types or the same event repeatedly. The latter type of data are called *recurrent-event* data. Several decisions must be made when analyzing such multiple-event data. For instance, is a subject at risk of the kth event before experiencing the $(k-1)$th event? Does the risk of an event depend on whether other events have already occurred? The answers to these questions determine the definition of analysis time and other aspects of the analysis. These issues for recurrent-event data are discussed in section 15.12 of the chapter on continuous-time survival. The same issues apply to discrete-time survival.

14 Discrete-time survival

14.1 Introduction

As discussed in *Introduction to models for survival or duration data (part VII)*, in discrete-time survival data there are just a few possible survival times shared by many subjects. This can be due to *interval-censoring*, where an event actually occurs in continuous time but the researcher only knows the time interval within which the event occurred. For instance, employment status may be known only at panel waves in longitudinal surveys. Another form of interval censoring is using coarse time-scales, for instance, expressing time to events in years. Alternatively, time scales can be *inherently or intrinsically discrete*. Examples include the number of menstrual cycles to conception and the number of elections to a change of government.

Discrete-time survival models are specified in terms of the discrete-time *hazard*, defined as the conditional probability of the event occurring at a time point given that it has not already occurred. A convenient feature of these survival models is that they become models for dichotomous responses when the data have been expanded to so-called person–period data. Standard logit and probit models can then be used, as well as complementary log-log models that have not been introduced yet. It is a good idea to read chapter 10 before embarking on this chapter.

We discuss multilevel discrete-time survival models where random effects, often called *frailties* in this context, are included to handle unobserved heterogeneity between clusters and within-cluster dependence. Discrete-time frailty models are applied to data on time from birth to death (or censoring) for different children nested in mothers.

14.2 Single-level models for discrete-time survival data

14.2.1 Discrete-time hazard and discrete-time survival

We now consider data used by Long, Allison, and McGinnis (1993) and provided with the book by Allison (1995). Three hundred one male biochemists who received their doctorates in 1956 or 1963 and were assistant professors at research universities in the United States some time in their careers were followed up for 10 years from the beginning of their assistant professorships.

The event of interest is promotion to associate professor, which usually corresponds to receiving tenure (a permanent position). Promotions typically take effect at a specific date of the year, making the survival time to promotion inherently discrete. Censoring occurs when assistant professors leave their research university for a job outside academia or for a position at a college or university in which teaching is the primary mission. If some of these transitions occur because of concerns about not being able to pass tenure, the survival and censoring times are not independent given the covariates and censoring becomes informative, unless these concerns are captured by the covariates in the model. Here we assume independent censoring.

The promotions data can be read in by typing

    . use http://www.stata-press.com/data/mlmus3/promotion

Initially, we will not include covariates and only use the following variables:

- id: person identifier
- dur: number of years from beginning of assistant professorship to promotion or censoring
- event: dummy variable for promotion to associate professor (1: promoted; 0: censored)

We will use the notation T for the time in years to promotion, which can take on integer values $t = 1, 2, \ldots, 10$. If the variable event equals 1, we know that T equals dur, and if event equals 0, we know that T is greater than dur.

Discrete-time survival models are specified in terms of the *discrete-time hazard*, defined as the conditional probability that the event occurs at time t, given that it has not yet occurred:

$$h_t \equiv \Pr(T = t | T > t - 1) = \Pr(T = t | T \geq t)$$

Stata's ltable command for life tables provides us with all the information we need to estimate these hazards:

    . ltable dur event, noadjust

Interval		Beg. Total	Deaths	Lost	Survival	Std. Error	[95% Conf. Int.]	
1	2	301	1	1	0.9967	0.0033	0.9767	0.9995
2	3	299	1	6	0.9933	0.0047	0.9737	0.9983
3	4	292	17	12	0.9355	0.0143	0.9008	0.9584
4	5	263	42	10	0.7861	0.0243	0.7337	0.8294
5	6	211	53	9	0.5887	0.0297	0.5280	0.6442
6	7	149	46	7	0.4069	0.0303	0.3473	0.4656
7	8	96	31	6	0.2755	0.0282	0.2217	0.3319
8	9	59	15	2	0.2055	0.0262	0.1567	0.2590
9	10	42	7	6	0.1712	0.0248	0.1258	0.2227
10	11	29	4	25	0.1476	0.0241	0.1043	0.1981

(See section 14.4 for an explanation of the noadjust option.)

We see that there are 301 individuals at the beginning of the first year, one of whom gets promoted (somewhat inappropriately for this particular application denoted as Deaths in the output) and one of whom is censored (Lost). The estimated hazard for this interval is $\widehat{h}_1 = 1/301 = 0.0033$. At the beginning of year 2, there are 299 individuals left in the sample who have not yet experienced the event (301 minus 1 promoted minus 1 censored) and are hence *at risk*. One of these individuals gets promoted in year 2, so $\widehat{h}_2 = 1/299 = 0.0033$. For the following years, we have $\widehat{h}_3 = 17/292 = 0.0582$, $\widehat{h}_4 = 42/263 = 0.1597$, $\widehat{h}_5 = 53/211 = 0.2512$, $\widehat{h}_6 = 46/149 = 0.3087$, $\widehat{h}_7 = 31/96 = 0.3229$, $\widehat{h}_8 = 15/59 = 0.2542$, $\widehat{h}_9 = 7/42 = 0.1667$, and $\widehat{h}_{10} = 4/29 = 0.1379$. The individuals who are at risk at a given interval and contribute to the denominator for the estimated hazard are called the *risk set* for that interval.

We can obtain these estimated hazards directly using the ltable command with the hazard and noadjust options:

```
. ltable dur event, hazard noadjust
                   Beg.    Cum.    Std.            Std.
     Interval      Total   Failure Error   Hazard  Error   [95% Conf. Int.]

         1    2     301    0.0033  0.0033  0.0033  0.0033  0.0001   0.0123
         2    3     299    0.0067  0.0047  0.0033  0.0033  0.0001   0.0123
         3    4     292    0.0645  0.0143  0.0582  0.0141  0.0339   0.0890
         4    5     263    0.2139  0.0243  0.1597  0.0246  0.1151   0.2115
         5    6     211    0.4113  0.0297  0.2512  0.0345  0.1882   0.3232
         6    7     149    0.5931  0.0303  0.3087  0.0455  0.2260   0.4041
         7    8      96    0.7245  0.0282  0.3229  0.0580  0.2194   0.4461
         8    9      59    0.7945  0.0262  0.2542  0.0656  0.1423   0.3981
         9   10      42    0.8288  0.0248  0.1667  0.0630  0.0670   0.3109
        10   11      29    0.8524  0.0241  0.1379  0.0690  0.0376   0.3023
```

We see that the estimated hazard reaches a maximum in year 7 and then declines; those who have not been promoted by the end of year 9 have only a 14% chance of being promoted in year 10. Perhaps some of these assistant professors will never be promoted.

The discrete-time survival function is the probability of not experiencing the event by time t,

$$S_t \equiv \Pr(T > t)$$

For $t = 1$, this is simply 1 minus the probability that the event occurs at time 1 (given that it has not yet occurred, which is impossible), $S_1 = (1 - h_1)$. For $t = 2$, S_2 is the probability that the event did not occur at time 1 and that it did not occur at time 2, given that it did not occur at time 1, and can be expressed as $S_2 = \Pr(T > 2) = \Pr(T > 2 | T > 1)\Pr(T > 1) = (1 - h_2)(1 - h_1)$. In general, we have

$$S_t = \prod_{s=1}^{t}(1 - h_s) \tag{14.1}$$

Substituting the estimated hazards gives the estimated survival function given under Survival in the output of the ltable command when the hazard option was not

used, as shown on page 750. The cumulative failure function $1 - S_t$ is given under
Cum. Failure in the output of the ltable command with the hazard option. The
ltable command provides estimated standard errors for the survival function, the cumulative failure function, and the hazard function.

14.2.2   Data expansion for discrete-time survival analysis

By expanding the data appropriately and defining a binary variable, y, taking the values
0 and 1, we can obtain the estimated hazards as the proportions of 1s observed each
year. As we will see later, this data expansion is necessary for conducting the most
common type of discrete-time survival analysis.

Consider the first two individuals in the data:

```
. list id dur event if id<3, noobs
```

id	dur	event
1	10	0
2	4	1

The person with id=1 was censored at $t = 10$ because event takes the value 0 when
dur is 10. This person therefore represents 1 of the 301 individuals at risk in year 1
(that is, is part of the risk set in year 1), 1 of the 299 individuals at risk in year 2,
and so forth up to and including year 10. In the expanded dataset, this person should
contribute one observation for each of the 10 years, with y equal to zero, so that the
observation contributes to the denominator but not to the numerator of the estimated
hazard. Similarly, the person with id=2 should be represented by four observations, for
years 1–4, because the professor experiences the event at $t = 4$. For the first 3 years,
y should be 0. For the fourth year, y should be 1 because the event occurs that year,
and the observation should contribute to the numerator as well as the denominator
of the estimated hazard. After having experienced the event, the professor no longer
belongs to the risk set for promotion and therefore does not contribute more than four
observations.

In general, each person should be represented by a row of data for each year the
person was at risk. We therefore expand the data to obtain dur rows per person:

```
. expand dur
```

We then create a new variable, year, that labels the years,

```
. by id, sort: generate year = _n
```

and list the expanded data for the first two individuals:

```
. list id dur year event if id<3, sepby(id) noobs
```

id	dur	year	event
1	10	1	0
1	10	2	0
1	10	3	0
1	10	4	0
1	10	5	0
1	10	6	0
1	10	7	0
1	10	8	0
1	10	9	0
1	10	10	0
2	4	1	1
2	4	2	1
2	4	3	1
2	4	4	1

The response variable y should be 1 if a promotion occurs for the person that year and 0 otherwise.

```
. generate y = 0
. replace y = event if year==dur
```

The expansion of the data for the first two professors is also shown in figure 14.1.

Person–year data

Original data

id	dur	event		id	year	y
1	10	0	$\longrightarrow$	1	1	0
2	4	1		1	2	0
				1	3	0
				1	4	0
				1	5	0
				1	6	0
				1	7	0
				1	8	0
				1	9	0
				1	10	0
				2	1	0
				2	2	0
				2	3	0
				2	4	1

Figure 14.1: Expansion of original data to person–year data for first two assistant professors (first one right-censored in year 10, second one promoted in year 4)

The data are now in person–year or person–period form, where each year of observation for each assistant professor has one record or observation for each period he or she is at risk of being promoted.

14.2.3 Estimation via regression models for dichotomous responses

We can obtain the sample hazards or estimated hazards $\widehat{h}_t$ for each year t by finding the proportion of 1s each year, for instance, using the `tabulate` command:

```
. tabulate year y, row
```

Key
frequency
row percentage

year	y 0	1	Total
1	300 99.67	1 0.33	301 100.00
2	298 99.67	1 0.33	299 100.00
3	275 94.18	17 5.82	292 100.00
4	221 84.03	42 15.97	263 100.00
5	158 74.88	53 25.12	211 100.00
6	103 69.13	46 30.87	149 100.00
7	65 67.71	31 32.29	96 100.00
8	44 74.58	15 25.42	59 100.00
9	35 83.33	7 16.67	42 100.00
10	25 86.21	4 13.79	29 100.00
Total	1,524 87.54	217 12.46	1,741 100.00

The penultimate column in the output above gives the estimated hazards as percentages, and these agree with the estimated hazards on page 751 produced by the `ltable` command.

Alternatively, we can obtain estimated hazards as predicted probabilities by using a logistic regression model where the covariates are dummy variables for each year,

$$\text{logit}\{\Pr(y_{si} = 1|\mathbf{d}_{si})\} = \alpha_1 + \alpha_2 d_{2si} + \cdots + \alpha_{10} d_{10,si}$$

Here y_{si} is an indicator for the event occurring at time s for person i; $d_{2si}, \ldots, d_{10,si}$ are dummy variables for years 2–10; and $\mathbf{d}_{si} = (d_{2si}, \ldots, d_{10,si})'$ is a vector containing all the dummy variables for professor i.

The Stata command to fit this model by maximum likelihood is

```
. logit y i.year
Logistic regression                              Number of obs   =       1741
                                                 LR chi2(9)      =     251.17
                                                 Prob > chi2     =     0.0000
Log likelihood = -529.15641                      Pseudo R2       =     0.1918
```

y	Coef.	Std. Err.	z	P>|z|	[95% Conf. Interval]	
year						
2	.006689	1.416576	0.00	0.996	-2.76975	2.783128
3	2.920224	1.032372	2.83	0.005	.8968116	4.943637
4	4.043289	1.01571	3.98	0.000	2.052533	6.034045
5	4.611479	1.014165	4.55	0.000	2.623753	6.599205
6	4.897695	1.017242	4.81	0.000	2.903937	6.891452
7	4.963382	1.025171	4.84	0.000	2.954084	6.97268
8	4.627643	1.045336	4.43	0.000	2.578822	6.676463
9	4.094344	1.083864	3.78	0.000	1.970009	6.218679
10	3.871201	1.137248	3.40	0.001	1.642236	6.100166
_cons	-5.703782	1.001665	-5.69	0.000	-7.66701	-3.740555

where the `i.` preceding the covariate `year` produces dummy variables for `year` and includes them in the model, omitting the dummy variable for year 1. Predicted probabilities, which here correspond to estimated discrete-time hazards, can be obtained using the `predict` command with the `pr` option:

```
. predict haz, pr
```

Finally, we list the estimated hazards, the time-specific event indicator y_{is} and the variable event, for the years where each of the first two assistant professors in the dataset are at risk of promotion:

```
. list id year haz y event if id<3, sepby(id) noobs
```

id	year	haz	y	event
1	1	.0033223	0	0
1	2	.0033445	0	0
1	3	.0582192	0	0
1	4	.1596958	0	0
1	5	.2511848	0	0
1	6	.3087248	0	0
1	7	.3229167	0	0
1	8	.2542373	0	0
1	9	.1666667	0	0
1	10	.137931	0	0
2	1	.0033223	0	1
2	2	.0033445	0	1
2	3	.0582192	0	1
2	4	.1596958	1	1

The estimated hazards are identical to those obtained using the ltable command. We could use any other link function, such as the probit link or the complementary log-log link introduced in section 14.6, to obtain the same predicted hazards.

We can plot the discrete-time hazards by selecting an assistant professor who has observations for all periods, such as professor 1 (the same results would be obtained here if we picked another professor), and using the twoway command,

```
. twoway (line haz year if id==1, connect(stairstep)), legend(off)
> xtitle(Year) ytitle(Discrete-time hazard)
```

which produces the graph in figure 14.2.

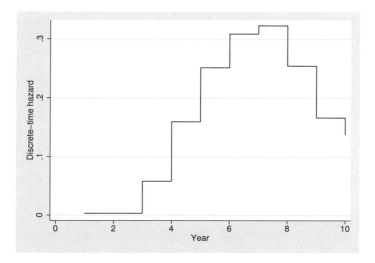

Figure 14.2: Discrete-time hazard (conditional probability of promotion given that promotion has not yet occurred)

Here we have used the `connect(stairstep)` option to plot a step function (but you can omit this option).

The likelihood from the binary regression models based on the expanded data is just the required likelihood for discrete-time survival data, and the resulting predicted hazards are maximum likelihood estimates. To see this, consider the required likelihood contribution for a person who was censored at time t, which is just the corresponding probability $S_{ti} \equiv \Pr(T_i > t)$. From (14.1), this probability is given by

$$S_{ti} = \prod_{s=1}^{t}(1 - h_{si})$$

where h_{si} is the discrete-time hazard at time s for person i. In the expanded dataset, someone who is censored at time t is represented by a row of data for $s = 1, 2, \ldots, t$ with $\mathbf{y} = 0$. The corresponding likelihood contributions from binary regression are the model-implied probabilities of the observed responses, $\Pr(y_{si} = 0) = (1 - h_s)$, and these are simply multiplied together because the observations are taken as independent, giving the required likelihood contribution for discrete-time survival.

For an assistant professor who was promoted at time t, the likelihood contribution should be

$$\Pr(T_i = t) = \Pr(T_i = t | T_i > t - 1)\Pr(T_i > t - 1)$$
$$= h_{ti} \prod_{s=1}^{t-1}(1 - h_{si})$$

Such a person is represented by a row of data for $s = 1, 2, \ldots, t$ with y equal to zero for $s \leq t - 1$ and y equal to 1 for $s = t$. The likelihood contribution from binary regression is therefore $\Pr(y_{1i} = 0) \times \cdots \times \Pr(y_{t-1,i} = 0) \times \Pr(y_{ti} = 1)$ as required.

We can interpret the logistic regression model as a linear model for the logit of the discrete-time hazard,

$$\text{logit}\{\Pr(y_{si} = 1|\mathbf{d}_{si})\} \;=\; \alpha_1 + \alpha_2 d_{2si} + \cdots + \alpha_{10} d_{10,si} \;=\; \text{logit}\{\underbrace{\Pr(T_i = s|T_i \geq s, \mathbf{d}_{si})}_{h_{si}}\}$$

No functional form is imposed on the relationship between discrete-time hazard and year because dummy variables are used for years 2–10 (the intercept represents year 1).

14.2.4 Including covariates

Although it is interesting to investigate how the population-averaged or marginal hazard evolves over time, the main purpose of survival analysis is usually to estimate the effects of covariates on the hazard. Here regression models become useful.

Time-constant covariates

The following time-constant or person-specific covariates are available:

- `undgrad`: selectivity of undergraduate institution (scored from 1–7) (x_{2i})
- `phdmed`: dummy variable for having a PhD from a medical school (1: yes; 0: no) (x_{3i})
- `phdprest`: a measure of prestige of the PhD institution (ranges from 0.92 to 4.62) (x_{4i})

We specify a logistic regression for the expanded data with these three covariates added to the dummy variables for years 2–10,

$$\text{logit}\{\Pr(y_{si} = 1|\mathbf{d}_{si}, \mathbf{x}_i)\} \;=\; \alpha_1 + \alpha_2 d_{2si} + \cdots + \alpha_{10} d_{10,si} + \beta_2 x_{2i} + \beta_3 x_{3i} + \beta_4 x_{4i}$$

where $\mathbf{x}_i = (x_{2i}, x_{3i}, x_{4i})'$ is the vector of covariates. The first part of the linear predictor, from α_1 to $\alpha_{10} d_{10,si}$ determines the so-called *baseline hazard*, the hazard when the covariates $\mathbf{x}_i$ are all zero. The model could be described as semiparametric because no assumptions are made regarding the functional form for the baseline hazard, whereas the effects of covariates are assumed to be linear and additive on the logit scale.

We can fit this model by using the `logit` command:

```
. logit y i.year undgrad phdmed phdprest
Logistic regression                              Number of obs   =       1741
                                                 LR chi2(12)     =     260.79
                                                 Prob > chi2     =     0.0000
Log likelihood = -524.34273                      Pseudo R2       =     0.1992
```

y	Coef.	Std. Err.	z	P>|z|	[95% Conf.	Interval]
year						
2	.0063043	1.416703	0.00	0.996	-2.770383	2.782991
3	2.923036	1.032552	2.83	0.005	.899272	4.946801
4	4.051509	1.015925	3.99	0.000	2.060334	6.042685
5	4.619307	1.01441	4.55	0.000	2.631099	6.607514
6	4.924625	1.017614	4.84	0.000	2.930139	6.919111
7	4.992068	1.025703	4.87	0.000	2.981728	7.002409
8	4.69053	1.046322	4.48	0.000	2.639776	6.741284
9	4.167769	1.085192	3.84	0.000	2.040833	6.294706
10	3.964635	1.139095	3.48	0.001	1.73205	6.197219
undgrad	.1576609	.06007	2.62	0.009	.039926	.2753959
phdmed	-.0950034	.1665181	-0.57	0.568	-.4213728	.2313661
phdprest	.0650372	.0854488	0.76	0.447	-.1024394	.2325138
_cons	-6.67189	1.081551	-6.17	0.000	-8.791692	-4.552088

This model assumes that the difference in log odds between individuals with different covariates is constant over time. For instance, the predicted log odds of individuals 1 and 4 (both of whom have 10 observations and who differ in all three covariates) can be obtained and plotted by typing

```
. predict lo, xb
. twoway (line lo year if id==1, connect(stairstep) lpatt(solid))
> (line lo year if id==4, connect(stairstep) lpatt(dash)),
> legend(off) xtitle(Year) ytitle(Log odds)
```

giving the graph in figure 14.3.

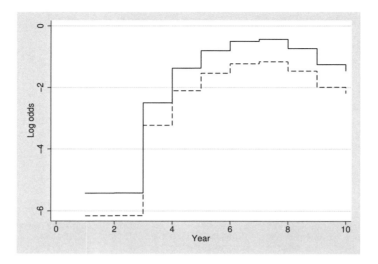

Figure 14.3: Predicted log odds of promotion given that promotion has not yet occurred for professors 1 (solid) and 4 (dashed)

A constant difference in the log odds corresponds to a constant ratio of the odds. For this reason, the model is often called a *proportional odds model*, not to be confused with the ordinal logistic regression model of the same name discussed in chapter 11. We will see in section 14.6 how time-varying covariates can be used to relax the proportionality assumption.

The model is also referred to as a *continuation-ratio logit* model or sequential logit model because the logit can be written as

$$\text{logit}\{\Pr(T_i = s | T_i \geq s, \mathbf{d}_i, \mathbf{x}_i)\} \quad = \quad \ln\underbrace{\left\{\frac{\Pr(T_i = s | T_i \geq s, \mathbf{d}_i, \mathbf{x}_i)}{\Pr(T_i > s | T_i \geq s, \mathbf{d}_i, \mathbf{x}_i)}\right\}}_{\text{Odds}(T_i = s | T_i \geq s, \mathbf{d}_i, \mathbf{x}_i)}$$

$$= \quad \ln\left\{\frac{\Pr(T_i = s, T_i \geq s, \mathbf{d}_i, \mathbf{x}_i)/\Pr(T_i \geq s, \mathbf{d}_i, \mathbf{x}_i)}{\Pr(T_i > s, T_i \geq s, \mathbf{d}_i, \mathbf{x}_i)/\Pr(T_i \geq s, \mathbf{d}_i, \mathbf{x}_i)}\right\}$$

$$= \quad \ln\left\{\frac{\Pr(T_i = s, T_i \geq s, \mathbf{d}_i, \mathbf{x}_i)}{\Pr(T_i > s, T_i \geq s, \mathbf{d}_i, \mathbf{x}_i)}\right\}$$

$$= \quad \ln\left\{\frac{\Pr(T_i = s | \mathbf{d}_i, \mathbf{x}_i)}{\Pr(T_i > s | \mathbf{d}_i, \mathbf{x}_i)}\right\}$$

which is the log of the odds (or ratio of probabilities) of stopping at time s versus continuing beyond time s. This becomes a model for the odds of continuing versus stopping if we replace 0 with 1 and 1 with 0 in the response variable (see exercises 14.5

and 14.8). The relevant continuation-ratio odds for a four-interval example, $S = 4$, is shown in figure 14.4.

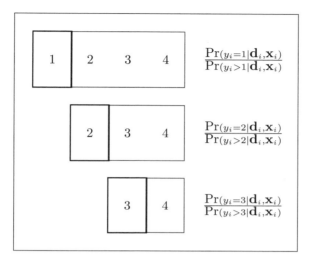

Figure 14.4: Relevant odds $\Pr(y_i = s|\mathbf{d}_i, \mathbf{x}_i)/\Pr(y_i > s|\mathbf{d}_i, \mathbf{x}_i)$ for $(s = 1, 2, 3)$ in a continuation-ratio logit model with four time intervals. Odds is ratio of probabilities of events; events included in numerator probability are in thick frames and events included in denominator probability are in thin frames. [This diagram is very similar to those in Brendan Halpin's web notes on "Models for ordered categories (ii)".]

We can obtain the estimated odds ratios associated with the three covariates by using the `or` option with the `logit` command or by exponentiating the coefficients. For `undergrad`, the odds ratio is estimated as $\exp(0.157) = 1.17$, implying that the odds of being promoted in any given year (given that promotion has not already occurred) increase 17% for every unit increase in the selectivity of the undergraduate institution, controlling for the other covariates.

Log-odds curves are not easy to interpret, and usually survival curves are presented instead. The estimated hazards $\widehat{h}_{si}$ can be obtained from the log odds `lo` by using the `invlogit()` function. To get the cumulative products in (14.1) for each person, we use the `sum()` function combined with the `by` prefix and apply this to the logarithms, using the relationship

$$\widehat{S}_{ti} = \prod_{s=1}^{t}(1 - \widehat{h}_{si}) = \exp\left\{\sum_{s=1}^{t}\ln(1 - \widehat{h}_{si})\right\}$$

```
. generate ln_one_m_haz = ln(1-invlogit(lo))
. by id (year), sort: generate ln_surv = sum(ln_one_m_haz)
. generate surv = exp(ln_surv)
. twoway (line surv year if id==1, connect(stairstep) lpatt(solid))
> (line surv year if id==4, connect(stairstep) lpatt(dash)),
> legend(off) xtitle(Year) ytitle(Survival)
```

The graph is given in figure 14.5.

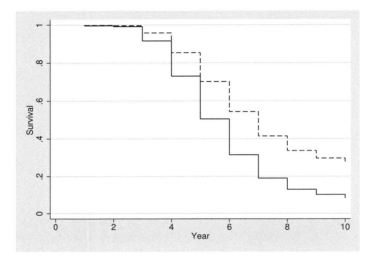

Figure 14.5: Predicted probability of remaining an assistant professor for professors 1 (solid) and 4 (dashed)

We see that professors with the covariate values of professor 1 (undgrad = 7, phdmed = 0, and phdprest = 2.21) have a 50% chance of being promoted by year 5.

Time-varying covariates

The dataset contains four types of time-varying information: the cumulative number of papers published each year, the cumulative number of citations each year to all papers previously published, whether the employer is the same as at the start of the assistant professorship, and the prestige of the current employer. This information is coded in the following variables:

- art1–art10: the cumulative number of articles published in each of the 10 years
- cit1–cit10: the cumulative number of citations to all previous articles in each of the 10 years
- jobtime: number of years until change of employer from start of assistant professorship, coded as missing for those who did not change employer

- **prest1**: a measure of prestige of the assistant professor's first employing institution (ranges from 0.65 to 4.6)

- **prest2**: prestige of the assistant professor's second employing institution, coded as missing for those who did not change employers (no one had more than two employers)

Because the data are in person–year form with one observation for each year each person is at risk, it is straightforward to include time-varying covariates by creating variables that take on the appropriate values for each person in each year. For the first two variables above, this is most easily achieved by reading in the data in their original form (with one row per person),

```
. use http://www.stata-press.com/data/mlmus3/promotion, clear
```

and using the **reshape** command to produce person–year data:

```
. reshape long art cit, i(id) j(year)
(note: j = 1 2 3 4 5 6 7 8 9 10)
Data                              wide   ->   long

Number of obs.                     301   ->   3010
Number of variables                 29   ->     12
j variable (10 values)                   ->   year
xij variables:
                        art1 art2 ... art10   ->   art
                        cit1 cit2 ... cit10   ->   cit
```

We can then list the data:

```
. list id year art cit dur event if id<3, sepby(id) noobs
```

id	year	art	cit	dur	event
1	1	0	0	10	0
1	2	0	0	10	0
1	3	2	1	10	0
1	4	2	1	10	0
1	5	2	1	10	0
1	6	2	1	10	0
1	7	2	1	10	0
1	8	2	1	10	0
1	9	2	1	10	0
1	10	2	1	10	0
2	1	8	27	4	1
2	2	10	44	4	1
2	3	14	57	4	1
2	4	18	63	4	1
2	5	.	.	4	1
2	6	.	.	4	1
2	7	.	.	4	1
2	8	.	.	4	1
2	9	.	.	4	1
2	10	.	.	4	1

We see that the first assistant professor stopped publishing after year 3, was cited only once in 10 years, and did not get promoted within 10 years, whereas the second professor published 18 papers and had 63 citations by the fourth year when he or she got promoted.

The `reshape` command has produced 10 rows of data for each person although only `dur` rows are required. We therefore delete the excess rows of data and create the variables `year` and `y` as before:

```
. drop if year>dur
(1269 observations deleted)
. generate y = 0
. replace y = event if year==dur
(217 real changes made)
```

The time-varying covariates `art` and `cit` are now ready for inclusion in the model. Following Allison (1995), we will use the prestige of the current employer as the third time-varying covariate. We generate a new variable, `prestige`, which initially equals `prest1` for everyone and then changes to `prest2` for those who changed employers in the year given in `jobtime`:

```
. generate prestige = prest1
. replace prestige = prest2 if year>=jobtime
(181 real changes made)
```

The variable `jobtime` is missing for those who never changed employers. Because Stata interprets missing values as very large numbers, the logical expression `year>=jobtime` is never true for those who did not change employers, and therefore `prestige` remains equal to `prest1` for those individuals as required. The data for the first two assistant professors are given in table 14.1.

Table 14.1: Expanded data with time-constant and time-varying covariates for first two assistant professors

		Covariates						Response
		Time-constant			Time-varying			
id	year	undgrad	phdmed	phdprest	art	cit	prestige	y
i	s	x_{2i}	x_{3i}	x_{4i}	x_{5si}	x_{6si}	x_{7si}	y_{si}
1	1	7	0	2.21	0	0	2.36	0
1	2	7	0	2.21	0	0	2.36	0
1	3	7	0	2.21	2	1	2.36	0
1	4	7	0	2.21	2	1	2.36	0
1	5	7	0	2.21	2	1	2.36	0
1	6	7	0	2.21	2	1	2.36	0
1	7	7	0	2.21	2	1	2.36	0
1	8	7	0	2.21	2	1	2.36	0
1	9	7	0	2.21	2	1	2.36	0
1	10	7	0	2.21	2	1	2.36	0
2	1	6	0	2.21	8	27	1.84	0
2	2	6	0	2.21	10	44	1.84	0
2	3	6	0	2.21	14	57	2.88	0
2	4	6	0	2.21	18	63	2.88	1

As in figure 14.1, the expanded data has 10 records for the first assistant professor and 4 records for the second. For the first professor, the response y_{si} is equal to 0 throughout the 10 years ($s = 1, \ldots, 10$); for the second professor, the response is 0 for the first 3 years ($s = 1, 2, 3$) and then 1 in year $s = 4$. There are three time-constant covariates, undgrad, phdmed, and phdprest (x_{2i}, x_{3i}, and x_{4i}), and three time-varying covariates, art, cit, and prestige (x_{5si}, x_{6si}, and x_{7si}).

Using expanded data of the above form, we can fit a logistic discrete-time hazards model including both time-constant and time-varying covariates,

$$\text{logit}\{\Pr(y_{si} = 1 | \mathbf{d}_{si}, \mathbf{x}_{si})\} = \alpha_1 + \alpha_2 d_{2si} + \cdots + \alpha_{10} d_{10,si}$$
$$+ \beta_2 x_{2i} + \beta_3 x_{3i} + \beta_4 x_{4i} + \beta_5 x_{5si} + \beta_6 x_{6si} + \beta_7 x_{7si}$$

where $\mathbf{x}_{si} = (x_{2i}, x_{3i}, x_{4i}, x_{5si}, x_{6si}, x_{7si})'$.

This model can be fit using the command

```
. logit y i.year undgrad phdmed phdprest art cit prestige, or

Logistic regression                              Number of obs   =        1741
                                                 LR chi2(15)     =      303.80
                                                 Prob > chi2     =      0.0000
Log likelihood = -502.83976                      Pseudo R2       =      0.2320
```

y	Odds Ratio	Std. Err.	z	P>|z|	[95% Conf.	Interval]
year						
2	.9126348	1.29337	-0.06	0.949	.056753	14.67591
3	15.10083	15.61356	2.63	0.009	1.990204	114.5787
4	44.34795	45.12593	3.73	0.000	6.035867	325.8423
5	73.16483	74.3748	4.22	0.000	9.977552	536.5136
6	93.20208	95.14476	4.44	0.000	12.60323	689.2383
7	97.98613	100.932	4.45	0.000	13.01284	737.8311
8	69.18332	72.81854	4.03	0.000	8.79172	544.4137
9	37.2479	40.90447	3.29	0.001	4.328515	320.527
10	32.10152	36.91194	3.02	0.003	3.371095	305.6892
undgrad	1.21355	.0767456	3.06	0.002	1.07208	1.373688
phdmed	.7916123	.1357166	-1.36	0.173	.5656917	1.107759
phdprest	1.022572	.0951069	0.24	0.810	.8521685	1.227049
art	1.076152	.0194367	4.06	0.000	1.038723	1.114929
cit	1.000092	.0013031	0.07	0.944	.997541	1.002649
prestige	.7816668	.0887699	-2.17	0.030	.6256841	.976536
_cons	.0017972	.0019711	-5.76	0.000	.0002094	.0154221

The maximum likelihood estimates of the odds ratios for this model are also presented in table 14.2.

Table 14.2: Maximum likelihood estimates for logistic discrete-time hazards model for promotions of assistant professors

	Odds ratio	(95% CI)
$\exp(\alpha_2)$ [year 2]	0.91	(0.1, 14.7)
$\exp(\alpha_3)$ [year 3]	15.10	(2.0, 114.6)
$\exp(\alpha_4)$ [year 4]	44.35	(6.0, 325.8)
$\exp(\alpha_5)$ [year 5]	73.17	(10.0, 536.5)
$\exp(\alpha_6)$ [year 6]	93.20	(12.6, 689.3)
$\exp(\alpha_7)$ [year 7]	97.99	(13.0, 737.8)
$\exp(\alpha_8)$ [year 8]	69.18	(8.8, 544.4)
$\exp(\alpha_9)$ [year 9]	37.25	(4.3, 320.5)
$\exp(\alpha_{10})$ [year 10]	32.10	(3.4, 305.7)
$\exp(\beta_2)$ [undgrad]	1.21	(1.1, 1.4)
$\exp(\beta_3)$ [phdmed]	0.79	(0.6, 1.1)
$\exp(\beta_4)$ [phdprest]	1.02	(0.9, 1.2)
$\exp(\beta_5)$ [art]	1.08	(1.0, 1.1)
$\exp(\beta_6)$ [cit]	1.00	(1.0, 1.0)
$\exp(\beta_7)$ [prestige]	0.78	(0.6, 1.0)
Log likelihood	-502.84	

At the 5% level, only the odds ratios for **undergrad**, **art**, and **prestige** are significant. As might be expected, a more selective undergraduate institution and more published articles are associated with a greater odds of promotion (given that promotion has not yet occurred), whereas being employed at a more prestigious institution is associated with a decreased odds of promotion. Regarding the coefficients of the dummy variables, their exponentials represent the odds ratio of promotion in a given year (given that promotion did not occur earlier) compared with year 1, when $\mathbf{x}_{si} = \mathbf{0}$. So, for instance, the odds of promotion are 98 times as great in year 7 as they are in year 1 when all covariates take the value zero. The reason for these large odds ratios and wide confidence intervals is that the estimated odds of promotion in year 1 (reported in the output next to _cons) are merely 0.0018 and poorly estimated. It would perhaps be better to use a different year as reference category.

It is also interesting to consider a parametric form for the discrete-time baseline hazard as a function of time instead of using dummy variables. Allison (1995) specifies a polynomial relationship for these data (see also exercise 14.1).

14.2.5   Multiple absorbing events and competing risks

There are sometimes multiple types of absorbing events or, in other words, multiple modes of failure. A classical example is different causes of death in mortality studies.

In the promotions data, we might, for instance, distinguish between two types of events that could be experienced by an assistant professor: 1) promotion at the first university he works at and 2) leaving the first university before being promoted there. A separate discrete-time hazard can then be defined for each type of absorbing event. This situation is referred to as *competing risks* because the events are competing in the sense that a person experiencing one of the events is removed from the risk sets of all other events.

Discrete-time hazard modeling for multiple absorbing events can proceed in a similar manner as for single events with the difference that the responses in the expanded data are nominal (with different categories of events) and no longer binary. The data are expanded so that professors have a 0 for every year where neither event happened and a 1 or 2 for the year where the professor is promoted or leaves the university, respectively. We can then fit a logistic discrete-time hazards model for multiple absorbing events by using the multinomial logit model for nominal responses discussed in section 12.2.1.

We start by rereading in the promotions data:

```
. use http://www.stata-press.com/data/mlmus3/promotion, clear
```

We then define the two events with dummy variables `event1` and `event2` for being promoted and leaving, respectively. The first type of event is defined by using

```
. generate event1 = 0
. replace event1 = 1 if dur<jobtime & event==1
(163 real changes made)
```

This produces a 1 if a promotion was experienced (`event==1`) *and* the promotion occurred before leaving the first university (`dur<jobtime`), and a 0 otherwise. We see that 163 of the professors are promoted at the first university they work at. The second kind of event is defined by using

```
. generate event2 = 0
. replace event2 = 1 if dur>=jobtime
(74 real changes made)
```

producing a 1 if the assistant professor leaves his first university before having been promoted (`dur>=jobtime`) and a 0 otherwise. It is assumed here that a professor who is promoted in the same year as he leaves his first university is promoted at his new university. We see that 74 of the professors leave their first university without having been promoted there. For this event, we also need to redefine the time to event, which is now the time of leaving the first university (`jobtime`):

```
. replace dur = jobtime if event2==1
(31 real changes made)
```

Repeating the commands we previously used for single-event data, we expand the original data to person–year data,

```
. reshape long art cit, i(id) j(year)
(note: j = 1 2 3 4 5 6 7 8 9 10)
```

Data		wide	->	long
Number of obs.		301	->	3010
Number of variables		31	->	14
j variable (10 values)			->	year
xij variables:				
	art1 art2 ... art10		->	art
	cit1 cit2 ... cit10		->	cit

and get rid of redundant rows in the data:

```
. drop if year>dur
(1376 observations deleted)
```

We can then list some of the person–year data for three of the assistant professors, including both `event1` and `event2`:

```
. list id year jobtime dur event1 event2 if id==1|id==2|id==161, sepby(id) noobs
```

id	year	jobtime	dur	event1	event2
1	1	.	10	0	0
1	2	.	10	0	0
1	3	.	10	0	0
1	4	.	10	0	0
1	5	.	10	0	0
1	6	.	10	0	0
1	7	.	10	0	0
1	8	.	10	0	0
1	9	.	10	0	0
1	10	.	10	0	0
2	1	3	3	0	1
2	2	3	3	0	1
2	3	3	3	0	1
161	1	.	2	1	0
161	2	.	2	1	0

We see that the first professor experienced neither of the events over the 10-year observation period. In contrast, the second professor experienced the second event, leaving the first university he was employed at in year 3 without having been promoted there (he was, however, promoted at his new university in year 4, as can be seen in table 14.1). Professor 161 was promoted in year 2 at her original university.

In preparation for modeling multiple types of events, we then define a nominal variable, `yy`, taking the value 0 until an event is experienced, 1 if the professor is promoted at his first university, and 2 if the professor is leaving without having been promoted:

```
. generate yy = 0
. replace yy = 1 if year==dur & event1==1
(163 real changes made)
. replace yy = 2 if year==dur & event2==1
(74 real changes made)
```

We define value labels for the nominal response:

```
. label define eventlabel 1 "promfirst" 2 "leave"
. label values yy eventlabel
```

The data for the three professors now look like this:

```
. list id year jobtime dur event1 event2 yy if id==1|id==2|id==161,
> sepby(id) noobs
```

id	year	jobtime	dur	event1	event2	yy
1	1	.	10	0	0	0
1	2	.	10	0	0	0
1	3	.	10	0	0	0
1	4	.	10	0	0	0
1	5	.	10	0	0	0
1	6	.	10	0	0	0
1	7	.	10	0	0	0
1	8	.	10	0	0	0
1	9	.	10	0	0	0
1	10	.	10	0	0	0
2	1	3	3	0	1	0
2	2	3	3	0	1	0
2	3	3	3	0	1	leave
161	1	.	2	1	0	0
161	2	.	2	1	0	promfirst

We are now ready to fit a logistic discrete-time hazard model for multiple absorbing events. The same covariates are used as previously for the single-event case (we use `prest1` instead of `prestige` because it is the prestige of the first institution that matters here). Letting the superscripts [1] and [2] denote the two types of events, we specify a model with separate baseline hazards for each type of event:

$$\ln\left\{\frac{\Pr(y_{si} = 1 | \mathbf{d}_{si}, \mathbf{x}_{si})}{\Pr(y_{si} = 0 | \mathbf{d}_{si}, \mathbf{x}_{si})}\right\} = \alpha_1^{[1]} + \alpha_2^{[1]} d_{2si} + \cdots + \alpha_{10}^{[1]} d_{10,si}$$
$$+ \beta_2^{[1]} x_{2i} + \beta_3^{[1]} x_{3i} + \beta_4^{[1]} x_{4i}$$
$$+ \beta_5^{[1]} x_{5si} + \beta_6^{[1]} x_{6si} + \beta_7^{[1]} x_{7si}$$

$$\ln\left\{\frac{\Pr(y_{si} = 2 | \mathbf{d}_{si}, \mathbf{x}_{si})}{\Pr(y_{si} = 0 | \mathbf{d}_{si}, \mathbf{x}_{si})}\right\} = \alpha_1^{[2]} + \alpha_2^{[2]} d_{2si} + \cdots + \alpha_{10}^{[2]} d_{10,si}$$
$$+ \beta_2^{[2]} x_{2i} + \beta_3^{[2]} x_{3i} + \beta_4^{[2]} x_{4i}$$
$$+ \beta_5^{[2]} x_{5si} + \beta_6^{[2]} x_{6si} + \beta_7^{[2]} x_{7si}$$

This model is identical to the multinomial logit model for nominal responses discussed in section 12.2.1.

We can use `mlogit` with the `rrr` option to fit the model by maximum likelihood and obtain estimated odds ratios. We specify `ib6.year` to use year 6 as reference category for the categorical variable `year` (instead of the default year 1).

```
. mlogit yy ib6.year undgrad phdmed phdprest art cit prest1, rrr
Multinomial logistic regression                 Number of obs   =       1634
                                                LR chi2(30)     =     304.51
                                                Prob > chi2     =     0.0000
Log likelihood = -671.38843                     Pseudo R2       =     0.1849
```

yy	RRR	Std. Err.	z	P>|z|	[95% Conf. Interval]	
0	(base outcome)					
promfirst						
year						
1	.0128989	.0132414	-4.24	0.000	.0017248	.0964621
2	.0122117	.0125222	-4.30	0.000	.0016366	.0911208
3	.1534968	.056125	-5.13	0.000	.074966	.3142931
4	.5828765	.1631651	-1.93	0.054	.3367433	1.008914
5	.8152931	.2278044	-0.73	0.465	.4714939	1.409781
7	1.341087	.4405855	0.89	0.372	.7043897	2.553296
8	.5111122	.2438342	-1.41	0.159	.200649	1.301954
9	.4703872	.2643801	-1.34	0.180	.156329	1.415375
10	.2883422	.2297488	-1.56	0.119	.0604892	1.374481
undgrad	1.144861	.0832787	1.86	0.063	.992739	1.320292
phdmed	.8185376	.1616294	-1.01	0.311	.5558523	1.205363
phdprest	.9981063	.1047621	-0.02	0.986	.8125198	1.226082
art	1.073669	.0213433	3.58	0.000	1.032641	1.116327
cit	.9999352	.0014397	-0.05	0.964	.9971174	1.002761
prest1	.7938277	.1035595	-1.77	0.077	.6147265	1.02511
_cons	.1967245	.10549	-3.03	0.002	.0687728	.5627301
leave						
year						
1	1.02e-07	.0000599	-0.03	0.978	0	.
2	.249396	.1322296	-2.62	0.009	.0882233	.705011
3	.7734526	.3347741	-0.59	0.553	.3311368	1.806591
4	.5758507	.2730709	-1.16	0.244	.227335	1.458658
5	.7946052	.3772739	-0.48	0.628	.3133328	2.015102
7	1.639593	.8569544	0.95	0.344	.5886331	4.566963
8	.897801	.6303747	-0.15	0.878	.2267381	3.554969
9	1.633939	1.067844	0.75	0.452	.4538807	5.882065
10	1.241043	1.041922	0.26	0.797	.2394185	6.433038
undgrad	1.226955	.1219456	2.06	0.040	1.009785	1.490832
phdmed	.5137594	.1324057	-2.58	0.010	.3100193	.8513944
phdprest	.8188393	.1205438	-1.36	0.175	.6136076	1.092714
art	1.011262	.0306538	0.37	0.712	.9529311	1.073163
cit	1.001728	.0020015	0.86	0.388	.9978124	1.005658
prest1	1.264341	.2231535	1.33	0.184	.8946	1.786897
_cons	.0412265	.031836	-4.13	0.000	.0090753	.1872807

The estimated odds ratios for being promoted at the first university (given in the upper panel of the output under `promfirst`) are generally not that different from those previously reported for promotion at any research university in the single-event case. Regarding the estimated odds ratios for leaving without having been promoted (given in the lower panel under `leave`), a one-unit increase in the selectivity of the undergraduate institution corresponds to an increase in the odds of leaving by about 23%, whereas having a PhD from a medical school reduces the odds by 49%, controlling for other covariates. None of the other adjusted odds ratios are significantly different from 1 at the 5% level.

14.2.6 Handling left-truncated data

In the promotion data, all individuals were sampled at the beginning of their assistant professorships, that is, when they became at risk for promotion to associate professor. Such a sampling design, where the start of the observation period coincides with the time individuals become at risk, is sometimes called *sampling the inflow* (here the inflow to the state of assistant professor). This design is typical in experimental studies where, for instance, duration to death following surgery is investigated. When sampling the inflow, the start of observation is also the origin of analysis time.

However, in observational studies, we often have more complex sample designs where the times individuals become at risk do not necessarily coincide with the start of the observation period. *Delayed entry* means that individuals become at risk before entering the study. In such a design, *left-truncation* occurs if those who have experienced the event are not included in the study. For instance, when investigating the duration of unemployment, a *stock sample* includes individuals from the stock of unemployed at a given time and their times to employment (or censoring) are recorded. Those not unemployed at the start of observation are not eligible for the study, and the sample is in this sense truncated. If we analyze unemployment durations among those selected into the sample, without any correction, we are likely to underestimate the hazard of employment in the general population. The sampling is said to be *length biased* because the probability of selection at a given point in time depends on the time spent at risk.

In truncated samples, we require the conditional hazard of survival, given that the subject is selected into the study. Consider an individual who we know has already been at risk for a time t_b when he enters the study. This situation was illustrated by case 5 in the right panel of the figure on page 745, where B denotes the beginning of the study. Because the individual would not have been included in the study if the event had occurred at a time $t \leq t_b$, we must condition on the fact that $t > t_b$. The likelihood contribution for this person if he is right-censored at time $t > t_b$ therefore becomes

$$\Pr(T_i > t | T_i > t_b) \;=\; \frac{\Pr(T_i > t, T_i > t_b)}{\Pr(T_i > t_b)} \;=\; \frac{\prod_{s=1}^{t}(1 - h_{si})}{\prod_{s=1}^{t_b}(1 - h_{si})} \;=\; \prod_{s=t_b+1}^{t}(1 - h_{si})$$

and the likelihood contribution if he experiences the event at time $t > t_b$ is

$$\Pr(T_i = t | T_i > t_b) \;=\; \frac{\Pr(T_i = t, T_i > t_b)}{\Pr(T_i > t_b)} \;=\; \frac{h_{ti} \prod_{s=1}^{t-1}(1 - h_{si})}{\prod_{s=1}^{t_b}(1 - h_{si})} \;=\; h_{ti} \prod_{s=t_b+1}^{t-1}(1 - h_{si})$$

The correct likelihood contribution under delayed entry is thus simply obtained by letting the individual start contributing observations from entering the study at $t_b + 1$ and discarding the preceding periods, $t = 1, \ldots, t_b$. For instance, if professor 10 had already completed 3 years as assistant professor when he entered the study, we would simply delete the data for years 1–3 for him and let his initial hazard in the study at year 4 be $h_{4,10}$.

The above correction cannot be used if the time at risk before entering the study t_b is unknown. Fortunately, bias can be avoided here by simply discarding subjects with late entry; however, this can incur a substantial loss in efficiency.

Different kinds of sampling designs for survival analysis are discussed in, for instance, Lancaster (1990), Hamerle (1991), Guo (1993), Jenkins (1995), and Klein and Moeschberger (2003).

14.3 How does birth history affect child mortality?

Pebley and Stupp (1987) and Guo and Rodríguez (1992) analyzed data on child mortality in Guatemala. The data come from a retrospective survey conducted in 1974–76 by the Instituto Nutrición de Centroamérica y Panamá (UNCAP) and the think tank RAND, and have been made available by Germán Rodríguez.

Most of the data come from a female life history survey that covered all women aged 15–49 who lived in six villages and towns. The survey questionnaire asked about the complete birth history, maternal education, and related subjects. The data include all children except multiple births (twins or triplets), stepchildren, and children born more than 15 years before the survey. Observations with missing socioeconomic data or inconsistent dates of events have been deleted.

The outcome of interest is the children's length of life. Because most of the mothers have several children, this is an example of multilevel or clustered survival data. The variables we will use here are

- `kidid`: child identifier (i)
- `momid`: mother identifier (j)
- `time`: time in months from birth to death or censoring
- `death`: dummy variable for death (1: death; 0: censoring)
- `mage`: mother's age at time of birth of child
- `border`: birth order of child

- p0014, p1523, p2435, p36up: dummy variables for time between birth of index child and birth of previous child being 0–14, 15–23, 24–35, and 36 or more months, respectively (the reference category is no previous birth)

- pdead: dummy variable for previous child having died before conception of the index child

- f0011 and f1223: dummy variables for next child being born within 0–11 and 12–23 months after the birth of the index child, respectively, and before the death of the index child (the reference category is no subsequent birth within 23 months and before the death of the index child)

We read in the data by typing

```
. use http://www.stata-press.com/data/mlmus3/mortality, clear
```

14.4 Data expansion

Pebley and Stupp (1987) treat these data as continuous-time survival data and use a piecewise exponential model (see section 15.4.1 and exercise 15.5) with constant hazards in the intervals <1, 1–5, 6–11, 12–23, and >23 months. Here we will treat these same intervals as discrete-time intervals and use discrete-time survival models. When the exact timing is known, the piecewise exponential model described in section 15.4.1 may be more appropriate because it takes into account that individuals were at risk for only parts of the time interval within which they either died or were censored. In contrast, discrete-time survival analysis treats these individuals as being at risk for the entire interval, an issue we will return to shortly. On the other hand, the piecewise exponential model assumes that the hazard is constant throughout each interval, whereas discrete-time models make no such assumption.

First, we use the egen function cut() to categorize the variable time, and then we use the table command to make sure that we have done this correctly:

```
. egen discrete = cut(time), at(0 1 6 12 24 61) icodes
. table discrete, contents(min time max time)
```

discrete	min(time)	max(time)
0	.25	.25
1	1	5
2	6	11
3	12	23
4	24	60

The variable discrete takes on consecutive integer values from 0 to 4 because we used the icodes option in the egen command. However, for the data expansion to work, we require that discrete starts from 1, so we add 1:

```
. replace discrete = discrete + 1
```

Before expanding the data, we summarize them in a life table using the `ltable` command so that we can check that our data expansion is correct:

```
. ltable discrete death, noadjust
```

Interval		Beg. Total	Deaths	Lost	Survival	Std. Error	[95% Conf. Int.]	
1	2	3120	109	9	0.9651	0.0033	0.9580	0.9710
2	3	3002	92	94	0.9355	0.0044	0.9263	0.9436
3	4	2816	66	126	0.9136	0.0051	0.9031	0.9230
4	5	2624	94	238	0.8808	0.0059	0.8687	0.8919
5	6	2292	42	2250	0.8647	0.0063	0.8518	0.8765

Returning to the issue of how to deal with individuals who died or were censored within an interval, `ltable` with the `noadjust` option treats these individuals as being at risk during the entire interval (assuming that the event and censoring take place at the end of the interval). This gives estimated hazards of $\widehat{h}_1 = 109/3120 = 0.0349$ and so on. The data expansion method described in section 14.2.2 yields these same estimates. If time is truly discrete, this is the correct method for estimating the hazards.

However, if time is treated as interval-censored as it is here, we do not know when during the first interval the $109 + 9 = 118$ individuals died or were censored. To allow for partial contributions to the risk set for those who were censored or died during the interval, the *actuarial adjustment* consists of excluding half the 118 individuals from the denominator, giving $\widehat{h}_1 = 109/(3120 - 118/2) = 0.0356$. This can be motivated by assuming that the hazard of removal from the risk set follows a uniform distribution over the interval. The estimated hazards from the actuarial method are provided by `ltable` without the `noadjust` option. In this dataset, the adjustment makes a considerable difference in the last interval where 2,250 individuals were censored.

We expand the data and produce a table for comparison with the life table above:

```
. expand discrete
(10734 observations created)
. by kidid, sort: generate interval = _n
. generate y = 0
. replace y = death if interval == discrete
(403 real changes made)
```

```
. tabulate interval y, row
```

```
┌─────────────────┐
│ Key             │
├─────────────────┤
│      frequency  │
│  row percentage │
└─────────────────┘
```

interval	y 0	1	Total
1	3,011	109	3,120
	96.51	3.49	100.00
2	2,910	92	3,002
	96.94	3.06	100.00
3	2,750	66	2,816
	97.66	2.34	100.00
4	2,530	94	2,624
	96.42	3.58	100.00
5	2,250	42	2,292
	98.17	1.83	100.00
Total	13,451	403	13,854
	97.09	2.91	100.00

The row totals and numbers of 1s agree with the `Beg. Total` and `Deaths` in the life table, so we can trust our data expansion.

14.5 ❖ Proportional hazards and interval-censoring

The most popular methods for modeling covariate effects for continuous-time survival data (discussed in detail in section 15.4) assume that the continuous-time hazards are proportional,

$$h(z|\mathbf{x}_{ij}) = h_0(z)\exp(\beta_2 x_{2ij} + \cdots + \beta_p x_{pij})$$

where $h(z|\mathbf{x}_{ij})$ is the continuous-time hazard function for subject i in cluster j at time z and $h_0(z)$ is the *baseline hazard function* (the hazard when $x_{2ij} = \cdots = x_{pij} = 0$). It follows that continuous-time hazards are proportional in the sense that the hazard ratio

$$\frac{h(z|\mathbf{x}_{ij})}{h(z|\mathbf{x}_{i'j'})} = \exp\{\beta_2(x_{2ij} - x_{2i'j'}) + \cdots + \beta_p(x_{pi} - x_{pi'j'})\}$$

does not depend on time. We see that an exponentiated coefficient, say, $\exp(\beta_2)$, represents the *hazard ratio* for a one-unit change in x_{2ij}, controlling for the other covariates.

It can be shown (see display 15.3 in the next chapter) that the proportionality assumption translates to the following relationship for the survival function,

$$S(z|\mathbf{x}_{ij}) \equiv \Pr(Z_{ij} > z|\mathbf{x}_{ij}) = S_0(z)^{\exp(\beta_2 x_{2ij} + \cdots + \beta_p x_{pij})}$$

where Z_{ij} is the continuous survival time for subject i in cluster j and $S_0(z)$ is the *baseline survival function*, the survival function when the covariates are all zero.

If the survival times are interval-censored, so we only observe integer values $T_{ij} = t$ if $z_{t-1} < Z_{ij} \leq z_t$ $(t=1, 2, \ldots)$, then the discrete-time hazard is given by

$$h_{tij} \equiv \Pr(T_{ij} = t | \mathbf{x}_{ij}, T_{ij} > t - 1) \quad = \quad \frac{\Pr(Z_{ij} > z_{t-1} | \mathbf{x}_{ij}) - \Pr(Z_{ij} > z_t | \mathbf{x}_{ij})}{\Pr(Z_{ij} > z_{t-1} | \mathbf{x}_{ij})}$$

$$= \quad 1 - \frac{S(z_t | \mathbf{x}_{ij})}{S(z_{t-1} | \mathbf{x}_{ij})}$$

It follows that

$$1 - h_{tij} \quad = \quad \frac{S(z_t | \mathbf{x}_{ij})}{S(z_{t-1} | \mathbf{x}_{ij})} \quad = \quad \left\{ \frac{S_0(z_t)}{S_0(z_{t-1})} \right\}^{\exp(\beta_2 x_{2ij} + \cdots + \beta_p x_{pij})}$$

and

$$\ln(1 - h_{tij}) \quad = \quad \exp(\beta_2 x_{2ij} + \cdots + \beta_p x_{pij}) \{\ln S_0(z_t) - \ln S_0(z_{t-1})\}$$

The right-hand side is negative, so we reverse the signs and then take the logarithms and obtain

$$\ln\{-\ln(1 - h_{tij})\} \quad = \quad \beta_2 x_{2ij} + \cdots + \beta_p x_{pij} + \underbrace{\ln\{\ln S_0(z_{t-1}) - \ln S_0(z_t)\}}_{\alpha_t}$$

This model, which is linear in the covariates, contains the same regression parameters as the continuous-time proportional-hazards model and contains time-specific constants α_t. The model is a generalized linear model with a complementary log-log link $g(h_{tij}) = \ln\{-\ln(1 - h_{tij})\}$.

By estimating the parameters α_t freely for each time point t, we are making no assumption regarding the shape of the baseline survival function $S_0(z_t)$ or the corresponding baseline hazard function within the time intervals $z_{t-1} < Z_{ij} \leq z_t$. This is in contrast to the piecewise exponential model discussed in section 15.4.1, which assumes constant baseline hazards within time intervals.

14.6   Complementary log-log models

We will use the complementary log-log link here instead of the logit link because, as shown in section 14.5, this link function follows if a proportional hazards model holds in continuous time and the survival times are interval-censored. Hence, for child i of mother j, we consider the complementary log-log discrete-time survival model,

$$\text{cloglog}(h_{sij}) \equiv \ln\{-\ln(1 - h_{sij})\} \quad = \quad \alpha_1 d_{1sij} + \cdots + \alpha_5 d_{5sij}$$

where we have included one dummy variable for each period but no constant (instead of a constant and dummy variables for all but the first period as in section 14.2). The model can alternatively be written as

$$h_{sij} = \text{cloglog}^{-1}(\alpha_1 d_{1sij} + \cdots + \alpha_5 d_{5sij}) \equiv 1 - \exp\{-\exp(\alpha_1 d_{1sij} + \cdots + \alpha_5 d_{5sij})\}$$

As mentioned in section 14.5, the exponentiated regression coefficients can be interpreted as hazard ratios in continuous time.

Unlike probit and logit models that are symmetric with

$$\text{probit}(h_{sij}) \;=\; \Phi^{-1}(h_{sij}) = -\Phi^{-1}(1 - h_{sij})$$

$$\text{logit}(h_{sij}) \;=\; -\text{logit}(1 - h_{sij})$$

the complementary log-log link is not symmetric,

$$\text{cloglog}(h_{sij}) \;\neq\; -\text{cloglog}(1 - h_{sij})$$

This means that switching the 0s and 1s in the response variable will not merely reverse the sign of the estimated coefficients as for logit and probit models. It is worth noting that the complementary log-log and logit models are very similar when the time intervals are small.

By using dummy variables for the time intervals, we are not making any assumption regarding the shape of the discrete-time hazard function, and in the model without further covariates, we will get the same hazard estimates whatever link we use.

We construct the dummies for the time intervals,

```
. quietly tabulate interval, generate(int)
```

and use the `cloglog` command to fit a model with a complementary log-log link,

```
. cloglog y int1-int5, noconstant
```

Complementary log-log regression				Number of obs	=	13854
				Zero outcomes	=	13451
				Nonzero outcomes	=	403
				Wald chi2(5)	=	4943.80
Log likelihood = -1811.6791				Prob > chi2	=	0.0000

y	Coef.	Std. Err.	z	P>|z|	[95% Conf. Interval]	
int1	-3.336513	.0957877	-34.83	0.000	-3.524253	-3.148772
int2	-3.469723	.1042614	-33.28	0.000	-3.674072	-3.265374
int3	-3.741583	.1230944	-30.40	0.000	-3.982844	-3.500323
int4	-3.310976	.1031478	-32.10	0.000	-3.513142	-3.108809
int5	-3.990277	.1543055	-25.86	0.000	-4.292711	-3.687844

We can estimate the hazards for the children in each time interval by typing

```
. predict haz, pr
```

and list the estimated hazards (which are the same for all children) for child 101 by typing

```
. sort kidid interval
. list kidid interval haz if kidid==101, noobs
```

kidid	interval	haz
101	1	.0349359
101	2	.03064624
101	3	.0234375
101	4	.03582317
101	5	.01832461

The marginal (over the covariates) hazard function estimated above is of some interest, but we are more interested in the relationships between covariates and the death hazard. Understanding these relationships might provide insight into the etiology of child mortality and could potentially guide targeted interventions to reduce mortality.

Mother's age at birth is known to be a risk factor for child mortality. We thus allow for linear and quadratic functions of age by including the covariates mage (x_{2ij}) and mage2 (x_{3ij}), with the latter variable constructed as

```
. generate mage2 = mage^2
```

Because birth order (often called *parity* in epidemiology) is a known risk factor, we also include border (x_{4ij}).

The other covariates relate to the timing of other births in the family. As discussed by Pebley and Stupp (1987), birth spacing may be important for several reasons; for instance, several children of similar ages may compete for scarce resources (food, clothing, and parental time). We therefore include dummy variables for the time intervals between the births of the index child and the previous child: p0014 (x_{5ij}), p1523 (x_{6ij}), p2435 (x_{7ij}), and p36up (x_{8ij}).

However, Pebley and Stupp point out that there may be a spurious relationship between the previous birth interval and the death of the index child if the previous birth interval is shortened by the previous child's death and if the risks of dying are correlated between children in the same family. For this reason, the survival status of the previous child, pdead (x_{9ij}), is included as a covariate.

Also, the birth interval to the next child may be shortened by the index child's death. Therefore, the dummy variables f0011 and f1223 for the next birth interval being 0–11 months and 12–23 months, respectively, have been set to zero if the index child died within the corresponding time intervals. The birth of a subsequent child can affect the index child only after it has occurred, for a short birth interval (0–11 months) when the index child is at least 12 months old (interval 4 onward) and for a long birth interval (12–23 months) when the index child is at least 24 months old (interval 5). We therefore follow Guo and Rodríguez (1992) in defining the time-varying dummy variables comp12 $(x_{10,sij})$, comp24e $(x_{11,sij})$, and comp24l $(x_{12,sij})$ for competition with the subsequent child when the index child is 12–23 months old (interval 4) and over 24 months old (interval 5), differentiating in the latter interval between competition with children born earlier and later.

```
. generate comp12 = f0011*(interval==4)

. generate comp24e = f0011*(interval==5)

. generate comp24l = f1223*(interval==5)
```

We now include all the above covariates in the complementary log-log model:

$$\ln\{-\ln(1 - h_{sij})\} = \alpha_1 d_{1sij} + \cdots + \alpha_5 d_{5sij} + \beta_2 x_{2ij} + \cdots + \beta_{12} x_{12,sij}$$

Estimates with robust standard errors for clustered data can be obtained using

```
. cloglog y int1-int5 mage mage2 border p0014 p1523 p2435 p36up pdead
> comp12 comp24e comp24l, noconstant eform vce(cluster momid)
```

Complementary log-log regression			Number of obs	=	13854
			Zero outcomes	=	13451
			Nonzero outcomes	=	403
			Wald chi2(16)	=	4512.16
Log pseudolikelihood = -1784.1899			Prob > chi2	=	0.0000

(Std. Err. adjusted for 851 clusters in momid)

y	exp(b)	Robust Std. Err.	z	P>|z|	[95% Conf. Interval]	
int1	.2318333	.17652	-1.92	0.055	.052127	1.031071
int2	.2042155	.1572742	-2.06	0.039	.0451383	.9239144
int3	.1560266	.121621	-2.38	0.017	.0338607	.7189539
int4	.2300307	.17649	-1.92	0.055	.0511335	1.034824
int5	.1221787	.0922591	-2.78	0.005	.0278125	.5367243
mage	.8617077	.0499925	-2.57	0.010	.7690897	.9654793
mage2	1.002562	.0010004	2.56	0.010	1.000603	1.004525
border	1.063391	.0364623	1.79	0.073	.9942743	1.137311
p0014	1.73225	.3591742	2.65	0.008	1.153766	2.600779
p1523	.8764676	.1549602	-0.75	0.456	.6197873	1.23945
p2435	.7697006	.1391446	-1.45	0.148	.5400621	1.096983
p36up	.6681857	.1366279	-1.97	0.049	.4475556	.9975793
pdead	1.118051	.178947	0.70	0.486	.8170047	1.530026
comp12	4.909925	2.279635	3.43	0.001	1.976375	12.19777
comp24e	4.436945	3.158989	2.09	0.036	1.099135	17.91089
comp24l	.8476108	.3038621	-0.46	0.645	.4198045	1.711378

We see that mother's age at birth (mage and mage2), being born within 14 months of the previous child (p0014), and getting a new sibling within the first 11 months of life (comp12 and comp24e) have statistically significant coefficients at the 5% level. For example, the estimated hazard of death increases 73% if the index child was born within 14 months of the previous child, controlling for the other covariates. Having a new sibling within the first 11 months of life increases the estimated hazard 4.91-fold at age 12 to 23 months and 4.44-fold at age 24–60 months, controlling for the other covariates. It appears that competition with children born shortly before, but even more so shortly after, the index child is an important risk factor. The estimates are also given under "No random intercept" in table 14.3. The exponentiated coefficients of the dummy variables (not given in the table) do not represent hazard ratios here because the overall constant was omitted.

Table 14.3: Maximum likelihood estimates for complementary log-log models with and without random intercept for Guatemalan child mortality data

	No random intercept		Random intercept	
	Hazard ratio	(95% CI)*	Hazard ratio	(95% CI)
Fixed part				
$\exp(\beta_2)$ [mage]	0.86	(0.77, 0.97)	0.86	(0.76, 0.96)
$\exp(\beta_3)$ [mage2]	1.00	(1.00, 1.00)	1.00	(1.00, 1.00)
$\exp(\beta_4)$ [border]	1.06	(0.99, 1.14)	1.06	(0.99, 1.14)
$\exp(\beta_5)$ [p0014]	1.73	(1.15, 2.60)	1.79	(1.17, 2.74)
$\exp(\beta_6)$ [p1523]	0.88	(0.62, 1.24)	0.90	(0.62, 1.30)
$\exp(\beta_7)$ [p2435]	0.77	(0.54, 1.10)	0.79	(0.55, 1.15)
$\exp(\beta_8)$ [p36up]	0.67	(0.45, 1.00)	0.68	(0.45, 1.03)
$\exp(\beta_9)$ [pdead]	1.12	(0.82, 1.53)	0.95	(0.67, 1.35)
$\exp(\beta_{10})$ [comp12]	4.91	(1.98, 12.20)	4.94	(1.98, 12.34)
$\exp(\beta_{11})$ [comp24e]	4.44	(1.10, 17.91)	4.53	(1.07, 19.17)
$\exp(\beta_{12})$ [comp24l]	0.85	(0.42, 1.71)	0.85	(0.41, 1.73)
Random part				
$\sqrt{\psi}$			0.43	
Log likelihood	$-1,784.19$		$-1,782.70$	

*Using robust standard errors for clustered data

14.7   A random-intercept complementary log-log model

14.7.1   Model specification

To accommodate dependence among the survival times of different children of the same woman, after conditioning on the observed covariates, we include a random intercept ζ_j for mother j in the complementary log-log model,

$$\ln\{-\ln(1 - h_{sij})\} = \alpha_1 d_{1sij} + \cdots + \alpha_5 d_{5sij} + \beta_2 x_{2ij} + \cdots + \beta_{12} x_{12,sij} + \zeta_j$$

where $\zeta_j \sim N(0, \psi)$ and the random intercept is independent from the covariates.

The model can also be written in terms of a continuous latent response,

$$y_{sij}^* = \alpha_1 d_{1sij} + \cdots + \alpha_5 d_{5sij} + \beta_2 x_{2ij} + \cdots + \beta_{12} x_{12,sij} + \zeta_j + \epsilon_{sij}$$

where ϵ_{sij} has a standard extreme-value type-1 or Gumbel distribution, given the covariates and the random intercept ζ_j. The standard Gumbel distribution has a mean of about 0.577 (called Euler's constant) and a variance of $\pi^2/6$, and is asymmetric. Recall that such a Gumbel distribution was also used for the multinomial and conditional logit models discussed in chapter 12. It is assumed that the ϵ_{sij} are mutually independent and independent of ζ_j and the covariates. The latent response y^*_{sij} determines the observed binary response y_{sij} via the threshold model

$$
y_{sij} = \begin{cases} 1 & \text{if } y^*_{sij} > 0 \\ 0 & \text{otherwise} \end{cases}
$$

The exponentiated random intercept, $\exp(\zeta_j)$, is called a *shared frailty* because it is a mother-specific disposition or *frailty* that is shared among children nested in a mother. Frailties are sometimes also included in single-level survival models, a practice that we do not generally recommend because such models are often not well identified.

14.7.2 Estimation using Stata

Two commands are available for fitting the random-intercept complementary log-log model: `xtcloglog` and `gllamm` with the `link(cll)` option. Whereas `xtcloglog` can be used only for a two-level random-intercept model, `gllamm` can also be used for random-coefficient and higher-level models. However, `xtcloglog` is considerably faster than `gllamm`.

The syntax for `xtcloglog` is analogous to that for `xtlogit`:

```
. quietly xtset momid
. xtcloglog y int1-int5 mage mage2 border p0014 p1523 p2435 p36up pdead
> comp12 comp24e comp24l, noconstant eform
```

```
Random-effects complementary log-log model    Number of obs      =     13854
Group variable: momid                          Number of groups   =       851

Random effects u_i ~ Gaussian                  Obs per group: min =         1
                                                              avg =      16.3
                                                              max =        40

                                               Wald chi2(16)      =   2582.92
Log likelihood  = -1782.7021                   Prob > chi2        =    0.0000
```

y	exp(b)	Std. Err.	z	P>|z|	[95% Conf. Interval]	
int1	.2333078	.1806049	-1.88	0.060	.0511693	1.063774
int2	.2071299	.1606492	-2.03	0.042	.0452954	.9471781
int3	.1591151	.1239453	-2.36	0.018	.0345664	.7324346
int4	.2358834	.1833431	-1.86	0.063	.0514151	1.082191
int5	.1261628	.0998945	-2.61	0.009	.0267275	.5955317
mage	.8559014	.0513772	-2.59	0.010	.7609018	.9627618
mage2	1.00268	.0010671	2.52	0.012	1.000591	1.004774
border	1.059005	.0375963	1.61	0.106	.9878226	1.135316
p0014	1.790715	.3899374	2.68	0.007	1.168619	2.743974
p1523	.8967051	.1696206	-0.58	0.564	.6189227	1.299161
p2435	.7923456	.1489941	-1.24	0.216	.5480915	1.14545
p36up	.6810089	.1444591	-1.81	0.070	.449357	1.032082
pdead	.948696	.1702308	-0.29	0.769	.6674064	1.34854
comp12	4.941595	2.307624	3.42	0.001	1.978671	12.3413
comp24e	4.534147	3.335828	2.05	0.040	1.072151	19.17499
comp24l	.8459522	.3100246	-0.46	0.648	.4124748	1.734979
/lnsig2u	-1.672034	.6326815			-2.912067	-.4320015
sigma_u	.4334333	.1371126			.2331592	.8057347
rho	.1025014	.0582035			.0319916	.2829852

```
Likelihood-ratio test of rho=0: chibar2(01) =     2.98 Prob >= chibar2 = 0.042
```

The estimated subject-specific or conditional hazard ratios and their corresponding 95% confidence intervals were also shown under "Random intercept" in table 14.3.

The estimated residual correlation among the latent responses for two children of the same mother is only

$$\widehat{\rho} = \frac{\widehat{\psi}}{\widehat{\psi} + \pi^2/6} = \frac{0.433^2}{0.433^2 + \pi^2/6} = 0.10$$

(given as `rho` in the output), suggesting that there is not much dependence among the survival times for children of the same mother after controlling for the observed covariates.

Because of the small within-mother correlation, the estimated hazard ratios are close to those from the model without a random intercept. Not surprisingly, we see from the output that the test of the null hypothesis of zero between-mother variance, $H_0: \psi = 0$, is barely significant at the 5% level. However, `pdead`, a dummy variable for the previous child having died before conception of the index child is included in the model, so some of the dependence among the siblings is captured by the fixed part of the model.

Another way of assessing the degree of unobserved heterogeneity is in terms of the median hazard ratio. Imagine randomly drawing children with the same covariate values but having different mothers j and j'. The hazard ratio comparing the child whose mother has the larger random intercept with the other child is given by $\exp(|\zeta_j - \zeta_{j'}|)$. Following the derivation in section 10.9.2, the median hazard ratio is given by

$$\mathrm{HR}_{\mathrm{median}} = \exp\left\{\sqrt{2\psi}\,\Phi^{-1}(3/4)\right\}$$

Plugging in the parameter estimates, we obtain $\widehat{\mathrm{HR}}_{\mathrm{median}}$:

```
. display exp(sqrt(2*.433433^2)*invnormal(3/4))
1.5120099
```

When two children with the same covariate values but different mothers are randomly sampled, the hazard ratio comparing the child who has the larger hazard with the child who has the smaller hazard will exceed 1.51 in 50% of the samples, which is of moderate magnitude compared with the estimated hazard ratios for some of the covariates.

14.8 ❖ Population-averaged or marginal vs. subject-specific or conditional survival probabilities

We now consider the model-implied subject-specific or conditional survival probabilities, given that $\zeta_j = 0$, as well as the corresponding population-averaged or marginal survival probabilities, integrating over the random-intercept distribution.

The predicted conditional or mother-specific survival probabilities for $\zeta_j = 0$ are given by

$$S_{tij}^{C} = \prod_{s=1}^{t}\{1 - \mathrm{cloglog}^{-1}(\widehat{\alpha}_1 d_{1sij} + \cdots + \widehat{\beta}_{12}x_{12,sij} + \underbrace{0}_{\zeta_j})\}$$

and the predicted marginal or population-averaged survival probabilities are given by

$$S_{tij}^{M} = \int\left[\prod_{s=1}^{t}\{1 - \mathrm{cloglog}^{-1}(\widehat{\alpha}_1 d_{1sij} + \cdots + \widehat{\beta}_{12}x_{12,sij} + \zeta_j)\}\right]\phi(\zeta_j; 0, \widehat{\psi})\,d\zeta_j$$

The product inside the square brackets is the probability that a child with given covariates and a given value of the random intercept ζ_j survives past time t. Integrating this over the random-intercept distribution gives the corresponding survival probability, averaged over all children with the same covariate values. As explained in section 10.8, marginal probabilities are different from conditional probabilities evaluated at the population mean of the random intercept. Indeed, the marginal survival curve need not correspond to any subject-specific survival curve because at each point in time it represents the average over the selected population of subjects who are still at risk at that time.

For a mother with one child in the data who was censored at time t, the contribution to the marginal likelihood has exactly the same form as S_{tij}^M. By constructing an appropriate prediction dataset, we can therefore obtain S_{tij}^M by calculating the marginal likelihood contributions using `gllapred` with the `ll` option, but first we must fit the model using `gllamm`:

```
. gllamm y int1-int5 mage mage2 border p0014 p1523 p2435 p36up pdead
> comp12 comp24e comp24l, noconstant eform i(momid) link(cll) family(binomial)
> adapt

number of level 1 units = 13854
number of level 2 units = 851

Condition Number = 27451.829

gllamm model

log likelihood = -1782.702
```

y	exp(b)	Std. Err.	z	P>|z|	[95% Conf. Interval]	
int1	.2333055	.1806242	-1.88	0.060	.0511597	1.063951
int2	.2071278	.1606663	-2.03	0.042	.0452869	.947336
int3	.1591135	.1239584	-2.36	0.018	.03456	.7325563
int4	.2358809	.1833625	-1.86	0.063	.0514054	1.082371
int5	.1261616	.0999049	-2.61	0.009	.0267225	.5956308
mage	.8559022	.0513834	-2.59	0.010	.7608919	.9627761
mage2	1.00268	.0010672	2.51	0.012	1.000591	1.004774
border	1.059005	.0375963	1.61	0.106	.9878227	1.135317
p0014	1.790712	.3899412	2.68	0.007	1.168611	2.743983
p1523	.8967037	.1696228	-0.58	0.564	.6189185	1.299165
p2435	.7923441	.1489964	-1.24	0.216	.548087	1.145455
p36up	.6810076	.1444607	-1.81	0.070	.4493538	1.032085
pdead	.9487002	.1702355	-0.29	0.769	.6674039	1.348557
comp12	4.941589	2.307621	3.42	0.001	1.978668	12.34128
comp24e	4.534112	3.335808	2.05	0.040	1.07214	19.17489
comp24l	.8459486	.310024	-0.46	0.648	.4124724	1.734974

```
Variances and covariances of random effects
------------------------------------------------------------------------------

***level 2 (momid)

    var(1): .18785803 (.11886731)
------------------------------------------------------------------------------
```

Now we will produce a small prediction dataset from the data for child 101 because this child has a row of data for each interval (we will not save the data first, but sometimes this may be a good idea):

```
. keep if kidid==101
```

It is required that y be zero for each interval, which is already true. We can change the covariates to whatever values we want and then save the data under the name `junk.dta`

```
. replace pdead = 0
. replace p2435 = 0
. save junk, replace
```

We now replicate the data five times for this child, creating unique identifiers 1–5 in `kidid` for the replications. To obtain S_{tij}^{M} for $t = 1, \dots, 5$, we let child 1 have data for interval 1 only (corresponding to censoring at $t = 1$), child 2 for intervals 1 and 2, etc. The likelihood contributions for children 1–5 will then correspond to the marginal survival probabilities for times 1–5, respectively.

We first produce data for the five children by appending the data for child 101 to itself four times, changing `kidid` appropriately each time,

```
. replace kidid = 1
. append using junk
. replace kidid = 2 if kidid==101
. append using junk
. replace kidid = 3 if kidid==101
. append using junk
. replace kidid = 4 if kidid==101
. append using junk
. replace kidid = 5 if kidid==101
```

and then delete the unnecessary rows of data:

```
. drop if interval > kidid
```

The data now look like the following:

```
. sort kidid interval
. list kidid interval y, sepby(kidid) noobs
```

kidid	interval	y
1	1	0
2	1	0
2	2	0
3	1	0
3	2	0
3	3	0
4	1	0
4	2	0
4	3	0
4	4	0
5	1	0
5	2	0
5	3	0
5	4	0
5	5	0

For the log-likelihood contributions to correspond to $\ln S_{tij}^M$, we must have one child per mother (otherwise we would get joint survival probabilities for all children):

```
. replace momid = kidid
```

The log-likelihood contributions are obtained using `gllapred` with the `ll` option, and with the `fsample` option to get predictions for the full sample (not just the estimation sample but everyone in the data):

```
. gllapred loglik, ll fsample
```

The population-averaged or marginal survival probabilities S_{tij}^M, called `msurv`, are then obtained by exponentiation:

```
. generate msurv = exp(loglik)
```

The mother-specific or conditional hazards can be obtained by defining a variable, `zeta1`, for the random-intercept values and then using the `mu` (and not `marginal`) option, together with the `us(zeta)` option:

```
. generate zeta1 = 0
. gllapred chaz, mu us(zeta) fsample
(mu will be stored in chaz)
```

To get the corresponding conditional survival probabilities S_{tij}^C, we use commands similar to those on page 762:

```
. generate ln1mchaz = ln(1-chaz)
. by kidid (interval), sort: generate sln1mchaz = sum(ln1mchaz)
. generate csurv = exp(sln1mchaz)
```

We can now plot the survival probabilities as a function of `interval`:

```
. twoway (line msurv interval if kidid==interval, sort lpatt(solid))
> (line csurv interval if kidid==5, sort lpatt(dash)),
> legend(order(1 "marginal" 2 "conditional"))
> ytitle(Survival probability)
```

This command produces figure 14.6.

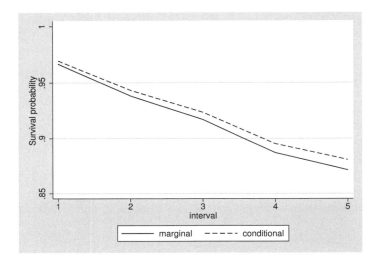

Figure 14.6: Predicted subject-specific or conditional survival functions and population-averaged or marginal survival functions

We see that the conditional and marginal survival curves are similar because of the relatively small random-intercept variance.

14.9   Summary and further reading

We have demonstrated how discrete-time survival analysis can proceed by using standard models for dichotomous responses once the data have been expanded appropriately. Using a logit link gives the continuation-ratio logit model in which the conditional odds of the event occurring given that it has not yet occurred are proportional. Using a complementary log-log link corresponds to proportional hazards in continuous time. In both models, proportionality can be relaxed by constructing appropriate time-varying covariates by multiplying the covariates with dummy variables for time intervals (or other functions of time); see also exercises 14.4, 14.5, and 14.6.

Discrete-time survival analysis can also be performed using versions of the cumulative model for ordinal responses discussed in chapter 11 that allow for right-censoring. These models are sometimes called grouped continuous models, whereas the models discussed in this chapter are called continuation-ratio models. A major drawback of the ordinal approach is that time-varying covariates cannot be handled.

Good introductory books on survival analysis in general include Collett (2003b) and Hosmer, Lemeshow, and May (2008) for medicine and Allison (1984, 1995), Singer and Willett (2003), and Box-Steffensmeier and Jones (2004) for social science. Useful introductory articles on single-level discrete-time survival modeling include Allison (1982) and Singer and Willett (1993).

To accommodate clustered or multilevel survival data, we can include random effects in discrete-time survival models. Here we have included a random intercept, the exponential of which is usually called a *frailty* in the survival context. Random coefficients could of course also be included in the same manner as demonstrated in other chapters of the book, using `xtmelogit` or `gllamm`.

Multilevel discrete-time survival models are discussed for instance in the papers by Hedeker, Siddiqui, and Hu (2000), Barber et al. (2000), Rabe-Hesketh, Yang, and Pickles (2001b), and Grilli (2005). Models with discrete random effects, and many other issues in survival analysis, are discussed in the book by Vermunt (1997).

There are three exercises on single-level discrete-time survival analysis: exercise 14.1 based on the promotions data used in this chapter, exercise 14.2 on teacher turnover, and exercise 14.3 on the time to first sexual intercourse. Exercises on multilevel discrete-time survival analysis are based on the child mortality data used in this chapter (exercise 14.4), survey data on the length of schooling among brothers (exercise 14.5), a randomized trial on time to onset of blindness in patients with diabetic retinopathy (exercise 14.6), and a cluster-randomized intervention to reduce the risk of onset of smoking (exercise 14.7). Exercise 14.8 on early math proficiency is not about survival but uses the continuation-ratio logit model for the cognitive milestone reached.

14.10   Exercises

14.1 Assistant professor promotions data

In this exercise, we revisit the dataset `promotion.dta` on promotion to associate professor that was used to introduce discrete-time survival modeling.

1. Fit the logistic regression model considered on page 766.
2. Modify the model from step 1 by using a low-order polynomial of time instead of dummy variables for time. Start with a linear relationship and include successively higher-order terms until the highest-order term is not significant at the 5% level.
3. Produce graphs of the model-implied discrete-time hazards for professors 1 and 4 using the models from steps 1 and 2.

4. Fit the model from step 1 but with a complementary log-log link instead of a logit link.

5. Interpret the estimated exponentiated regression coefficients from step 4 for `undgrad` and `art`.

6. For individuals 1 and 4, plot the model-implied survival functions for the models with a logit link (from step 1) and a complementary log-log link (from step 4). Compare the curves for the two different models for the same individual.

14.2 Teacher turnover data

Here we consider data from Singer (1993) made available by Singer and Willett (2003). The careers of 3,941 special educators newly hired in Michigan public schools between 1972 and 1978 were tracked for 13 years. The outcome of interest is the number of years the teachers worked at the school, from the date they were hired until they left the school.

The variables in the dataset `teachers.dta` are

- `id`: teacher identifier
- `t`: number of years teacher worked at the school, that is, time to leaving school (no censoring) or end of study (censoring)
- `censor`: censoring indicator, equal to 1 if censored and 0 otherwise

1. Expand the data to person–years.
2. Obtain predicted hazards using a regression model of your choice with dummy variables for the periods.
3. Calculate and plot the predicted survival function over time.
4. Fit a continuation-ratio logit discrete-time survival model with a polynomial in time. What order of the polynomial seems to be required?
5. Produce a graph of the model-implied survival functions for the models in steps 2 and 4.

14.3 First sexual intercourse data

We now analyze the age at first intercourse data from Capaldi, Crosby, and Stoolmiller (1996) that is supplied with the book by Singer and Willett (2003). One hundred eighty at-risk boys (who had not had intercourse before) were followed from grades 6–12 and the date of first intercourse was recorded.

The variables in `firstsex.dta` are

- `id`: boy identifier (j)
- `time`: grade (school-year) in which either intercourse occurred (no censoring) or the boy dropped out of the study (censoring). Children are about 6 years old in first grade, so their approximate age in a given grade is the grade plus 5 years.

- censor: censoring indicator, equal to 1 if censored and 0 otherwise
- pt: a dummy variable for parental transition having occurred before seventh grade (0: lived with both parents; 1: experienced one or more parental transitions)
- pas: parental antisocial behavior when the boy was in fourth grade (based on arrests and driver license suspensions, drug-use, subscales 2 and 9 of the Minnesota Multiphasic Personality Inventory, and the mother's age at birth)

1. Expand the data to person–years, remembering that boys in this study could not experience the event before grade 7 (an example of a stock sample; see section 14.2.6).
2. Fit a logistic regression model with a dummy variable for each grade and pt as the only covariate, and interpret the exponentiated estimated regression coefficient of pt.
3. Plot the predicted survival functions separately for boys who have and have not had parental transitions (variable pt).
4. Extend the model by including pas as a further covariate, and interpret the exponentiated estimated coefficients of pt and pas.
5. Is there evidence for an interaction between pas and pt?

14.4 Child mortality data

Consider some further analyses of the child mortality data from Pebley and Stupp (1987) and Guo and Rodríguez (1992).

In addition to the variables listed in section 14.3, the following variables are included in mortality.dta:

- sex: sex of child (0: female; 1: male)
- home: dummy variable for child being born at home (1: home; 0: clinic or hospital)
- edyears: mother's years of education
- income: family annual income in quetzales in 1974–1975; the quetzal was pegged to the U.S.$ with a fixed exchange rate of 1 quetzal = 1 U.S.$. Missing responses are coded as −9.

1. Repeat the analyses presented in section 14.7.2.
2. Add the variables given above to the analysis.
3. Interpret the estimated hazard ratios for these new variables.
4. Relax the proportional-hazards assumption for home birth. Specifically, include an interaction between the first time interval and home birth to test whether home birth has a greater effect on the hazard of death shortly after, or during, birth than later on.

14.5 Brothers' school transition data

Mare (1994) analyzed data from the 1973 Occupational Changes in a Generation II survey (OCG) to investigate the relationship between the educational attainment of fathers and their sons. He selected male respondents who had at least one brother and considered the years of schooling of the respondent, his oldest brother, and his father.

The variables in `brothers.dta` are

- `id`: family identifier
- `brother`: brother identifier (0: respondent; 1: respondent's brother)
- `time`: length of schooling (1: <12 years; 2: 12 years; 3: ≥13 years)
- `feduc`: length of schooling for father (1: 0–8 years; 2: 9–11 years; 3: 12 years; 4: 13–15 years; 5: ≥16 years)
- `freq`: number of families

The data are in collapsed form, with `freq` indicating the number of families having the given levels of education for the respondent, his brother, and his father.

1. Expand the data so that there is one row of data for the transition to the 12th year of schooling for every subject and another row of data for the transition to the 13th year of schooling for those brothers who had 13 years or more of schooling. Define a variable, `transition`, equal to 1 and 2 for the two transitions and a dummy variable, `event`, equal to 1 if the transition occurred and 0 otherwise. This is just the expansion to person–period data discussed in this chapter except that the response variable is now a dummy variable for continuing in education ("censoring"), not for leaving.

2. For the respondents (and not their brothers), fit a continuation-ratio logit model with dummy variables `tr1` and `tr2` for the two transitions and dummy variables `ed2` to `ed5` for father's education levels 2 to 5. Use `freq` as frequency weights (see `help weights`).

3. Now relax the proportional odds assumption by allowing the effect of father's education to be different for the first and second transitions. Use a likelihood-ratio test (at the 5% level) to choose the more appropriate model.

4. Fit the model selected in step 2 for both brothers, assuming that the covariate effects are the same for both brothers, and including a random intercept for families. Use `gllamm`, specifying `freq` as level-2 weights by first renaming `freq` to `wt2` and then using the `gllamm` option `weight(wt)`.

5. Interpret the parameter estimates for the random-intercept model.

14.6 Blindness data

Diabetic retinopathy is a complication associated with diabetes mellitus and is a major cause of visual loss. Huster, Brookmeyer, and Self (1989) describe data from the Diabetic Retinopathy Study (DRS) conducted in 1971 to investigate the effectiveness of laser photocoagulation in delaying the onset of blindness in patients with diabetic retinopathy. One eye of each patient was randomly selected for treatment, and patients were assessed at 4-month intervals. The endpoint used to assess the treatment effect was the occurrence of visual acuity less than 5/200 at two consecutive assessments. The data considered by Huster, Brookmeyer, and Self (1989) and provided by Ross and Moore (1999) are a 50% sample of the high-risk patients as defined by DRS criteria. Following Ross and Moore (1999), we treat the data as discrete-time survival data using the intervals [6,10), [10,14), ..., [50,54), [54,58), [58,66), [66,83).

The variables in `blindness.dta` are

- `id`: patient identifier
- `eye`: eye identifier
- `treat`: dummy variable for eye being treated by photocoagulation
- `time`: onset of blindness in months
- `discrete`: discrete-time intervals 1: [6,10); 2: [10,14); ...; 12: [50,54); 13: [54,58); 14: [58,66); 15: [66,83)
- `censor`: dummy variable for blindness occurring
- `late`: dummy variable for late onset, or adult diabetes (onset at age 20 or later)

1. Expand the data to eye periods, with periods defined by the variable `discrete`.

2. Fit a continuation-ratio logit discrete-time survival model with dummy variables for time intervals; `treat`, `late`, and their interaction as covariates; and a random intercept for patients.

3. Test the proportional odds assumption for `treat` by including an interaction between `treat` and a new variable, `midp`, defined as the middle of the discrete-time intervals. (Hint: Except for the last two intervals, $midp = 4 + 4*discrete$.)

4. ❖ For the selected model, plot the model-implied marginal survival functions for the two eyes for patients with late onset diabetes.

See also exercise 15.6 for continuous-time analysis of these data.

14.7   Cigarette data   Solutions

Hedeker, Siddiqui, and Hu (2000) analyzed data from a subset of the Television School and Family Smoking Prevention and Cessation Project (TVSFP; see Flay et al. [1988]). The data are supplied with the SuperMix program (Hedeker et al. 2008).

In 1986, schools in Los Angeles were randomized to one of four conditions given by different combinations of two factors: (1) TV: a media (television) intervention (1: present; 0: absent) and (2) CC: a social-resistance classroom curriculum

(1: present; 0: absent). The intervention took place when the students were in seventh grade (when the children were about 12 years old). The students were assessed pre- and postintervention in 1986 and again a year later and two years later. At each of the four time points, the students were asked: "Have you ever smoked a cigarette?" To estimate the effect of the intervention, only those who answered "no" to the question before the intervention were included in the study. The first time the students answered "yes" after the intervention will be the event of interest in this exercise.

In addition to the clustering of students in classes, the classes are clustered in schools. We will consider two-level models in this exercise but will revisit the data for three-level modeling in exercise 16.5.

The variables in `cigarette.dta` are

- `school`: school identifier
- `class`: class identifier
- `time`: postintervention time point (1: 7th grade; 2: 8th grade; 3: 9th grade)
- `event`: dummy variable for student responding "yes" to the question about smoking
- `cc`: social-resistance classroom curriculum (dummy variable)
- `tv`: television intervention (dummy variable)

1. Expand the data to person–period data.
2. Estimate the discrete-time model that assumes the continuous-time hazards to be proportional. Include `cc`, `tv`, and their interaction as explanatory variables and specify a random intercept for classes. Use dummy variables for periods.
3. Interpret the exponentials of the estimated regression coefficients.
4. Obtain the estimated residual intraclass correlation of the latent responses.

See also exercise 16.5 for three-level models of these data.

14.8 Early childhood math proficiency data

In this exercise, we analyze the data described in exercise 11.9. There are two datasets: `ecls_child.dta` contains the student-level data and `ecls_school.dta` contains the school-level data.

As described in exercise 11.9, the response variable, math proficiency, is the highest of five developmental milestones reached, where it is assumed that a given milestone can be reached only after reaching the previous milestone. Although these data are not survival data, it makes sense to fit a continuation-ratio logit model because children are assumed to progress through the stages sequentially. However, instead of modeling the odds of stopping at a given milestone given that it has been reached, as in discrete-time survival, it makes more sense to model the odds of progressing to the next milestone given that the milestone has been reached. The data expansion will be the same as discussed in this chapter, but the response variable will be 1 for every milestone where the child did *not* stop (analogous to censoring) and 0 for the highest milestone the child reached (analogous to death). Because 5 is the highest level that can be reached, everyone with data for that level in the expanded dataset will have a 0 response. In the expanded data, there should therefore not be a row of data for level 5.

The first three steps of the exercise below are identical to the first three steps of exercise 11.9.

1. Use the `merge` command to combine the two datasets.
2. Create a numeric school identifier from the string variable `s4_id`.
3. Summarize the student-level and school-level variables, treating school as the unit of analysis when summarizing school-level variables.
4. Fit a continuation-ratio logit model to the math proficiency data with `numrisks`, `nbhoodcl`, and `pubpriv2` as covariates. Because `profmath` takes the value 0 for only nine children, merge the lowest two categories before fitting the model. See text above for how to expand the data and define the response variable.
5. Interpret the estimates.

15 Continuous-time survival

15.1 Introduction

In this chapter, we consider continuous survival data or duration data where an event (such as death) can occur at any instant in continuous time and survival times can be viewed as recorded on a continuous scale. In contrast, survival data are viewed as discrete when the event can occur only at specific time points (as in the number of menstrual cycles until conception) or when the time units are coarse (as in the number of semesters until dropout from university). In theory, a continuous scale implies that there is a zero probability of ties, where several units have identical survival times. In practice, the survival times are of course rounded, for instance, to seconds, hours, or days, depending on the application. However, the time units must be sufficiently small for the probability of ties to be small. If ties are common, the data are typically treated as discrete-time survival data and the methods discussed in the previous chapter are used.

Basic features of survival data, such as censoring, truncation, delayed entry, and time-varying covariates, were discussed in the *Introduction to models for survival or duration data (part VII)*. We assume that you have read this introduction before embarking on this chapter.

15.2 What makes marriages fail?

We will use data from the 1968–1990 waves of the Panel Study of Income Dynamics (PSID) to introduce survival modeling. The PSID is the world's longest running household panel survey, starting in 1968 with a sample of approximately 5,500 U.S. individuals and their families. The household members and their children were followed up even if they moved out of the original household. New households formed by sample members were also followed and interviewed in the same way as the original households. The sample members were surveyed annually.

The dataset is from the aML manual (Lillard and Panis 2003) and is part of the data used by Lillard and Panis (1996) that included all respondents who survived and remained in the panel through 1984. In 1985, the PSID collected special retrospective marital and fertility histories of both the household heads and their spouses and continually updated the information since then. This information was available on 95% of the sample. For the remaining respondents, Lillard and Panis (1996) constructed marriage histories using 1968–1990 panel information, including more highly detailed questions in 1968 (for the male head) and 1976 (for the wife).

The variable of main interest for us is the duration of the respondents' first marriage until divorce or censoring. Censoring can occur because individuals drop out of the study, because of death of a spouse, or because the first marriage is intact at the last interview. If divorced, the respondents were asked in which month of which year they were divorced. However, in some cases, the recall was less precise and the variables `lower` and `upper` provide the lower and upper bounds of the time interval in which divorce occurred. For right-censored durations, the lower and upper time bounds are identical. To produce a continuous time variable when the bounds were different, we simulated random numbers from uniform distributions on the intervals.

The sample includes only individuals who were married at least once. The unit of analysis is the couple, but this need not be a couple that is cohabiting at any of the panel waves. For divorcees, the couple would be themselves and their spouse from their first marriage.

The file `divorce.dta` contains the following variables for 3,371 couples:

- `id`: identification number for couple
- `lower`: lower bound for time from wedding to divorce in years
- `upper`: upper bound for time from wedding to divorce in years
- `dur`: time from wedding to divorce or censoring in years; made continuous if `lower` ≠ `upper` by simulating from a uniform distribution on [`lower`,`upper`)
- `divorce`: dummy variable for marriage ending in divorce (1: divorce; 0: censored)
- `hiseduc`: husband's education in years of schooling
- `hereduc`: wife's education in years of schooling
- `heblack`: dummy variable for husband being black (1: black; 0: nonblack)
- `sheblack`: dummy variable for wife being black (1: black; 0: nonblack)
- `agediff`: age difference between husband and wife in years (positive value means husband is older)

We start by reading in the divorce data:

```
. use http://www.stata-press.com/data/mlmus3/divorce
```

Following Lillard and Panis (2003), we then construct several dummy variables that will be used as covariates: `mixrace` for marriage being between spouses of different races,

```
. generate mixrace = (heblack & !sheblack) | (sheblack & !heblack)
```

`hedropout` for husband having less than 12 years of education,

```
. generate hedropout = hiseduc<12
```

`hecollege` for husband having 16 years or more of education,

> . generate hecollege = hiseduc>=16 if hiseduc<.

`heolder` for husband being 10 years or more older than his wife,

> . generate heolder = agediff > 10 if agediff<.

and `sheolder` for wife being 10 years or more older than her husband,

> . generate sheolder = agediff< -10

15.3   Hazards and survival

We assume that the events are *absorbing* in the sense that once an event has occurred for a unit, the unit is no longer at risk for this event. For instance, after break up of a first marriage, the first marriage can no longer break up (although subsequent marriages can of course break up).

The time or duration from becoming at risk of experiencing the event is called the *survival time* and is denoted t. The survival time t may be regarded as a realization of a continuous random variable T with a cumulative density function $F(t)$ and probability density function $f(t)$. In the divorce dataset, the observed time from the first wedding to divorce for a particular couple i is denoted t_i (its value is unknown when `divorce=0`).

The start of the observation period coincides with the time of becoming at risk of divorce (time of wedding) because respondents report the wedding and divorce dates of their first marriage retrospectively. If, instead of considering the break up of the first marriage, the survey had focused on break up of the marriage that couples were in at the first panel wave in 1968, we would have delayed entry. This means that all couples who married before 1968 have been at risk of divorce before entering the study. Couples who divorced before 1968 would not have been included in the dataset (unless they have remarried). As discussed in section 14.2.6, such a design is known to suffer from length-biased sampling.

An obvious quantity of interest in survival analysis is the probability of surviving to time t or beyond, the *survival function* $S(t)$, which is given by

$$S(t) \equiv \Pr(T \geq t) \ = \ 1 - F(t) \tag{15.1}$$

where $F(\cdot)$ is a cumulative density function.

A further function of interest for survival data is the *hazard function, intensity function*, or *incidence rate* $h(t)$. This represents the instantaneous risk of the event per unit of time, given that the event has not yet occurred.

To obtain a formal definition of the hazard, consider an event occurring in a time interval from t to $t + \Delta$ (where Δ is positive), that is, $t \leq T < t + \Delta$. If we want the probability of the event occurring in the interval among those still at risk of experiencing the event, we must condition on the event not yet having occurred at time t, that is,

$T \geq t$. The conditional probability of the event occurring in the time interval given that the event has not yet occurred is $\Pr(t \leq T < t + \Delta | T \geq t)$. The hazard is this probability divided by the length of the time interval Δ when the time interval becomes tiny:

$$h(t) \equiv \lim_{\Delta \to 0} \left\{ \frac{\Pr(t \leq T < t + \Delta | T \geq t)}{\Delta} \right\} \tag{15.2}$$

Note that the hazard is a *rate*, expressed in units of 1 over time, for example, per second or per month.

The hazard function (15.2) can alternatively be expressed as the density function $f(t)$ divided by the survival function $S(t)$:

$$h(t) = \frac{f(t)}{S(t)}$$

The survival function (15.1) can alternatively be expressed as

$$S(t) = \exp\{-H(t)\} = \exp\{-\underbrace{\int_0^t h(u)\,du}_{H(t)}\} \tag{15.3}$$

where $H(t)$ is the *integrated hazard function* or *cumulative hazard function*. Display 15.1 derives these relationships, but we invite you to skip these derivations because they involve calculus and are somewhat more technical than most of the book.

1. Demonstrating that $h(t) = f(t)/S(t)$:

Because $\Pr(A|B) = \Pr(A, B)/\Pr(B)$, we obtain

$$\Pr(t \leq T < t + \Delta | T \geq t) = \frac{\Pr(t \leq T < t + \Delta, T \geq t)}{\Pr(T \geq t)} = \frac{\Pr(t \leq T < t + \Delta)}{\Pr(T \geq t)}$$

Using the definitions of cumulative density function and survival function, we see that the numerator and denominator become

$$\Pr(t \leq T < t + \Delta) = F(t + \Delta) - F(t)$$

and

$$\Pr(T \geq t) = S(t)$$

Then from (15.2),

$$
\begin{aligned}
h(t) &= \lim_{\Delta \to 0} \left\{ \frac{\Pr(t \leq T < t + \Delta | T \geq t)}{\Delta} \right\} \\
&= \lim_{\Delta \to 0} \left[\frac{\{F(t + \Delta) - F(t)\}/S(t)}{\Delta} \right] \\
&= \lim_{\Delta \to 0} \left\{ \frac{F(t + \Delta) - F(t)}{\Delta} \right\} \frac{1}{S(t)}
\end{aligned}
$$

Because the density function is the derivative of the cumulative density function,

$$f(t) = \frac{\partial F(t)}{\partial t} \equiv \lim_{\Delta \to 0} \left\{ \frac{F(t + \Delta) - F(t)}{\Delta} \right\}$$

it follows that

$$h(t) = \frac{f(t)}{S(t)}$$

2. Demonstrating that $S(t) = \exp\{-H(t)\}$:

By the chain rule, we obtain

$$\frac{\partial \ln\{S(t)\}}{\partial t} = \frac{1}{S(t)} \frac{\partial S(t)}{\partial t} = \frac{1}{S(t)} \frac{\partial [1 - F(t)]}{\partial t} = -\frac{f(t)}{S(t)} = -h(t)$$

Integrating both sides, we get

$$\ln\{S(t)\} = \int_0^t -h(u) \, du = -H(t)$$

and exponentiating both sides, it follows that

$$S(t) = \exp\{-H(t)\}$$

Display 15.1: Demonstration of two central relations in survival analysis: $h(t) = f(t)/S(t)$ and $S(t) = \exp\{-H(t)\}$

Returning to the divorce data, we start by listing the variables `id`, `dur`, and `divorce` for the first four couples:

```
. list id dur divorce in 1/4, clean noobs
    id        dur    divorce
     9     10.546          0
    11     34.943          0
    13   2.870226          1
    15   18.18403          1
```

We see that the marriage of couple 13 lasted 2.87 years before it ended in divorce (`divorce` takes the value 1), whereas the marriage of couple 9 was still intact after 10.55 years when the couple was censored (`divorce` takes the value 0).

To allow us to use Stata's survival time commands (which have the prefix `st` for "survival time"), we now declare or *set* the data to be survival data by using the `stset` command:

```
. stset dur, failure(divorce) id(id)
                 id:   id
      failure event:   divorce != 0 & divorce < .
obs. time interval:    (dur[_n-1], dur]
   exit on or before:  failure
_____

        3371   total obs.
           0   exclusions
_____

        3371   obs. remaining, representing
        3371   subjects
        1032   failures in single failure-per-subject data
    62098.25   total analysis time at risk, at risk from t =          0
                             earliest observed entry t =              0
                               last observed exit t =         73.068
```

The first argument, `dur`, declares that this variable contains the time from wedding to divorce or censoring. The Stata documentation calls time scales that are 0 at the time a unit becomes at risk of the event, here the time of the wedding, *analysis time*. If we had used the calendar time at the divorce, we would have to specify the wedding date in the `origin()` option. The observation period is assumed to start when `dur` is zero. If there had been delayed entry, we would need to use the `enter()` option to specify the time observation began on the `dur` time scale. For example, if the survey questions had been about the marriage couples were in at the first panel wave, the `enter()` option would specify a variable containing the number of years couples were already married at the time of the first panel wave.

The option `failure(divorce)` designates `divorce` as the indicator for divorce (and not censoring), and the option `id(id)` designates `id` as the couple identifier. In the output, `failure event:   divorce != 0 & divorce < .` means that the event is assumed to occur when the variable `divorce` does not take the value 0 (hence taking the value 1 for divorce) and is not missing. We see that there are 3,371 couples in the dataset and that the marriages of 1,032 of these couples ended in divorce (rather aptly called "failures" in the output). The total time at risk for the couples is 62,098.25 years, and the longest duration (to divorce or censoring) observed is 73.068 years.

We can estimate the survival function $S(t)$ for the divorce data by using the *Kaplan–Meier estimator* or *product limit estimator* described in display 15.2.

Order the n survival times (all times that end in the event or failure instead of censoring) such that $t_{(1)} \leq t_{(2)} \cdots \leq t_{(n)}$ where $t_{(k)}$ is the kth largest unique survival time. This defines $n-1$ intervals $(t_{(2)} - t_{(1)}), \ldots, (t_{(n)} - t_{(n-1)})$.

The Kaplan–Meier estimator $\widehat{S}(t)$ of the survival function $S(t)$ is defined as

$$\widehat{S}(t) \;=\; \prod_{k \,|\, t_{(k)} \leq t} \left(1 - \frac{d_k}{r_k} \right)$$

where the product is over all intervals k that end before time t. Here r_k is the number of subjects at risk just before $t_{(k)}$, and d_k is the number who experience the event at time $t_{(k)}$. The ratio d_k/r_k can therefore be interpreted as the estimated probability of experiencing the event at the end of the kth interval, given that the subject is still at risk.

For example, the survival function at the second event time $t_{(2)}$ is equal to the estimated probability of not experiencing the event at time $t_{(1)}$ times the estimated probability of experiencing the event at time $t_{(2)}$, given that the subject is still at risk at $t_{(2)}$.

The form of the Kaplan–Meier estimator is the same as for the life-table estimator for the discrete-time survival function in equation (14.1). The difference is that the intervals are now defined by splitting the analysis time at each of the unique event or failure times in the data.

Display 15.2:   Kaplan–Meier estimator of survival function

It is convenient to use the `sts graph` command to produce a plot of the Kaplan–Meier estimates $\widehat{S}(t)$ of the survival function, called a *Kaplan–Meier plot*:

```
. sts graph

        failure _d:  divorce
  analysis time _t:  dur
               id:  id
```

The resulting graph is shown in figure 15.1, where we see that about 10% of the marriages break up within the first 5 years (survival function is at about 0.9), 20% of the marriages break up within the first 10 years, and 35% of the marriages break up within the first 20 years. It appears that about 55% of the marriages are never dissolved in these data.

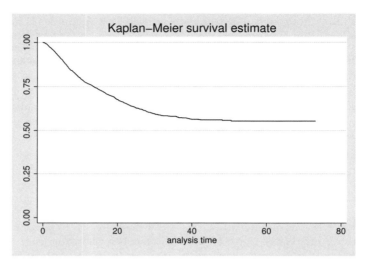

Figure 15.1: Kaplan–Meier survival plot of $\widehat{S}(t)$ for divorce data

We can produce a graph of the estimated hazard function $\widehat{h}(t)$ (using kernel smooth-ing) by including the **hazard** option in the **sts graph** command:

```
. sts graph, hazard
        failure _d:  divorce
   analysis time _t:  dur
              id:  id
```

The resulting graph is shown in figure 15.2.

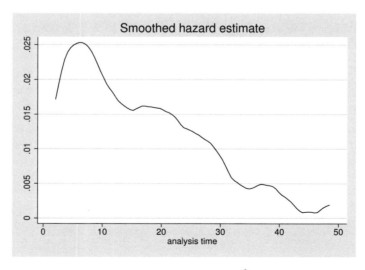

Figure 15.2: Smoothed hazard estimate $\widehat{h}(t)$ for divorce data

The hazard is quite high at the beginning and increases until it reaches a peak at about 7 years of marriage. It is tempting to see this as evidence for the "seven-year itch", famous from a 1955 movie with that title featuring Marilyn Monroe, because the hazard of a given marriage failing apparently increases until 7 years, after which the hazard decreases. However, this hazard function represents the *marginal* or *population-averaged* hazard and does not necessarily reflect the couple-specific hazard for any couple. Later in the chapter (sections 15.9 and 15.10), we discuss models for subject-specific hazards.

15.4   Proportional hazards models

In proportional hazards (PH) models with one covariate x_i, the hazard given the covariate $h(t|x_i)$ is specified as

$$h(t|x_i) = h_0(t)\exp(\beta x_i) \tag{15.4}$$

where $h_0(t)$ is the *baseline hazard*, the hazard when $x_i = 0$.

The *hazard ratio* or *incidence-rate ratio* associated with a unit increase in x_i, from a value a to $a + 1$, becomes

$$\frac{h(t|x_i = a + 1)}{h(t|x_i = a)} = \frac{h_0(t)\exp\{\beta(a+1)\}}{h_0(t)\exp\{\beta(a)\}} = \exp(\beta)$$

Because the hazard ratio is constant over time, the hazard functions of different individuals are proportional, and that is the reason why the model is called a proportional hazards model.

Taking the logarithm on both sides of (15.4), the PH model can alternatively be written as a log-linear model for the hazard

$$\ln\{h(t|x_i)\} = \ln\{h_0(t)\} + \beta x_i$$

The survival function for the PH model is

$$S(t|x_i) \equiv \Pr(T_i > t|x_i) = S_0(t)^{\exp(\beta x_i)}$$

Display 15.3 derives this relationship, but you can skip this if you like.

We start by restating the relation between the survival function $S(t)$ and the cumulative hazards function $H(t)$ given in (15.3)

$$S(t) = \exp\{-H(t)\} = \exp\{-\int_0^t h(u)\,du\}$$

Substituting the proportional hazards model (15.4) for the hazard $h(u)$, we get

$$S(t) = \exp\{-\int_0^t h_0(u)\exp(\beta x_i)\,du\}$$

$$= \exp[\{-\int_0^t h_0(u)\,du\}\exp(\beta x_i)]$$

$$= \exp\{-\int_0^t h_0(u)\,du\}^{\exp(\beta x_i)}$$

Using the definition of the cumulative hazards function and the relation between the survival and cumulative hazards function in (15.3), which also apply to baseline hazards, it follows that

$$S(t) = \exp\{-H_0(t)\}^{\exp(\beta x_i)}$$

$$= S_0(t)^{\exp(\beta x_i)}$$

Display 15.3: Demonstration of $S(t|x_i) = S_0(t)^{\exp(\beta x_i)}$ for proportional hazards model

The simplest and most restrictive assumption for the baseline hazard $h_0(t)$ is that it is constant, $h_0(t) = \lambda$. The corresponding distribution for the survival time is the exponential distribution with mean $1/\lambda$. For example, if the hazard rate is two per day, the average time to the event is half a day. Because the hazard rate λ must be nonnegative, it is convenient to write it as $\lambda = \exp(\alpha_0)$ where there is no constraint on α_0. Note that we use α's instead of β's to refer to coefficients that determine the baseline hazard function $h_0(t)$.

Table 15.1 shows the form of the baseline hazard for the exponential model as well as two more general proportional-hazards models: the Weibull and Gompertz models.

Table 15.1: Parametric proportional hazards models: Name of density $f(t)$, form of baseline hazard function $h_0(t)$, and parameters

$f(t)$	$h_0(t)$	Parameters
Exponential	$\exp(\alpha_0)$	α_0
Weibull	$pt^{p-1}\exp(\alpha_0)$	α_0, p "shape"
Gompertz	$\exp(\gamma t)\exp(\alpha_0)$	α_0, γ "shape"

It is clear that the exponential model is a special case of the Weibull model with $p = 1$ and a special case of the Gompertz model with $\gamma = 0$. The exponential model has a constant baseline hazard, and the Weibull and Gompertz models have monotonic baseline hazards. The hazard of the Weibull model increases with time if $p > 1$ and decreases if $p < 1$, and the hazard of the Gompertz model increases with time if $\gamma > 0$ and decreases if $\gamma < 0$. None of the models therefore appear to be suitable for the divorce data because the smoothed hazard function in figure 15.2 has a peak.

One way to make the parametric proportional hazards models more flexible is to divide the time scale into intervals, each of which has different parameters for the baseline hazard. The most common approach is to construct a piecewise constant baseline hazard $h_0(t)$ by assuming an exponential distribution with different values of α_s for different time intervals. Such a piecewise exponential model is discussed in the next section.

15.4.1   Piecewise exponential model

When using a piecewise exponential model, we have to decide for which intervals or "pieces" we are going to assume that the hazard is constant. Somewhat informed by the shape of the estimated hazard function in figure 15.2, we will use the six intervals (0,1], (1,5], (5,9], (9,15], (15,25], and (25,). Here "(" means that the left-hand-side number is not included in the interval, whereas "[" means that it is included and similarly for the right-hand-side number ")" and "]". So the interval (0,1] does not include 0 (it does however include numbers slightly larger than 0) but does include 1.

We first use the `stptime` (for "person-time") command to estimate the hazard for each of the chosen intervals:

```
. stptime, at(0,1,5,9,15,25)
            failure _d:  divorce
      analysis time _t:  dur
                   id:   id

    Cohort      |  person-time    failures        rate    [95% Conf. Interval]
    ------------+--------------------------------------------------------------
    (0   -   1] |    3337.4472          30    .00898891    .0062849    .0128562
    (1   -   5] |   12023.083          270    .0224568     .0199318    .0253017
    (5   -   9] |    9816.07           251    .02557031    .0225948    .0289377
    (9   -  15] |   11438             210    .01835985    .0160373    .0210188
    (15  -  25] |   12509.839          193    .01542786    .0133978    .0177655
          >  25 |   12973.809           78    .00601211    .0048156    .007506
    ------------+--------------------------------------------------------------
         total  |   62098.248         1032    .01661883    .0156352    .0176643
```

The total number of couples at risk at year 0 was 3,371 (the number of couples in the dataset). However, we see from the column called `person-time` that the first-year interval (0,1] contains only 3,337.45 person-years. The difference is due to couples being removed from the risk set because their marriages lasted less than a year or due to censoring within the first year of marriage.

For each time interval, the number of divorces during the time interval is given under `failures`, and the estimated hazard is given under `rate`. For instance, the hazard of divorce in the interval (5,9] is estimated as the number of divorces divided by the number of person-years within the interval, $0.02557031 = 251/9816.07$. We see that the estimated hazards of divorce follow the same general pattern as seen in the plot of the estimated hazard function in figure 15.2, peaking in the interval (5,9]. The last two columns contain the lower and upper limits for the estimated 95% confidence intervals for the hazards. If we had instead used the `ltable` command (with the `hazard` and `noadjust` options) discussed for discrete-time survival in chapter 14, we would have obtained estimates of the hazards that do not take into account that the risk set is gradually depleted within each interval.

To fit a piecewise exponential survival model, we use the `stsplit` command to expand the data according to the chosen intervals, producing one row of data or one record for each interval a couple is at risk for divorce. Thereafter, we sort and list the data for the first four couples as before.

```
. stsplit interval, at(1,5,9,15,25)
(10754 observations (episodes) created)
. sort id _t
. list id interval _t0 _t _d if id<34, sepby(id) noobs
```

id	interval	_t0	_t	_d
9	0	0	1	0
9	1	1	5	0
9	5	5	9	0
9	9	9	10.546	0
11	0	0	1	0
11	1	1	5	0
11	5	5	9	0
11	9	9	15	0
11	15	15	25	0
11	25	25	34.943001	0
13	0	0	1	0
13	1	1	2.8702259	1
15	0	0	1	0
15	1	1	5	0
15	5	5	9	0
15	9	9	15	0
15	15	15	18.184032	1
33	0	0	1	0
33	1	1	1.418	0

We see that for each couple, there is one record for each interval they are at risk for divorce. For instance, couple 9 has four records because censoring occurred in the fourth interval, whereas couple 15 has five records because a divorce occurred in the fifth interval. Each record in the expanded dataset hence represents a part of the history for a couple, called an *episode* in Stata.

The `stsplit` command has split each couple's history into episodes and created variables _t_0 and _t for the start and end times of the episodes. All episodes except the final episode for a couple end at one of the predetermined cutpoints (1, 5, 9, 15, 25), and the couple can be viewed as censored at these times (because nothing happened at these times—we just split up the couple's history there). The variable _d takes the value 0 for censoring and 1 for divorce. The variable `interval` keeps track of the intervals and takes the value of the lower bound of each interval (here it is equal to _t0 because there is no delayed entry). The data expansion is the same as for discrete-time survival discussed in section 14.2.2 (where the intervals here correspond to the discrete times there) except that we keep track of the exact timing of divorces and censoring within each interval in the variable _t. Stata interprets the expanded data in exactly the same way as it interpreted the original data. All `st` commands will produce the same output before and after the data expansion.

To understand the expanded data better, we produce a summary table. For each interval, there should be as many records as there were couples at risk of divorce (that is, not yet divorced) at the start of the interval. The sum of the time segments _t−_t0 across all couples represented in a given interval should equal the person-years of observation for that interval, and _d should take the value 1 for each couple that got divorced during the interval. We can use the `table` command to tabulate these quantities:

```
. generate pers_yr = _t-_t0
. table interval, contents(freq sum pers_yr sum _d)
```

interval	Freq.	sum(pers_yr)	sum(_d)
0	3,371	3337.447	30
1	3,283	12023.08	270
5	2,728	9816.07	251
9	2,205	11438	210
15	1,627	12509.84	193
25	911	12973.81	78

As expected, the last two columns agree with the first two columns produced by the `stptime` command on page 807; namely, `sum(pers_yr)` corresponds to `person_time` and `sum(_d)` corresponds to `failures`.

The reason for expanding the data was to allow us to include dummy variables for the time bands or intervals in a piecewise exponential survival model. Letting the index $s = 1, \ldots, 6$ denote the intervals in the expanded data, we specify the following model:

$$\ln\{h(t|\mathbf{d}_{si})\} = \alpha_1 d_{1si} + \alpha_2 d_{2si} + \cdots + \alpha_6 d_{6si} \tag{15.5}$$

where $\mathbf{d}_{si} = (d_{1si}, \ldots, d_{6si})'$ are dummy variables for intervals 1 to 6, the intervals starting at 0, 1, 5, 9, 15, and 25, with corresponding coefficients α_1 to α_6.

Using `ibn.interval` to include dummy variables for all the intervals and the option `noconstant` to suppress the intercept, the `streg` command with the option `distribution(exponential)` fits the piecewise exponential model:

```
. streg ibn.interval, distribution(exponential) noconstant

        failure _d: divorce
  analysis time _t: dur
               id: id

Exponential regression -- log relative-hazard form

No. of subjects =           3371          Number of obs   =      14125
No. of failures =           1032
Time at risk    =     62098.24824
                                          Wald chi2(6)    =   16685.34
Log likelihood  =       -3077.405         Prob > chi2     =     0.0000
```

_t	Haz. Ratio	Std. Err.	z	P>|z|	[95% Conf. Interval]	
interval						
0	.0089889	.0016411	-25.81	0.000	.0062849	.0128562
1	.0224568	.0013667	-62.38	0.000	.0199318	.0253017
5	.0255703	.001614	-58.09	0.000	.0225948	.0289377
9	.0183599	.001267	-57.93	0.000	.0160373	.0210188
15	.0154279	.0011105	-57.95	0.000	.0133978	.0177655
25	.0060121	.0006807	-45.17	0.000	.0048156	.007506

```
note: no constant term was estimated in the main equation
```

We see that the estimated hazards and their confidence intervals are identical to those previously produced by the `stptime` command on page 807 (the label `Haz. Ratio` is misleading here because the exponentiated coefficients are hazards due to omission of an intercept).

The benefit of using `streg` is that it makes it straightforward to include covariates in the analysis. Here we include the covariates `heblack` (x_{2i}), `mixrace` (x_{3i}), `hedropout` (x_{4i}), `hecollege` (x_{5i}), `heolder` (x_{6i}), and `sheolder` (x_{7i}), and let the corresponding coefficients be denoted β_2 to β_7. The piecewise exponential model becomes

$$\ln\{h(t|\mathbf{d}_{si}, \mathbf{x}_i)\} = \underbrace{\alpha_1 d_{1si} + \alpha_2 d_{2si} + \cdots + \alpha_6 d_{6si}}_{\ln\{h_0(t)\}} + \beta_2 x_{2i} + \cdots + \beta_7 x_{7i} \qquad (15.6)$$

where the *baseline hazard* $h_0(t) \equiv \exp(\alpha_1 d_{1si} + \alpha_2 d_{2si} + \cdots + \alpha_6 d_{6si})$ is the hazard when the covariates take the value zero ($\mathbf{x}_i = \mathbf{0}$).

Fitting the model using `streg` with the `distribution(exponential)` option, we obtain

```
. streg ibn.interval heblack mixrace hedropout hecollege heolder sheolder,
> distribution(exponential) noconstant

        failure _d:  divorce
   analysis time _t:  dur
              id:  id
```

Exponential regression -- log relative-hazard form

No. of subjects =	3371	Number of obs	=	14125
No. of failures =	1032			
Time at risk =	62098.24824			
		Wald chi2(12)	=	16560.43
Log likelihood =	-3056.2251	Prob > chi2	=	0.0000

_t	Haz. Ratio	Std. Err.	z	P>|z|	[95% Conf. Interval]	
interval						
0	.0093972	.0017478	-25.09	0.000	.0065264	.0135306
1	.0235649	.0016577	-53.28	0.000	.0205299	.0270485
5	.027068	.0019591	-49.87	0.000	.0234881	.0311935
9	.0196236	.0015249	-50.59	0.000	.0168513	.0228519
15	.0168828	.0013671	-50.40	0.000	.0144051	.0197867
25	.0068689	.0008251	-41.46	0.000	.0054279	.0086923
heblack	1.19882	.0956472	2.27	0.023	1.025277	1.401737
mixrace	1.274274	.1015894	3.04	0.002	1.089939	1.489785
hedropout	.7388078	.0504425	-4.43	0.000	.6462718	.8445934
hecollege	.760166	.0801867	-2.60	0.009	.6181854	.9347557
heolder	.7578042	.1640962	-1.28	0.200	.4957199	1.158451
sheolder	1.633342	.4265919	1.88	0.060	.9789543	2.725158

note: no constant term was estimated in the main equation

The estimated hazard ratios for the six covariates with associated 95% confidence intervals are also reported under "Piecewise exponential" in table 15.2.

Table 15.2: Estimated hazard ratios (HR) for proportional hazards (PH) models and time ratio (TR) for accelerated failure-time (AFT) model with associated 95% confidence intervals

| | PH | | | | | AFT | |
| | Piecewise exponential | | Cox | | Poisson cubic spline | | Log normal | |
	HR	(95% CI)	HR	(95% CI)	HR	(95% CI)	TR	(95% CI)
heblack	1.20	(1.03, 1.40)	1.19	(1.02, 1.39)	1.19	(1.02, 1.39)	0.87	(0.72, 1.05)
mixrace	1.27	(1.09, 1.49)	1.27	(1.08, 1.48)	1.27	(1.08, 1.48)	0.73	(0.60, 0.89)
hedropout	0.74	(0.65, 0.84)	0.75	(0.65, 0.85)	0.75	(0.65, 0.85)	1.39	(1.18, 1.63)
hecollege	0.76	(0.62, 0.93)	0.76	(0.62, 0.93)	0.76	(0.62, 0.93)	1.41	(1.10, 1.79)
heolder	0.76	(0.50, 1.16)	0.77	(0.50, 1.18)	0.77	(0.50, 1.18)	1.14	(0.73, 1.79)
sheolder	1.63	(0.98, 2.73)	1.64	(0.98, 2.74)	1.64	(0.98, 2.73)	0.53	(0.28, 0.98)
Log likelihood	−3,056.23		−7,821.91†		−9,442.73		−3,077.61	

†Partial log likelihood

Controlling for the other covariates, the estimated hazard ratios suggest that the hazard of marriage dissolution is about 20% [= 100%(1.19882 − 1)] higher for black husbands than for white husbands and 27% [= 100%(1.274274 − 1)] higher for mixed couples than for nonmixed couples. The model implies that couples where the husband is black and the wife is white have a 53% [= 100%(1.19882 × 1.274274 − 1)] higher hazard than white couples (see table 15.3 for hazard ratios comparing each type of couple with white couples).

		mixrace	
		0	1
heblack	0	both white 1	he white & she black 1.27
	1	both black 1.20	he black & she white 1.20 × 1.27 = 1.53

Table 15.3: Estimated hazards ratios for combinations of the spouses' race (both spouses white as reference category and adjusted for other covariates)

For a husband having less than 12 years of education, the hazard of marriage dissolution is about 26% lower [−26% = 100%(0.7388078 − 1)] than for the reference category (12 to 16 years of education), and the hazard is 24% lower [−24% = 100%(0.760166 − 1)] for husband having more than 16 years of education compared with the reference, controlling for the other covariates. Still controlling for the other covariates, the hazard is estimated as 24% *lower* [−24% = 100%(0.7578042 − 1)] if the husband is more than 10 years older than the wife than if he is not (not significant at the 5% level), whereas the hazard is 63% [= 100%(1.633342 − 1)] *higher* if the wife is more than 10 years older than the husband than if she is not (not significant at the 5% level).

We now plot the estimated piecewise constant baseline hazard function, which was omitted from the model. The postestimation command stcurve cannot be used to accomplish this because Stata interprets the baseline hazard to be just the exponential of the intercept $\exp(\widehat{\alpha}_1)$. Therefore we calculate the estimated baseline hazard ourselves from the estimated coefficients:

```
. generate h0 = exp(_b[0.interval]*(interval==0) + _b[1.interval]*(interval==1)
> + _b[5.interval]*(interval==5) + _b[9.interval]*(interval==9)
> + _b[15.interval]*(interval==15) + _b[25.interval]*(interval==25))
```

(An alternative method for making predictions with a subset of coefficients that requires less typing uses the matrix score command as described on page 821.) We can then produce a graph using the twoway command:

```
. twoway line h0 _t, connect(J) sort xtitle(Time since wedding in years)
> ytitle(Baseline hazard)
```

Here the connect(J) option (synonymous with the connect(stairstep) option) connects the estimated hazards using a step function, because the hazard at the previous time point persists until the next time point. The resulting graph is shown in figure 15.3.

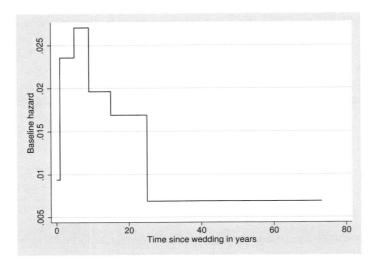

Figure 15.3: Piecewise constant baseline hazard curve

Because the baseline hazard represents the hazard when $\mathbf{x}_i = \mathbf{0}$, it is the hazard for a couple where the husband is white and has between 12 and 16 years of education, the spouses have the same race (that is, the wife is also white), and the age difference is less than 10 years. The estimated baseline hazard from the piecewise exponential model has a similar shape to the smoothed curve in figure 15.2 but that figure showed the marginal hazard, not conditioning on covariate values.

The piecewise exponential model can alternatively be fit using a Poisson regression model (see chapter 13) with the logarithm of the time at risk or exposure (the length of the interval) as an offset (a covariate without a coefficient). The model is

$$\ln(\mu_{si}) \;=\; \ln(t_{si}) + \alpha_1 + \alpha_2 d_{2si} + \cdots + \alpha_6 d_{6si} + \beta_2 x_{2i} + \cdots + \beta_7 x_{7i}$$

where μ_{si} is the mean parameter of the Poisson distribution and t_{si} is the time at risk in interval s for couple i. It may appear strange to use a model for counts $(0,1,\dots)$ in this context because the event can only happen once in each episode. However, we could count the number of events that occur for each combination of interval and covariate values, and Poisson regression on such aggregated data would yield the same results as shown in section 13.3. Using Poisson regression in this way to estimate a piecewise exponential survival model is useful when random effects are introduced in the survival model (because `streg` can fit random-intercept models only, whereas `xtmepoisson` and `gllamm` can fit models with random coefficients).

We first construct the offset containing the log time at risk,

```
. generate lexposure = ln(_t - _t0)
```

and fit the Poisson regression model with the offset() option to include the offset and the irr option to express parameter estimates for the covariates as incidence-rate ratios:

```
. poisson _d ibn.interval heblack mixrace hedropout hecollege heolder sheolder,
> offset(lexposure) noconstant irr
```

Poisson regression				Number of obs	=	14125
				Wald chi2(12)	=	16560.43
Log likelihood = -4541.2426				Prob > chi2	=	0.0000

_d	IRR	Std. Err.	z	P>|z|	[95% Conf. Interval]	
interval						
0	.0093972	.0017478	-25.09	0.000	.0065264	.0135306
1	.0235649	.0016577	-53.28	0.000	.0205299	.0270485
5	.027068	.0019591	-49.87	0.000	.0234881	.0311935
9	.0196236	.0015249	-50.59	0.000	.0168513	.0228519
15	.0168828	.0013671	-50.40	0.000	.0144051	.0197867
25	.0068689	.0008251	-41.46	0.000	.0054279	.0086923
heblack	1.19882	.0956472	2.27	0.023	1.025277	1.401737
mixrace	1.274274	.1015894	3.04	0.002	1.089939	1.489785
hedropout	.7388078	.0504425	-4.43	0.000	.6462718	.8445934
hecollege	.760166	.0801867	-2.60	0.009	.6181854	.9347557
heolder	.7578042	.1640962	-1.28	0.200	.4957199	1.158451
sheolder	1.633342	.4265919	1.88	0.060	.9789543	2.725158
lexposure	1	(offset)				

As you can see, the parameter estimates and the estimated standard errors are identical to those produced by the streg command above.

Before discussing Cox regression in the next section, we return to the original non-expanded data format by using the stjoin command, after first removing the variables interval, lexposure, and h0 that vary within couples between the episodes:

```
. drop interval pers_yr lexposure h0
. stjoin
(option censored(0) assumed)
(10754 obs. eliminated)
```

15.4.2   Cox regression model

A disadvantage of parametric proportional hazards models, such as the exponential, Weibull, or Gompertz models, is that a parametric form must be assumed for the baseline hazard $h_0(t)$.

We showed in the previous section that the parametric form can, at least partly, be relaxed by using a piecewise parametric model with different values of the parameters in each interval. The narrower the intervals, the more flexible the model becomes, and the less we are smoothing the data. Taking this idea to the extreme, we can let the intervals be so small that each interval contains only one event. This can be accomplished by defining intervals to start just after each failure that occurs in the data and end just after the next failure (see also display 15.2 on the Kaplan–Meier estimator). Such intervals capturing a single event are sometimes referred to as *clicks*.

Specifying a piecewise exponential model such as (15.6) with a dummy variable for each click s corresponds to the Cox regression model,

$$\ln\{h(t|\mathbf{x}_i)\} = \ln\{h_0(t)\} + \beta_2 x_{2i} + \cdots + \beta_7 x_{7i}$$

which can equivalently be expressed as

$$h(t|\mathbf{x}_i) = h_0(t) \exp(\beta_2 x_{2i} + \cdots + \beta_7 x_{7i})$$

The model is sometimes described as *semiparametric* because no parametric shape is assumed for the baseline hazard $h_0(t)$, whereas the covariate effects are parametric.

We first demonstrate the `stcox` command, which fits the model without any need to expand the data. We then show how to expand the data and fit the same model using Poisson regression. This is useful for fitting random-effects models that are not accommodated by the `stcox` command (see section 15.9). The appropriate `stcox` command is

```
. stcox heblack mixrace hedropout hecollege heolder sheolder

        failure _d:  divorce
  analysis time _t:  dur
             id:  id

Cox regression -- no ties

No. of subjects =          3371            Number of obs   =        3371
No. of failures =          1032
Time at risk    =   62098.24824
                                           LR chi2(6)      =       40.46
Log likelihood  =    -7821.9073            Prob > chi2     =      0.0000
```

_t	Haz. Ratio	Std. Err.	z	P>|z|	[95% Conf. Interval]	
heblack	1.192658	.095109	2.21	0.027	1.020085	1.394426
mixrace	1.266902	.1009307	2.97	0.003	1.083752	1.481003
hedropout	.7458869	.0509019	-4.30	0.000	.6525054	.8526325
hecollege	.7594514	.080115	-2.61	0.009	.6175985	.9338858
heolder	.769728	.166686	-1.21	0.227	.5035098	1.176702
sheolder	1.640293	.4284575	1.89	0.058	.9830622	2.73692

The estimated hazard ratios were also reported under "Cox" in table 15.2 on page 812. We see that the estimates are nearly identical to those for the piecewise exponential model.

The log likelihood in the output is actually a *partial log likelihood* (see display 15.4) where the coefficients of the dummy variables for the clicks are viewed as nuisance parameters and have been eliminated. The output also mentions that there are no ties because ties pose a challenge in Cox regression and can be handled in different ways as briefly described in display 15.4.

Partial likelihood

The partial likelihood that is maximized with respect to $\boldsymbol{\beta}$ is given by

$$\prod_f \frac{\exp(\mathbf{x}'_f\boldsymbol{\beta})}{\sum_{i \in R(f)} \exp(\mathbf{x}'_i\boldsymbol{\beta})}$$

where the product is over all failures f, and the summation in the denominator is over all subjects who are still at risk at the time of failure, the members of the risk set $R(f)$.

The term in the product represents the conditional probability that the event happens to a subject with covariates $\mathbf{x}_f$ given that it has happened to one of the subjects still at risk [a member of the risk set $R(f)$]. The partial likelihood is equivalent to the conditional likelihood for conditional logistic regression for discrete choice (see display 12.2), with risk sets corresponding to alternative sets and failures corresponding to the chosen alternatives.

The partial likelihood can also be viewed as a profile likelihood (that is, the likelihood in which the baseline hazard parameters have been replaced by functions of $\boldsymbol{\beta}$ that maximize the likelihood for fixed $\boldsymbol{\beta}$).

The partial likelihood estimator does not depend on the exact timing of the events but only on their ranking or chronological order (because time does not appear in the partial likelihood).

Ties

When several subjects experience the event at exactly the same time, we have ties.

By default, stcox uses *Breslow's method* for ties, where the risk sets $R(f)$ above contain all subjects who failed at or after the failure time of the subject contributing to the numerator. For a group of subjects with tied survival times, the contributions to the partial likelihood therefore all have the same denominator.

Because risk sets usually decrease by one after each failure, *Efron's method* downweights contributions to the risk set from the subjects with tied failure times in successive risk sets. Efron's method is requested by using the efron option of stcox.

In the *exact method*, the contribution to the partial likelihood from a group of tied survival times is the sum, over all possible orderings (or permutations) of the tied survival times, of the contributions to the partial likelihood corresponding to these orderings. The exact method is obtained by using the exactp option of stcox (the exactm option is an approximation to the truly exact method).

Display 15.4: Partial likelihood and ties in Cox regression

We can plot a kernel smoothed version of estimated hazard functions from Cox regression for different covariate values using the `stcurve` command with the `hazard` option. For instance, to plot separate curves for `sheolder` equal to 0 and 1, and with the other covariates evaluated at their mean, we use

```
. stcurve, hazard xtitle(Time since wedding in years) at1(sheolder=0)
> at2(sheolder=1)
```

The resulting graph is shown in figure 15.4.

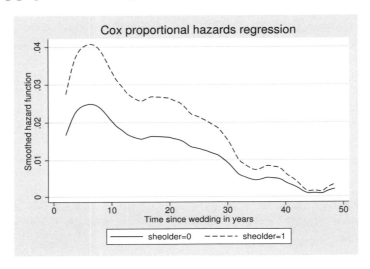

Figure 15.4: Kernel smoothed hazard curves from Cox regression

The model-implied curves are proportional, the curve for `sheolder=1` being a constant multiple of the curve for `sheolder=0`.

We also want to save the kernel-smoothed estimated baseline hazard for later use, which requires that we explicitly set all covariates to 0 in the `at()` option and that we save the predictions in a file by using the `outfile()` option (predictions cannot be saved in the data because the values for _t required to make a nice graph are not necessarily the same as the values of _t in the data),

```
. stcurve, hazard xtitle(Time since wedding in years)
> at(heblack=0 mixrace=0 hedropout=0 hecollege=0 heolder=0 sheolder=0)
> outfile(temp, replace)
```

(graph not shown).

15.4.3   Poisson regression with smooth baseline hazard

Now we demonstrate how the Cox regression model can be fit using Poisson regression by expanding the data with the `stsplit` command. This data expansion will allow us to model the baseline hazard using a smooth function such as a spline or polynomial.

Each divorce that occurred corresponds to an interval or click. For each click, the new dataset will contain a row of data (or episode) for the index couple that got divorced and all the couples that stayed married until the next click. This collection of records is called the *risk set* because it is the set of couples who were at risk of divorce at the time the index couple got divorced. We saw from the `stset` output that the dataset contains 1,032 divorces, so the expanded data will be huge. In the `stsplit` command, we specify that the limits between the intervals are the divorce times, called `failures`, and that the risk sets should have the identifier `click` (the command will take some time to run and, in Stata 11 or earlier, may require increasing the memory by using `set memory`):

```
. stsplit, at(failures) riskset(click)
(1032 failure times)
(2243134 observations (episodes) created)
```

We see that there are 1,032 unique times of divorce and that 2,243,134 new rows of data have been created!

We now list the data for the first four and last four clicks for the first couple in the dataset (`id=9`):

```
. list id click _t0 _t _d _st if id==9 & (click<5 | click>624), separator(4)
> noobs
```

id	click	_t0	_t	_d	_st
9	1	0	.10110053	0	1
9	2	.10110053	.11268183	0	1
9	3	.11268183	.13671894	0	1
9	4	.13671894	.35028338	0	1
9	625	10.41302	10.443006	0	1
9	626	10.443006	10.458218	0	1
9	627	10.458218	10.500483	0	1
9	.	10.500483	10.546	0	1

The last completed episode for the couple was click 627 with censoring occurring at time 10.546 during click 628. (This last contribution does not contribute to Cox regression.)

At this point, we could attempt to use maximum likelihood to fit a Poisson model with a dummy variable for each of the 1,032 risk sets. However, this is computationally very demanding, so we use an equivalent but less taxing approach based on maximizing the conditional likelihood, where the coefficients for the dummy variables are treated as nuisance parameters (see section 13.11.1).

As before, we first construct the offset, the logarithm of the time at risk for each risk set:

```
. generate lexposure = ln(_t - _t0)
```

We then fit a Poisson regression model by conditional maximum likelihood, using the `xtpoisson` command with the `fe` (for "fixed effects") option. We first `xtset` the data, declaring `click` as the grouping variable:

```
. quietly xtset click
```

In the `xtpoisson` command, we also specify `offset(lexposure)` to define the offset, and `irr` to display incidence-rate ratios:

```
. xtpoisson _d heblack mixrace hedropout hecollege heolder sheolder,
> fe offset(lexposure) irr
Conditional fixed-effects Poisson regression     Number of obs      =   2244166
Group variable: click                            Number of groups   =      1032

                                                 Obs per group: min =       139
                                                                avg =    2174.6
                                                                max =      3370

                                                 Wald chi2(6)       =     41.40
Log likelihood  = -7821.9073                     Prob > chi2        =    0.0000
```

_d	IRR	Std. Err.	z	P>|z|	[95% Conf. Interval]	
heblack	1.192658	.095109	2.21	0.027	1.020085	1.394426
mixrace	1.266902	.1009307	2.97	0.003	1.083752	1.481003
hedropout	.7458869	.0509019	-4.30	0.000	.6525054	.8526325
hecollege	.7594514	.080115	-2.61	0.009	.6175985	.9338858
heolder	.7697281	.1666861	-1.21	0.227	.5035099	1.176702
sheolder	1.640318	.4284608	1.89	0.058	.9830807	2.736951
lexposure	1	(offset)				

The estimates are identical to those produced by `stcox`. The offset is not needed here because it is constant within each risk set and drops out of the conditional likelihood. Identical estimates could also be obtained by using the following conditional logistic regression (see display 12.2):

```
clogit d heblack mixrace hedropout hecollege heolder sheolder, group(click)
```

As an alternative to this semiparametric approach, we can instead impose a smooth function on the baseline hazard, using, for instance, a polynomial or some kind of spline function. Here we adopt the latter strategy and use a cubic spline (see section 7.3.2 on linear splines). The necessary basis functions can be constructed using the `mkspline` command,

```
. mkspline sp = _t, cubic knots(1,5,7,9,12,15,20,25)
```

where the knots we have chosen for the spline are given in the `knots()` option. We then fit an ordinary Poisson model, including the offset (which is now needed), by maximum likelihood using the `poisson` command:

```
. poisson _d sp* heblack mixrace hedropout hecollege heolder sheolder,
> offset(lexposure) irr

Poisson regression                            Number of obs   =     2246505
                                              LR chi2(13)     =      279.14
                                              Prob > chi2     =      0.0000
Log likelihood = -9442.7291                   Pseudo R2       =      0.0146
```

_d	IRR	Std. Err.	z	P>|z|	[95% Conf. Interval]	
sp1	1.304973	.0751225	4.62	0.000	1.165738	1.460838
sp2	.044862	.0506301	-2.75	0.006	.0049117	.4097547
sp3	391459.9	3116978	1.62	0.106	.0653191	2.35e+12
sp4	1.88e-08	2.96e-07	-1.13	0.259	7.24e-22	486979.6
sp5	587734.7	8223518	0.95	0.342	7.23e-07	4.78e+17
sp6	.000776	.0064476	-0.86	0.389	6.57e-11	9160.43
sp7	2.955591	12.49459	0.26	0.798	.0007451	11723.31
heblack	1.192452	.0950753	2.21	0.027	1.019938	1.394145
mixrace	1.265812	.1008347	2.96	0.003	1.082836	1.479709
hedropout	.7464898	.0509393	-4.28	0.000	.6530391	.8533135
hecollege	.7587815	.0800423	-2.62	0.009	.6170569	.9330571
heolder	.7700903	.1667609	-1.21	0.228	.5037515	1.177245
sheolder	1.637608	.4277325	1.89	0.059	.9814807	2.732362
_cons	.0110737	.0016838	-29.62	0.000	.0082198	.0149184
lexposure	1	(offset)				

The estimated incidence-rate ratios or hazard ratios were also reported under "Poisson cubic spline" in table 15.2 on page 812. The estimates are almost identical to those reported for Cox regression, giving us some confidence that we are not over-smoothing the baseline hazard.

We now produce a graph of the baseline hazard, based on the cubic spline estimated above. The estimated baseline hazard could be calculated using the same kind of `generate` commands used on page 813, but here we demonstrate an approach that requires less typing, especially if the baseline hazard depends on many coefficients. First, we place the coefficients for the baseline hazard in a matrix, `coeff`,

```
. matrix a=e(b)
. matrix coeff = a[1,1..7],a[1,14..14]
```

where element 14 is the intercept. We then use the `matrix score` command to calculate the predicted log-baseline hazards, $\widehat{\alpha}_0 + \widehat{\alpha}_1 \texttt{sp1} + \cdots$,

```
. matrix score lhazard = coeff
```

and exponentiating this, we obtain the baseline hazard,

```
. generate h0=exp(lhazard)
```

Before plotting this baseline hazard, we append the data in `temp.dta` for the baseline hazard from Cox regression,

```
. append using temp
```

and plot both baseline hazards together using

```
. twoway (line h0 _t if divorce<., sort)  (line haz1 _t , sort),
> xtitle(Time since wedding in years)
> ytitle(Baseline hazard) legend(order(1 "Cubic spline" 2 "Cox"))
```

In the `line` subcommand used to plot the cubic spline version of the baseline hazard h0, we specified `if divorce<.` to plot the predictions only for the original 3,371 observations, which is much faster than plotting predictions for over 2 million expanded observations but produces an identical graph because the unique values of `_t` are just repeated in the expanded data. The resulting graph is presented in figure 15.5.

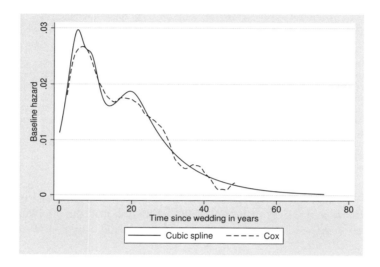

Figure 15.5: Estimated baseline hazard curve from piecewise exponential model with cubic spline and Cox model

We see that the cubic spline appears to be an appropriately smoothed version of the nonparametric baseline hazard from Cox regression (it is not too smooth and not too rough).

Before we proceed to a discussion of accelerated failure-time models, we restore the data to their unexpanded form. We first delete the observations from `temp.dta` and variables that vary within couples between the episodes created by `stsplit`,

```
. drop if _st >= .
(93 observations deleted)
. drop sp* h0 haz1 lexposure lhazard click
```

and then use `stjoin`,

```
. stjoin
(option censored(0) assumed)
(2243134 obs. eliminated)
```

Multilevel versions of proportional hazard models are discussed in section 15.9.

15.5   Accelerated failure-time models

Many parametric survival models can be written as log-linear models for the survival time T_i,

$$\ln(T_i) = \alpha_0 + \beta_2 x_i + \epsilon_i \tag{15.7}$$

or equivalently as multiplicative models of the form

$$T_i = \exp(\beta_2 x_i) \underbrace{\exp(\alpha_0 + \epsilon_i)}_{\tau_i} \tag{15.8}$$

The parametric distribution assumed for $\tau_i \equiv \exp(\alpha_0 + \epsilon_i)$ provides the name for the model. For example, in the log-normal survival model, τ_i is log normal (the log of τ_i has a normal distribution). The log-normal model is particularly interesting because if τ_i is log normal, then ϵ_i is normal, giving a standard linear regression model with a normally distributed residual for the log survival time in (15.7).

The parametric models for τ_i available in Stata are the following:

- *Log normal*, so ϵ_i is normal with variance σ^2
- *Log logistic*, so ϵ_i is logistic with variance $\gamma^2 \pi^2 / 3$ (the standard logistic distribution used in chapters 10 and 11 on dichotomous and ordinal responses is obtained when $\gamma = 1$)
- *Exponential*, so ϵ_i is standard extreme value type I or Gumbel distributed (this distribution was used in chapter 12 on nominal responses)
- *Weibull*, so ϵ_i is extreme value type I distributed with scale parameter p
- *Generalized gamma* with parameters α_0, κ, and σ (there is no simple distribution for ϵ_i in this case)

The exponential and Weibull models have both proportional hazards (PH) and accelerated failure time (AFT) parameterizations. The `time` option can be used in the `streg` command to switch from the default PH to the AFT parameterization for these models.

For model selection, it is useful that the Weibull, exponential, and log-normal models are all nested in the generalized gamma model. For other models, the Akaike information criterion (AIC) or the Bayesian information criterion (BIC) can be used for model selection (see section 6.5).

For log-normal and log-logistic models, the distribution of ϵ_i is symmetric around 0 so that the median log-survival time is simply $\alpha_0 + \beta_2 x_i$, and the median survival time is therefore $\exp(\alpha_0 + \beta_2 x_i)$. (For monotonic functions, such as the exponential, the quantile of the function of a random variable is the function of the quantile of the random variable). For the other distributions, the median of ϵ_i, denoted Median(ϵ_i), is not zero, but we can write the median survival time as

$$\text{Median}(T_i | x_i) = \exp(\beta_2 x_i) \text{Median}(\tau_i)$$

and similarly for any other quantiles or percentiles. The expected survival time is also multiplicative in the covariate:

$$E(T_i|x_i) = \exp(\beta_2 x_i)E(\tau_i)$$

When x_i changes from a to $a + 1$, the ratio of the median survival times and the ratio of the expected survival times both are

$$\frac{\exp\{\beta_2(a+1)\}}{\exp\{\beta_2(a)\}} = \exp(\beta_2)$$

Therefore, $\exp(\beta_2)$ is called the *time ratio* (TR) of x_i.

The models are called accelerated failure time (AFT) models because the survival function can be written as

$$S(t|x_i) = S_0\{t\exp(-\beta_2 x_i)\} \tag{15.9}$$

where $S_0(t)$ is the survival function when $x_i = 0$. For example, consider survival after onset of a terminal illness. If $\beta_2 = 0.69$ and some individual has $x_i = 1$ so that $\exp(-\beta_2 x_i) = 1/2$, then that individual reaches the same point in the survival curve, or the same probability of survival, in half the time as an individual with $x_i = 0$. The disease processes affecting survival are in that sense *accelerated*. When plotted against time, the survival curves for both individuals look the same, but the tick-mark on the time axis that represents 1 year since onset for the individual with $x_i = 1$ represents 2 years for the individual with $x_i = 0$.

To derive (15.9) from the AFT model, we first substitute for T_i from model (15.8) in the survival function and obtain

$$S(t|x_i) = \Pr(T_i > t|x_i) = \Pr\{\exp(\beta_2 x_i)\tau_i > t|x_i\} = \Pr\{\tau_i > t\exp(-\beta_2 x_i)|x_i\}$$

For the special case of $x_i = 0$, we then get

$$S_0(t) \equiv S(t|x_i = 0) = \Pr(T_i > t|x_i = 0) = \Pr(\tau_i > t)$$

because $\exp(-\beta_2 \times 0) = \exp(0) = 1$. Comparing the two expressions, we see that (15.9) follows.

15.5.1 Log-normal model

The log-normal survival model can be written as

$$\ln(T_i) = \alpha_0 + \beta_2 x_i + \epsilon_i, \quad \epsilon_i \sim N(0, \sigma^2) \tag{15.10}$$

where we assume that $E(\epsilon_i|x_i) = 0$. It follows that the mean survival time becomes

$$E(T_i|x_i) = \exp(\alpha_0 + \beta_2 x_i)\exp(\sigma^2/2)$$

We can fit the model using `streg` with the `distribution(lognormal)` option and the `time` option for the AFT parameterization:

```
. streg heblack mixrace hedropout hecollege heolder sheolder,
> distribution(lognormal) time

        failure _d:  divorce
  analysis time _t:  dur
               id:  id

Lognormal regression -- accelerated failure-time form

No. of subjects =          3371              Number of obs    =        3371
No. of failures =          1032
Time at risk    =  62098.24824
                                             LR chi2(6)       =       36.35
Log likelihood  =    -3077.6061              Prob > chi2      =      0.0000
```

_t	Coef.	Std. Err.	z	P>|z|	[95% Conf. Interval]	
heblack	-.137753	.0968523	-1.42	0.155	-.3275801	.0520742
mixrace	-.314885	.0988419	-3.19	0.001	-.5086115	-.1211584
hedropout	.3289508	.0818263	4.02	0.000	.1685742	.4893274
hecollege	.3404015	.123536	2.76	0.006	.0982754	.5825275
heolder	.1330227	.2304122	0.58	0.564	-.318577	.5846224
sheolder	-.6399424	.3181989	-2.01	0.044	-1.263601	-.016284
_cons	3.798159	.066096	57.46	0.000	3.668613	3.927705
/ln_sig	.5481884	.0241329	22.72	0.000	.5008887	.5954881
sigma	1.730116	.0417528			1.650187	1.813916

The exponentiated estimated coefficients or time ratios represents multiplicative factors for the time to reach a given survival probability. For instance, if the wife is more than 10 years older than the husband, the couple reaches a given survival probability in just over half the time $[0.53 = \exp(-0.6399424)]$ as couples with an age difference of less than 10 years. The mean or median survival time is correspondingly halved. Estimated coefficients greater than zero have time ratios greater than one corresponding to slower progression of the process. For example, if the husband completed fewer than 12 years of schooling, the process takes 39% longer $[1.39 = \exp(0.3289508)]$ to reach a given point, and the median or mean survival time is 39% longer.

The tr option can be used to display time ratios (TRs), which correspond to the exponentiated regression coefficients for the log-linear model. We can accomplish this by replaying the previous command with the tr option:

```
. streg, tr
Lognormal regression -- accelerated failure-time form
No. of subjects =         3371              Number of obs   =        3371
No. of failures =         1032
Time at risk    =   62098.24824
                                            LR chi2(6)      =       36.35
Log likelihood  =    -3077.6061             Prob > chi2     =      0.0000
```

_t	Tm. Ratio	Std. Err.	z	P>|z|	[95% Conf. Interval]	
heblack	.8713139	.0843888	-1.42	0.155	.7206656	1.053454
mixrace	.7298728	.072142	-3.19	0.001	.6013299	.8858936
hedropout	1.389509	.1136984	4.02	0.000	1.183616	1.631219
hecollege	1.405512	.1736313	2.76	0.006	1.103267	1.790558
heolder	1.142276	.2631943	0.58	0.564	.7271831	1.794313
sheolder	.5273228	.1677935	-2.01	0.044	.2826345	.9838479
_cons	44.61896	2.949137	57.46	0.000	39.1975	50.79027
/ln_sig	.5481884	.0241329	22.72	0.000	.5008887	.5954881
sigma	1.730116	.0417528			1.650187	1.813916

The estimated time ratios are given under `Tm. Ratio` in the output and were also reported together with estimates from other models in table 15.2 (under "log normal").

We can look at the model-implied hazard functions for `sheolder` equal to 0 and 1 when all other covariates are evaluated at their mean by using the postestimation command `stcurve` for `streg` with the `hazard` option,

```
. stcurve, hazard xtitle(Time since wedding in years) at1(sheolder=0)
> at2(sheolder=1)
```

which produces the curves presented in figure 15.6.

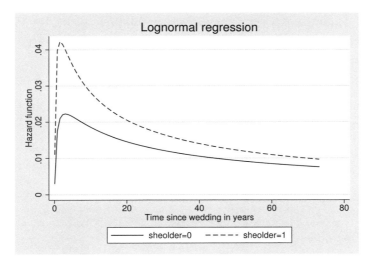

Figure 15.6: Hazards for log-normal survival model according to value taken by `sheolder`

The hazard curves in figure 15.6 have a similar shape as those based on Cox regression in figure 15.4, but they are much more smooth, as would be expected.

Model (15.10) is just a linear regression model for log survival times. If there were no censoring, we could simply log transform the survival times and fit a standard linear regression model by ordinary least squares (OLS) using the `regress` command. However, censoring is ubiquitous in survival data and also occurs frequently in the divorce data. For right-censored observations, all we know is that the survival time exceeds the censoring time. For such observations, the contribution to the likelihood is an integral from the censoring time to infinity. Such *censored regression models* are typically referred to as *tobit models* in economics (after the prominent economist James Tobin).

Stata's `tobit` command cannot handle censoring limits that are different across subjects, and we therefore use the `intreg` command instead (for "interval-censored regression"). The `intreg` command is very flexible, handling limits differing between couples and allowing for left-censoring, right-censoring, and interval-censoring. The command requires that two response variables be specified, representing the lower and upper limits of interval-censored data, respectively. If the lower limit is missing, we have left-censoring, and if the upper limit is missing, we have right-censoring, which is the only censoring we are concerned with in the divorce data.

Because we require a *log*-linear model for survival, we must log transform the durations,

```
. generate ldur = ln(dur)
```

before proceeding.

We now consider a log-normal survival model with *right-censoring*. In this case, the lower limit equals `ldur` and the upper limit should also be `ldur` unless right-censoring occurred, in which case it should be missing:

```
. generate ldur2 = ldur if divorce==1
(2339 missing values generated)
```

The model is fit using the `intreg` command:

```
. intreg ldur ldur2 heblack mixrace hedropout hecollege heolder sheolder

Interval regression                             Number of obs   =       3371
                                                LR chi2(6)      =      36.35
Log likelihood = -3077.6061                     Prob > chi2     =     0.0000
```

	Coef.	Std. Err.	z	P>|z|	[95% Conf.	Interval]
heblack	-.137753	.0968523	-1.42	0.155	-.3275801	.0520742
mixrace	-.314885	.0988419	-3.19	0.001	-.5086115	-.1211584
hedropout	.3289508	.0818263	4.02	0.000	.1685742	.4893274
hecollege	.3404015	.123536	2.76	0.006	.0982754	.5825276
heolder	.1330227	.2304122	0.58	0.564	-.318577	.5846224
sheolder	-.6399424	.3181989	-2.01	0.044	-1.263601	-.016284
_cons	3.798159	.066096	57.46	0.000	3.668613	3.927705
/lnsigma	.5481884	.0241329	22.72	0.000	.5008887	.5954881
sigma	1.730116	.0417528			1.650187	1.813916

```
Observation summary:       0  left-censored observations
                        1032      uncensored observations
                        2339 right-censored observations
                           0      interval observations
```

The estimates, *p*-values, and confidence intervals reported in the output are identical to those produced by the `streg` command.

We mentioned in section 15.2 that the time from wedding to divorce could not be determined exactly and that the variables `lower` and `upper` represented lower and upper bounds for the durations. We could use `intreg` to take this interval-censoring into account; see exercise 15.2.

Multilevel versions of accelerated failure-time models are discussed in section 15.10.

15.6 Time-varying covariates

We now want to include the number of children of each couple as a covariate. This covariate is time varying because it increases by one at every birth of a new child. Information about time-varying covariates can be stored in the data by splitting the survival time into episodes, from time 0 to the first birth, from the first to second birth, and so on, and finally until divorce or censoring.

We read in the dataset `divorce2.dta`, which is in this format,

```
. use http://www.stata-press.com/data/mlmus3/divorce2, clear
```

and list data for the first five couples:

```
. list id dur numkids divorce if id<34, sepby(id) noobs
```

id	dur	numkids	divorce
9	3.734	0	0
9	10.546	1	0
11	.767	0	0
11	32.512	1	0
11	34.943	2	0
13	2.585	0	0
13	2.870226	1	1
15	18.18403	0	1
33	1.418	0	0

The time-varying covariate, number of children, is called `numkids` in the dataset. We see that couple 9 had zero children during the first episode, from time 0 to 3.73 years when they had a child, then no more children before censoring occurred at 10.55 years. Couple 11 had two children by the time they were censored at 34.94 years, whereas couple 13 got divorced shortly after having a child. Couples 15 and 33 had no children by the time they got divorced and censored, respectively. Note the `divorce` takes the value 0 for each episode that does not end in divorce.

We define the survival time, failure indicator, and subject identifier by using the `stset` command with the `failure()` and `id()` options:

```
. stset dur, failure(divorce) id(id)
                 id:  id
      failure event:  divorce != 0 & divorce < .
obs. time interval:  (dur[_n-1], dur]
 exit on or before:  failure

        8728  total obs.
           0  exclusions

        8728  obs. remaining, representing
        3371  subjects
        1032  failures in single failure-per-subject data
    62098.25  total analysis time at risk, at risk from t =          0
                                    earliest observed entry t =          0
                                      last observed exit t =     73.068
```

This command created new variables _t0 and _t for the start and end times of each episode, as can be seen by listing the data:

```
. sort id dur
. list id _t0 _t numkids divorce if id<34, sepby(id) noobs
```

id	_t0	_t	numkids	divorce
9	0	3.734	0	0
9	3.734	10.546	1	0
11	0	.76700002	0	0
11	.76700002	32.512001	1	0
11	32.512001	34.943001	2	0
13	0	2.585	0	0
13	2.585	2.8702259	1	1
15	0	18.184032	0	1
33	0	1.418	0	0

Before we can fit the models described in the previous sections but now with numkids as an additional time-varying covariate, we must create several dummy variables as before:

```
. generate mixrace = (heblack & !sheblack) | (sheblack & !heblack)
. generate hedropout = hiseduc<12
. generate hecollege = hiseduc>=16 if hiseduc<.
. generate heolder = agediff > 10 if agediff<.
. generate sheolder = agediff< -10
```

We can use the stcox or streg commands, but not the intreg command, to fit all the models considered so far with the additional time-varying covariate numkids. For example, the log-normal accelerated failure-time model is fit like this:

```
. streg heblack mixrace hedropout hecollege heolder sheolder numkids,
> distribution(lognormal) time tr

       failure _d:  divorce
  analysis time _t:  dur
             id:  id

Lognormal regression -- accelerated failure-time form

No. of subjects =         3371           Number of obs   =        8728
No. of failures =         1032
Time at risk    =    62098.24824
                                         LR chi2(7)      =       73.70
Log likelihood  =     -3058.9311         Prob > chi2     =      0.0000
```

_t	Tm. Ratio	Std. Err.	z	P>|z|	[95% Conf. Interval]	
heblack	.92643	.0817663	-0.87	0.387	.7792663	1.101385
mixrace	.6900399	.061664	-4.15	0.000	.579173	.8221292
hedropout	1.350578	.0999625	4.06	0.000	1.168203	1.561425
hecollege	1.383798	.1545178	2.91	0.004	1.111798	1.722343
heolder	1.162471	.2440743	0.72	0.473	.7703046	1.754292
sheolder	.5572091	.1611488	-2.02	0.043	.3161147	.9821814
numkids	1.178646	.0287783	6.73	0.000	1.12357	1.236422
_cons	34.19375	2.400638	50.31	0.000	29.79795	39.23801
/ln_sig	.4662275	.0259057	18.00	0.000	.4154533	.5170017
sigma	1.59397	.0412928			1.515057	1.676992

Here the `tr` option was used to obtain time ratios. According to the fitted model, each extra child increases the mean and median survival time by 18%, controlling for the other covariates. However, the probability of having a child could be partly determined by the hazard of divorce. It is dangerous to interpret the effect as causal because the number of children is likely to be an endogenous covariate or "internal" covariate, as discussed by Lillard (1993).

Time-varying covariates can also be used to assess and relax proportionality assumptions in survival models. The models discussed so far assume that the covariates have multiplicative effects on the survival time (AFT models) or the hazards (PH models), and that these effects are constant over time. To relax these proportionality assumptions, we can introduce interactions between covariates and functions of analysis time.

For Cox regression, `stcox` provides two convenient options called `tvc()` and `texp()` for relaxing the proportional hazards assumption. The covariates for which the assumption is relaxed are specified in `tvc(varlist)`. If `texp()` is not specified, interactions with untransformed analysis time are included for all of these covariates. Alternatively, to include interactions between these covariates and a function of analysis time, the function is specified in `texp()`, for instance, `texp(log(_t))` represents the log of analysis time. For categorical explanatory variables, proportionality can also be relaxed by specifying category-specific (for example, group-specific) baseline hazards by using the `strata()` option in `stcox`; see section 15.8.1 for an example.

For other survival models, proportionality can be relaxed by splitting the observation period into episodes, as shown for the piecewise exponential model, and forming interactions between interval dummies and covariates (see exercise 15.6).

15.7   Does nitrate reduce the risk of angina pectoris?

Angina pectoris is severe chest pain due to insufficient blood and hence oxygen supply to the heart muscle. Severe angina attacks, sudden onset of angina at rest, and angina lasting more than 15 minutes are symptoms that may herald myocardial infarction or heart attack. In more than half of patients with angina, the severity of symptoms seriously limits their everyday activities, often leading to premature retirement. According to a meta-analysis reported by Hemingway et al. (2008), the prevalence of angina varied between about 6% and 9% for people aged 45 and above depending on the gender and age. The mean of the study-specific prevalences (weighted by population size) was 6.7% for women and 5.7% for men.

We will analyze a subset of the data published in Danahy et al. (1976) and previously analyzed by Pickles and Crouchley (1994, 1995) and Skrondal and Rabe-Hesketh (2004). Subjects with coronary heart disease participated in a randomized crossover trial comparing isosorbide dinitrate (ISDN) with placebo. The subjects were asked to exercise on exercise bikes until the onset of angina pectoris or, if angina did not occur, to exhaustion. The exercise time and outcome (angina or exhaustion) were recorded.

Each subject repeated the exercise test four times under each of two regimes: a placebo regime and a treatment regime (ISDN). For both regimes, the first exercise test of the day was a control test where the subjects were given neither placebo nor drug. Before the second test, which took place 1 hour after the initial test, the subjects were administered either ISDN or placebo orally. For each regime, the exercise test was repeated 1, 3, and 5 hours after drug or placebo administration. A crossover design was used where the regimes were implemented on 2 successive days and the order of the regimes was randomized. Unfortunately, we do not know the order for individual subjects because this was not reported in the original paper.

Each subject j therefore has repeated survival times to angina or exhaustion for eight exercise test occasions i in two regimes. Data on the timing of several events per subject are often called multivariate or multilevel survival data. The event could be of different types (for example, onset of different diseases) or of the same kind (as here). In the latter case, the data are typically called recurrent-event data (see section 15.12 for further discussion). The multiple events are not absorbing; otherwise, we would have competing risks.

Because the subjects started each of the four exercise tests at rest, so that a similar process leading to angina or exhaustion can be assumed to begin at the start of each test, we define the origin of analysis time to be at the beginning of each test. Such a time scale, that resets to 0 when subjects start being at risk for the next event, is called *gap time*.

The dataset `angina8.dta` contains the following variables for 42 subjects:

- `subject`: identification number for subject (j)
- `regime`: treatment regime (1: ISDN; 0: placebo)
- `occ`: exercise test number within each of the treatment regimes $(i,\ i = 1, 2, 3, 4)$
- `second`: time from start of exercise to angina or censoring in seconds
- `uncen`: dummy variable for type of event observed (1: angina; 0: censored)

We first read in the data and list the 16 observations for the first two subjects in the dataset:

```
. use http://www.stata-press.com/data/mlmus3/angina8, clear
. list subj regime occ second uncen in 1/16, sepby(subj) noobs
```

subj	regime	occ	second	uncen
1	0	1	150	1
1	0	2	172	1
1	0	3	118	1
1	0	4	143	1
1	1	1	136	1
1	1	2	445	0
1	1	3	393	0
1	1	4	226	1
2	0	1	205	1
2	0	2	287	1
2	0	3	211	1
2	0	4	207	1
2	1	1	250	1
2	1	2	306	1
2	1	3	206	1
2	1	4	224	1

The data are already in long form with eight records per subject representing the times to angina or exhaustion.

We now construct dummy variables x_{2i}, x_{3i}, and x_{4i}, called occ2, occ3, and occ4 in the dataset, for the second, third, and fourth exercise test occasion within each regime

```
. tabulate occ, generate(occ)
```

occ	Freq.	Percent	Cum.
1	42	25.00	25.00
2	42	25.00	50.00
3	42	25.00	75.00
4	42	25.00	100.00
Total	168	100.00	

and construct a treatment dummy x_{5ij}, called `treat`, that is equal to 1 after administration of the drug and equal to 0 otherwise:

```
. generate treat = regime==1 & occ>1
```

For the `id()` option of the `stset` command, we need an identifier for each survival time (or set of episodes constituting a survival time), not for each subject. We therefore construct an observation index, `id`, that labels all combinations of subjects and exercise test occasions,

```
. generate id=_n
```

The data now look like this:

```
. list id subj regime occ2 occ3 occ4 treat second uncen in 1/16, sepby(subj)
> noobs
```

id	subj	regime	occ2	occ3	occ4	treat	second	uncen
1	1	0	0	0	0	0	150	1
2	1	0	1	0	0	0	172	1
3	1	0	0	1	0	0	118	1
4	1	0	0	0	1	0	143	1
5	1	1	0	0	0	0	136	1
6	1	1	1	0	0	1	445	0
7	1	1	0	1	0	1	393	0
8	1	1	0	0	1	1	226	1
9	2	0	0	0	0	0	205	1
10	2	0	1	0	0	0	287	1
11	2	0	0	1	0	0	211	1
12	2	0	0	0	1	0	207	1
13	2	1	0	0	0	0	250	1
14	2	1	1	0	0	1	306	1
15	2	1	0	1	0	1	206	1
16	2	1	0	0	1	1	224	1

We are now ready to `stset` the dataset to define it as survival data:

```
. stset second, failure(uncen) id(id)

                id:  id
     failure event:  uncen != 0 & uncen < .
obs. time interval:  (second[_n-1], second]
 exit on or before:  failure
─────────────────────────────────────────────────────────
    168  total obs.
      0  exclusions
─────────────────────────────────────────────────────────
    168  obs. remaining, representing
    168  subjects
    155  failures in single failure-per-subject data
  47267  total analysis time at risk, at risk from t =         0
                          earliest observed entry t =         0
                            last observed exit t =          743
```

15.8 Marginal modeling

The basic idea of marginal or population-averaged modeling is to estimate the parameters as if the data were single level or nonclustered and take clustering into account when estimating standard errors.

15.8.1 Cox regression

We start by considering the following Cox regression model with a nonparametric baseline hazard function $h_0(t)$; occasion-specific dummy variables x_{2i}, x_{3i}, and x_{4i}; and treatment dummy x_{5ij}:

$$\ln\{h(t|\mathbf{x}_{ij})\} = \ln\{h_0(t)\} + \beta_2 x_{2i} + \beta_3 x_{3i} + \beta_4 x_{4i} + \beta_5 x_{5ij} \qquad (15.11)$$

The dummy variables for exercise test occasions are included to allow the hazard to change throughout the day as the drug concentration in the blood, the time of day, state of rest/tiredness, etc., change. The hazards for different test occasions are, however, assumed to be proportional (because the log-hazards are parallel). The hazards implied by the Cox model with occasion-specific dummy variables are shown under "Occ.spec. dummies" in table 15.4. In the literature on recurrent-event data, assuming a common form of the baseline hazard with different intercepts for different events is sometimes referred to as a semicommon baseline hazard.

The baseline hazard $h_0(t)$ is the hazard at the first or control exercise test occasion, where neither placebo nor ISDN is administered in either the placebo or the treatment regimes. At subsequent occasions, $i = 2, 3, 4$, the hazard for the treatment regime is $\exp(\beta_5)$ times the hazard for the placebo regime. It follows that $\exp(\beta_5)$ can be interpreted as a *hazard ratio*, comparing the hazards for subjects who are given the treatment with subjects who are given the placebo at each test occasion.

Table 15.4: Hazards implied by Cox models. "Occ.spec. dummies" refers to model (15.11) with occasion-specific dummy variables, and "Occ.spec. baselines" refers to model (15.12) with occasion-specific baseline hazards.

		Hazards	
Regime	Test occasion	Occ.spec. dummies	Occ.spec. baselines
Placebo	1 (control)	$h_0(t)$	$h_{01}(t)$
Placebo	2	$\{h_0(t)\exp(\beta_2)\}$	$h_{02}(t)$
Placebo	3	$\{h_0(t)\exp(\beta_3)\}$	$h_{03}(t)$
Placebo	4	$\{h_0(t)\exp(\beta_4)\}$	$h_{04}(t)$
Treatment	1 (control)	$h_0(t)$	$h_{01}(t)$
Treatment	2	$\{h_0(t)\exp(\beta_2)\}\exp(\beta_5)$	$h_{02}(t)\exp(\beta_5)$
Treatment	3	$\{h_0(t)\exp(\beta_3)\}\exp(\beta_5)$	$h_{03}(t)\exp(\beta_5)$
Treatment	4	$\{h_0(t)\exp(\beta_4)\}\exp(\beta_5)$	$h_{04}(t)\exp(\beta_5)$

We fit the Cox regression model with occasion-specific dummy variables by using `stcox` with the `vce(cluster subj)` option to obtain robust standard errors for clustered data:

```
. stcox occ2 occ3 occ4 treat, vce(cluster subj)

         failure _d:  uncen
    analysis time _t:  second
                 id:  id

Cox regression -- Breslow method for ties

No. of subjects   =        168          Number of obs   =         168
No. of failures   =        155
Time at risk      =      47267
                                        Wald chi2(4)    =       31.21
Log pseudolikelihood =   -648.95214     Prob > chi2     =      0.0000

                            (Std. Err. adjusted for 21 clusters in subj)
```

_t	Haz. Ratio	Robust Std. Err.	z	P>|z|	[95% Conf. Interval]	
occ2	.7283773	.0957445	-2.41	0.016	.5629461	.9424233
occ3	.9171277	.0861531	-0.92	0.357	.7629036	1.102529
occ4	1.228249	.1236968	2.04	0.041	1.008236	1.496273
treat	.4031625	.079238	-4.62	0.000	.2742736	.59262

The estimates suggest that treatment with ISDN reduces the hazard of angina by 60% $[-60\% = 100\%(0.4031625 - 1)]$. This estimate is also given under "Marginal" in table 15.5 on page 844. We also see that the hazard of angina within a given regime decreases by 27% $[-27\% = 100\%(0.7283773 - 1)]$ from occasion 1 to 2, does not change significantly at the 5% level from occasion 1 to 3, and increases by 23% $[= 100\%(1.228249 - 1)]$ from occasion 1 to 4.

To visualize the treatment effect, we plot smoothed estimated hazard curves at test occasion 2 (after placebo or ISDN has been administered) for the treatment and placebo groups,

```
. stcurve, hazard at1(occ2=1 occ3=0 occ4=0 treat=1)
> at1(occ2=1 occ3=0 occ4=0 treat=0) legend(order(1 "Treatment" 2 "Placebo"))
> xtitle(Time in seconds) ytitle(Hazard function)
```

producing the graph shown in figure 15.7.

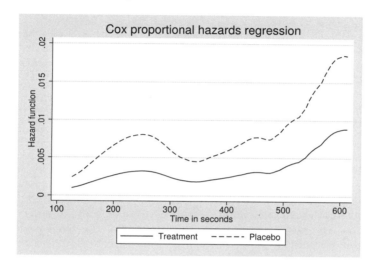

Figure 15.7: Smoothed estimated hazard functions at second exercise test occasion for treatment and placebo groups

Before proceeding, we store the smoothed estimated baseline hazard function (the hazard for the control test occasions) in `temp.dta` for later use,

```
. stcurve, hazard name(cox, replace) at(occ2=0 occ3=0 occ4=0 treat=0)
> outfile(temp, replace)
```

(graph not shown).

Instead of assuming that the log-baseline hazard functions at different exercise test occasions are parallel, we can let the baseline hazards in the Cox regression model be completely different at each occasion i (but identical across regimes for a given test occasion),

$$\ln\{h(t|\mathbf{x}_{ij})\} = \ln\{h_{0i}(t)\} + \beta_5 x_{5ij} \tag{15.12}$$

where $h_{0i}(t)$ is the occasion-specific baseline hazard for test occasion i. The hazards implied by this model are shown under "Occ.spec. baselines" in table 15.4.

The Cox model with occasion-specific baseline hazards can be fit using stcox with the strata() option:

```
. stcox treat, strata(occ) vce(cluster subj)

        failure _d:  uncen
  analysis time _t:  second
               id:  id

Stratified Cox regr. -- Breslow method for ties
No. of subjects     =          168         Number of obs   =         168
No. of failures     =          155
Time at risk        =        47267
                                            Wald chi2(1)    =       23.38
Log pseudolikelihood =   -441.86794         Prob > chi2     =      0.0000
                                  (Std. Err. adjusted for 21 clusters in subj)
```

_t	Haz. Ratio	Robust Std. Err.	z	P>|z|	[95% Conf. Interval]
treat	.4185219	.0753937	-4.84	0.000	.2940225 .5957389

```
                                            Stratified by occ
```

The strata() option causes the risk set for an event occurring at a particular exercise test occasion to be restricted to episodes for the same occasion. We see that the estimated treatment effect is close to that from model (15.11) with a nonparametric baseline hazard for the control condition and occasion-specific dummy variables. We will therefore use the latter and simpler specification of the fixed part of the model henceforth.

15.8.2 Poisson regression with smooth baseline hazard

Instead of a fully nonparametric specification of the baseline hazard function $h_0(t)$, we will now model the log-baseline hazard as a smooth function of time by using orthogonal polynomial terms (where the pth order term is a linear combination of t, t^2, ..., t^p such that the different order terms are mutually uncorrelated). An alternative would be to use a piecewise exponential model. That model is less smooth, but it also does not require the data to be expanded to episodes defined by clicks, only to episodes defined by the intervals within which the hazards are assumed to be constant. Yet another alternative would be to use splines as we did for the divorce data in section 15.4.2 or fractional polynomials.

Stata's stsplit command can be used to expand the data to episodes defined by clicks:

```
. stsplit, at(failures) riskset(click)
(118 failure times)
(10013 observations (episodes) created)
```

As explained in section 15.4.1, this command produces variables _t0, _t, and _d representing the start time, end time, and failure indicator for each episode. We then compute the lengths of the intervals between unique failure times so that we can use the log interval lengths as an offset in the Poisson regression.

```
. generate lny = ln(_t - _t0)
```

We could check whether the data expansion and the construction of the binary response variable and the offset is correct by fitting a Poisson regression model that corresponds to Cox regression by typing

```
xtset click
xtpoisson _d occ2 occ3 occ4 treat, fe offset(lny) irr
```

Both models give identical parameter estimates, confirming that our data have been set up correctly (at the time of writing this book, xtpoisson cannot be combined with vce(cluster subj) to obtain standard errors taking clustering into account).

Orthogonal polynomials of the desired degree, say, degree 4, can be created by using orthpoly:

```
. orthpoly _t, gen(t1-t4) degree(4)
```

Denoting the linear, quadratic, cubic, and quartic terms of the polynomial as p_{1sij} to p_{4sij} and the corresponding coefficients as δ_1 to δ_4, we specify a Poisson regression model with smooth baseline hazard as

$$\ln(\mu_{sij}) = \ln(t_{sij}) + \delta_0 + \delta_1 p_{1sij} + \cdots + \delta_4 p_{4sij} + \beta_2 x_{2i} + \beta_3 x_{3i} + \beta_4 x_{4i} + \beta_5 x_{5ij}$$

This model can be fit using poisson, and we obtain cluster-robust standard errors with the vce(cluster subj) option:

```
. poisson _d t1-t4 occ2 occ3 occ4 treat, offset(lny) irr vce(cluster subj)
Poisson regression                              Number of obs   =      10181
                                                Wald chi2(8)    =      48.21
                                                Prob > chi2     =     0.0000
Log pseudolikelihood = -811.77386               Pseudo R2       =     0.0839

                                   (Std. Err. adjusted for 21 clusters in subj)
```

		Robust				
_d	IRR	Std. Err.	z	P>|z|	[95% Conf.	Interval]
t1	1.759941	.334701	2.97	0.003	1.212324	2.554921
t2	.6195466	.101207	-2.93	0.003	.4498054	.8533423
t3	1.415135	.1151696	4.27	0.000	1.206489	1.659862
t4	.8220619	.03362	-4.79	0.000	.7587397	.8906688
occ2	.7357576	.0895868	-2.52	0.012	.5795506	.9340673
occ3	.9158704	.0907676	-0.89	0.375	.7541806	1.112225
occ4	1.213588	.1214511	1.93	0.053	.9974391	1.476577
treat	.4103732	.0797539	-4.58	0.000	.2803834	.6006281
_cons	.0054372	.0010889	-26.04	0.000	.003672	.0080508
lny	1	(offset)				

The estimated coefficients are quite close to those of the Cox model, and using higher-order polynomials does not improve the approximation appreciably.

We can also compare the orthogonal polynomial baseline hazard function with the nonparametric baseline hazard from Cox regression. We first predict the baseline hazard

function for the model just fit. The estimated coefficients for the baseline hazard are
placed in a matrix, coeff,

```
. matrix a=e(b)
. matrix coeff=a[1,1..4], a[1,9..9]
```

where element 9 is the intercept $\widehat{\delta}_0$. We then use the matrix score command to
calculate the predicted log-baseline hazards, $\widehat{\delta}_0 + \widehat{\delta}_1 p_{1sij} + \widehat{\delta}_2 p_{2sij} + \widehat{\delta}_3 p_{3sij} + \widehat{\delta}_4 p_{4sij}$,

```
. matrix score lhazard = coeff
```

and after exponentiating, we get the baseline hazard

```
. generate haz0 = exp(lhazard)
```

Appending the data in temp.dta for the smoothed predicted baseline hazard from
Cox regression,

```
. append using temp
```

we are now ready to plot both curves together in the same graph:

```
. twoway (line haz0 _t, sort) (line haz1 _t, sort),
> xtitle(Time in seconds) ytitle(Baseline hazard function)
> legend(order(1 "Polynomial" 2 "Cox"))
```

The resulting graph is shown in figure 15.8:

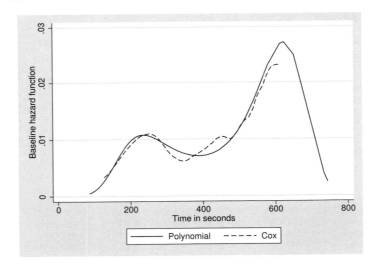

Figure 15.8: Estimated baseline hazard functions from Poisson regression with orthogonal polynomials and Cox regression

The dotted curve represents the baseline hazard from Cox regression, and the solid curve represents the baseline hazard from Poisson regression with orthogonal polynomials. We see that the baseline hazards are very similar. The nonparametric curve from Cox regression ends at about 600 seconds because only a few events happen later, whereas the curve based on the polynomials is parametric and ends at 743 seconds when the last event happened.

15.9   Multilevel proportional hazards models

We now discuss multilevel survival models where the dependence among multiple survival times is explicitly modeled using random effects.

15.9.1   Cox regression with gamma shared frailty

Consider a Cox regression model with a random intercept ζ_j for subject j:

$$
\begin{aligned}
\ln\{h(t|\mathbf{x}_{ij})\} &= \ln\{h_0(t)\} + \beta_2 x_{2i} + \beta_3 x_{3i} + \beta_4 x_{4i} + \beta_5 x_{5ij} + \zeta_j \\
&= [\ln\{h_0(t)\} + \zeta_j] + \beta_2 x_{2i} + \beta_3 x_{3i} + \beta_4 x_{4i} + \beta_5 x_{5ij}
\end{aligned}
$$

The random intercept ζ_j induces dependence among the hazards at different exercise test occasions for each subject. The exponential of the random intercept, $\exp(\zeta_j)$, is called a *shared frailty*, because it represents the unobserved frailty or proneness for angina shared across test occasions for a subject. It is assumed that the random intercept and frailty are independent of the covariates. We see that the subject-specific baseline hazard becomes $h_0(t)\exp(\zeta_j)$ in the model, so $h_0(t)$ is multiplied by the subject-specific frailty.

We now consider a Cox regression model with a frailty $\exp(\zeta_j)$ that is assumed to have a gamma distribution with expectation 1 and variance θ (scale=θ, shape=$1/\theta$). To fit the Cox model with a gamma frailty using `stcox`, we return to the unexpanded data by using the `stjoin` command (after deleting `click` and other variables we created after `stsplit`):

```
. drop click lny lhazard haz0 haz1 t1-t4
. stjoin
```

We can then fit the model by using the `stcox` command with the `shared()` option:

```
. stcox occ2 occ3 occ4 treat, shared(subj)

        failure _d:  uncen
   analysis time _t:  second
                 id:  id

Cox regression --
         Breslow method for ties            Number of obs    =       168
         Gamma shared frailty               Number of groups =        21
Group variable: subj

No. of subjects =        168                Obs per group: min =         8
No. of failures =        155                             avg =         8
Time at risk    =      47267                             max =         8

                                           Wald chi2(4)     =     68.45
Log likelihood  =  -580.70822              Prob > chi2      =    0.0000
```

_t	Haz. Ratio	Std. Err.	z	P>|z|	[95% Conf. Interval]
occ2	.5001758	.1354445	-2.56	0.011	.2941864 .850399
occ3	.9019859	.2320675	-0.40	0.688	.5447513 1.493486
occ4	1.697274	.4156496	2.16	0.031	1.050266 2.742867
treat	.2210261	.0514855	-6.48	0.000	.1400123 .3489162
theta	2.695165	.7875161			

```
Likelihood-ratio test of theta=0: chibar2(01) =    136.49 Prob>=chibar2 = 0.000
Note: standard errors of hazard ratios are conditional on theta.
```

The estimated hazard ratios (HRs) for the treatment along with corresponding standard errors and the estimate of the variance θ of the gamma distribution (not to be confused with our use of θ as the level-1 residual variance in linear multilevel models) are reported under "Gamma frailty" in table 15.5. We note that the estimated subject-specific or conditional hazard ratio is more different from 1 than the marginal or population-averaged counterpart. This is in accordance with the results in Gail et al. (1984), which can be used to contrast conditional and marginal effects in many survival models.

We can plot conditional or subject-specific hazard functions when the frailty is 1 (the mean) for the placebo and treatment groups at the second test occasion by using

```
. stcurve, at1(occ2=1 occ3=0 occ4=0 treat=1) at2(occ2=1 occ3=0 occ4=0 treat=0)
> hazard xtitle(Time in seconds) ytitle(Hazard function)
> legend(order(1 "Treatment" 2 "Placebo")) range(87 430)
```

giving the graph in figure 15.9. (We plotted the curves only up to 430 seconds because, for subjects with frailty equal to 1, survival goes to zero above 430 seconds, making the hazard irrelevant in that range.)

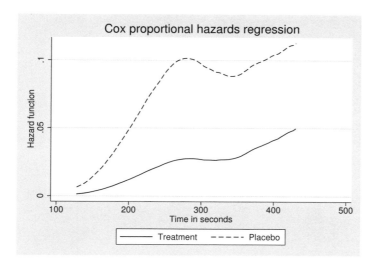

Figure 15.9: Smoothed estimated conditional hazard functions for the second exercise test occasion from Cox regression with gamma frailty evaluated at mean

We see that for subjects with mean frailty, the smoothed conditional or subject-specific hazard function increases almost linearly up to about 280 seconds, and because of the proportionality, the gap between treatment and placebo subjects increases markedly in that range. The shape of the subject-specific hazards from the shared frailty model are markedly different from the marginal hazards shown in figure 15.7. This illustrates the dangers of interpreting marginal hazards curves (and hence survival curves) as if they were subject specific.

Identical estimates to those produced by Cox regression with a gamma frailty can alternatively be obtained by expanding the data to clicks, as we did in the previous section, and fitting a Poisson regression model with a subject-specific random intercept ζ_j:

$$\ln(\mu_{sij}) = \ln(t_{sij}) + \alpha_s + \beta_2 x_{2i} + \beta_3 x_{3i} + \beta_4 x_{4i} + \beta_5 x_{5ij} + \zeta_j \qquad (15.13)$$

Here, μ_{sij} is the mean parameter of the Poisson distribution, t_{sij} is the time at risk in interval s for experiment i of subject j, α_s is a interval-specific parameter, and $\exp(\zeta_j)$ has a gamma distribution with expectation 1 and variance θ. The model can be fit by using the `xtpoisson` command,

```
xtset subj
xtpoisson _d i.click occ2 occ3 occ4 treat, re offset(lny) irr
```

which by default fits a model with a gamma-distributed frailty.

Table 15.5: Proportional hazards models for angina data. Estimated hazards ratios (HRs) for treatment ISDN versus placebo with corresponding 95% confidence intervals reported under "Fixed part" (other estimated regression coefficients not shown). Estimated variance parameters reported under "Random part".

| | Marginal | | Random effects | | | | Fixed effects | |
| | Cox regression | | Gamma frailty | | Log-normal frailty | | Normal RI & RC | | Subject-spec. baseline hazard | |
	Est	(95% CI)[†]	Est	(95% CI)	Est	(95% CI)	Est	(95% CI)	Est	(95% CI)
Fixed part										
$\exp(\beta_5)$ [treat]	0.40	(0.27, 0.59)	0.22	(0.14, 0.35)	0.23	(0.14, 0.36)	0.14	(0.05, 0.37)	0.26	(0.16, 0.42)
Random part										
$\theta^{\ddagger}$			2.70							
ψ_{11}					4.32		6.19			
ψ_{22}							3.54			
ρ_{21}							0.26			
Log likelihood	−648.95[•]		−580.71		−735.81		−717.56		−189.32[•]	

[†] Robust standard errors taking clustering into account
[‡] Variance of gamma distribution with expectation 1 for frailty
[•] Partial log likelihood

15.9.2   Poisson regression with normal random intercepts

We could assume that the frailty $\exp(\zeta_j)$ in (15.13) is distributed as log normal (and hence that the random intercept ζ_j is normal) instead of gamma. However, in contrast to the model based on a gamma frailty, this model is considerably more complicated to fit because the marginal likelihood cannot be expressed in closed form and the frailty must be integrated out using numerical integration or Monte Carlo integration. The reason for nevertheless considering models with normal random intercepts is that they are straightforward to extend to situations with *several* random effects. To reduce the computational burden, in particular for larger datasets than that considered here, we will smooth the baseline hazards. If appropriately implemented, this approach will considerably reduce the number of parameters to be estimated and yet produce estimates that are very similar to those produced without smoothing.

Because we have previously shown that a Poisson regression including a fourth-degree orthogonal polynomial for the risk sets produced parameter estimates that closely approximated those from Cox regression for the angina data, we now consider such a model with a normally distributed random intercept ζ_j or log-normal frailty $\exp(\zeta_j)$:

$$\ln(\mu_{sij}) = \ln(t_{sij}) + \delta_0 + \delta_1 p_{1sij} + \cdots + \delta_4 p_{4sij} + \beta_2 x_{2i} + \beta_3 x_{3i} + \beta_4 x_{4i} + \beta_5 x_{5ij} + \zeta_j$$

The random intercept is assumed to have zero mean, variance ψ, and to be independent of the covariates.

We first expand the data to clicks and re-create the orthogonal polynomial terms:

```
. stsplit, at(failures) riskset(click)
(118 failure times)
(10013 observations (episodes) created)
. generate lny = ln(_t - _t0)
(87 missing values generated)
. orthpoly _t, gen(t1-t4) degree(4)
```

We can then fit the model using **xtpoisson** with the **re** and **normal** options:

```
. quietly xtset subj

. xtpoisson _d t1-t4 occ2 occ3 occ4 treat, re normal offset(lny) irr

Random-effects Poisson regression              Number of obs      =      10181
Group variable: subj                           Number of groups   =         21

Random effects u_i ~ Gaussian                  Obs per group: min =         86
                                                              avg =      484.8
                                                              max =        873

                                               Wald chi2(8)       =     150.09
Log likelihood  = -735.80547                   Prob > chi2        =     0.0000
```

_d	IRR	Std. Err.	z	P>|z|	[95% Conf. Interval]	
t1	6.016739	.9847173	10.96	0.000	4.365671	8.29223
t2	.3748918	.0415088	-8.86	0.000	.3017582	.46575
t3	1.620945	.1169195	6.70	0.000	1.407248	1.867092
t4	.8047381	.0411539	-4.25	0.000	.7279884	.8895792
occ2	.5387437	.1450992	-2.30	0.022	.3177818	.913346
occ3	.9221486	.2367601	-0.32	0.752	.5575148	1.525266
occ4	1.731507	.4247124	2.24	0.025	1.070625	2.800342
treat	.2271963	.0525722	-6.40	0.000	.1443568	.3575735
_cons	.006078	.0029611	-10.47	0.000	.0023393	.0157924
lny	1	(offset)				
/lnsig2u	1.462888	.3473601	4.21	0.000	.7820747	2.143701
sigma_u	2.078079	.3609208			1.478514	2.92078

```
Likelihood-ratio test of sigma_u=0: chibar2(01) =    151.94 Pr>=chibar2 = 0.000
```

The estimated regression coefficients have been exponentiated in the output to give incidence rate-ratios (IRR) because we used the `irr` option. These IRRs can be interpreted as conditional hazard ratios (conditional on the random intercepts or frailties). The estimate for `treat`, also reported under "Log-normal frailty" in table 15.5, is close to the corresponding estimate for the model with a gamma frailty. Note that the random-intercept variance $\psi \equiv \mathrm{Var}(\zeta_j)$ cannot be compared directly with the frailty variance $\theta \equiv \mathrm{Var}\{\exp(\zeta_j)\}$.

There is a large random-intercept standard deviation. Comparing the log likelihood with that of the ordinary Poisson model suggests that the intercept varies significantly between subjects. We can interpret the heterogeneity by estimating the median incidence-rate ratio as discussed in section 13.7.2,

$$\mathrm{IRR}_{\mathrm{median}} = \exp\left\{\sqrt{2\psi}\,\Phi^{-1}(3/4)\right\}$$

Plugging in $\widehat{\psi}$, we obtain

```
. display exp(sqrt(2*2.078079^2)*invnormal(3/4))
7.258858
```

For two randomly sampled subjects with the same covariate values, the incidence-rate ratio (or hazard ratio) comparing the subject who has the larger random intercept with the other subject is as large or larger than 7.3 half the time.

15.9.3 Poisson regression with normal random intercept and random coefficient

In addition to including a random intercept ζ_{1j}, we now also allow the treatment effect to vary randomly between subjects by introducing a random coefficient ζ_{2j} for the dummy variable treat:

$$
\begin{aligned}
\ln(\mu_{sij}) &= \ln(t_{sij}) + \delta_0 + \delta_1 p_{1sij} + \cdots + \delta_4 p_{4sij} \\
&\quad + \beta_2 x_{2i} + \beta_3 x_{3i} + \beta_4 x_{4i} + \beta_5 x_{5ij} + \zeta_{1j} + \zeta_{2j} x_{5ij} \\
&= \ln(t_{sij}) + \delta_1 p_{1sij} + \cdots + \delta_4 p_{4sij} \\
&\quad + \beta_2 x_{2i} + \beta_3 x_{3i} + \beta_4 x_{4i} + (\delta_0 + \zeta_{1j}) + (\beta_5 + \zeta_{2j}) x_{5ij}
\end{aligned}
$$

Here ζ_{1j} and ζ_{2j} have a bivariate normal distribution with zero means, variances ψ_{11} and ψ_{22}, respectively, and covariance ψ_{21}. The random effects are assumed to be independent of the covariates.

Because the covariate x_{5ij} that has a random coefficient is binary in this model, the random intercept ζ_{1j} represents the (residual) between-subject variability in the log hazard when ISDN is *not* administered, whereas $\zeta_{1j} + \zeta_{2j}$ represents the between-subject variability when ISDN is administered. The conditional or subject-specific hazards implied by the model with a random intercept and a random treatment effect are shown in table 15.6.

Table 15.6: Conditional or subject-specific hazards implied by Cox model with random intercept and random treatment effect

Regime	Test occasion	Hazard
Placebo	1 (control)	$h_0(t) \exp(\zeta_{1j})$
Placebo	2	$\{h_0(t) \exp(\zeta_{1j})\} \exp(\beta_2)$
Placebo	3	$\{h_0(t) \exp(\zeta_{1j})\} \exp(\beta_3)$
Placebo	4	$\{h_0(t) \exp(\zeta_{1j})\} \exp(\beta_4)$
Treatment	1 (control)	$h_0(t) \exp(\zeta_{1j})$
Treatment	2	$\{h_0(t) \exp(\zeta_{1j})\} \exp(\beta_2)\{\exp(\beta_5) \exp(\zeta_{2j})\}$
Treatment	3	$\{h_0(t) \exp(\zeta_{1j})\} \exp(\beta_3)\{\exp(\beta_5) \exp(\zeta_{2j})\}$
Treatment	4	$\{h_0(t) \exp(\zeta_{1j})\} \exp(\beta_4)\{\exp(\beta_5) \exp(\zeta_{2j})\}$

We can use xtmepoisson to fit a Poisson regression model with both a random intercept and a random coefficient for treatment (models with random coefficients cannot be fit in xtpoisson):

```
. xtmepoisson _d t1-t4 occ2 occ3 occ4 treat || subj: treat,
> covariance(unstructured) offset(lny) irr
```

Mixed-effects Poisson regression   Number of obs   =   10181
Group variable: subj   Number of groups   =   21

                                                    Obs per group: min =   86
                                                                   avg =   484.8
                                                                   max =   873

Integration points =   7   Wald chi2(8)   =   138.45
Log likelihood =   -717.556   Prob > chi2   =   0.0000

_d	IRR	Std. Err.	z	P>|z|	[95% Conf. Interval]	
t1	10.31293	2.157379	11.15	0.000	6.844124	15.53983
t2	.3308228	.044848	-8.16	0.000	.2536311	.4315077
t3	1.688182	.1440924	6.14	0.000	1.428125	1.995595
t4	.7879996	.0466569	-4.02	0.000	.7016604	.8849627
occ2	.4359863	.1274784	-2.84	0.005	.2458042	.7733147
occ3	.8654263	.2323399	-0.54	0.590	.5113393	1.464708
occ4	1.780738	.4521346	2.27	0.023	1.082624	2.929022
treat	.1423461	.0695404	-3.99	0.000	.0546396	.3708372
_cons	.0061234	.0035093	-8.89	0.000	.0019914	.0188286
lny	1	(offset)				

Random-effects Parameters	Estimate	Std. Err.	[95% Conf. Interval]	
subj: Unstructured				
sd(treat)	1.880437	.4138663	1.221567	2.894677
sd(_cons)	2.488696	.4428286	1.755951	3.527209
corr(treat,_cons)	.2635586	.2416877	-.2346751	.652126

LR test vs. Poisson regression: chi2(3) =   188.44   Prob > chi2 = 0.0000
Note: LR test is conservative and provided only for reference.

Using a likelihood-ratio test, there is significant variability in the treatment effect
(twice the difference in log likelihoods is 36.50, which corresponds to a tiny *p*-value also
when taking into account that the null is on the boundary of the parameter space). The
estimated hazard-ratio for `treat`, also reported under "Normal RI & RC" in table 15.5, is
smaller than the corresponding estimate for the model with just a normally distributed
random intercept.

Before proceeding to multilevel accelerated failure-time models, we return to the
nonexpanded form of the data:

```
. drop lny click t1-t4

. stjoin
(option censored(0) assumed)
(10013 obs. eliminated)
```

15.10   Multilevel accelerated failure-time models

15.10.1   Log-normal model with gamma shared frailty

Stata's `streg` command allows gamma and inverse normal frailties but not a log-normal frailty. Here we illustrate a log-normal survival model with a shared frailty $\exp(\zeta_j)$ having a gamma distribution:

$$\ln(T_{ij}) = \beta_1 + \beta_2 x_{2i} + \beta_3 x_{3i} + \beta_4 x_{4i} + \beta_5 x_{5ij} + \zeta_j + \epsilon_{ij}, \qquad \epsilon_{ij} \sim N(0, \sigma^2)$$

This model may be fit using `streg` with the `distribution(lognormal)`, `shared()`, and `frailty(gamma)` options:

```
. streg occ2 occ3 occ4 treat, distribution(lognormal) shared(subj) frailty(gamma)

        failure _d:  uncen
  analysis time _t:  second
               id:  id

Lognormal regression --
        accelerated failure-time form          Number of obs    =       168
        Gamma shared frailty                    Number of groups =        21
Group variable: subj
No. of subjects =          168                  Obs per group: min =         8
No. of failures =          155                                 avg =         8
Time at risk    =        47267                                 max =         8
                                                LR chi2(4)       =     24.73
Log likelihood  =    -73.791288                 Prob > chi2      =    0.0001
```

_t	Coef.	Std. Err.	z	P>|z|	[95% Conf. Interval]	
occ2	.1145159	.0698114	1.64	0.101	-.0223118	.2513437
occ3	.0216564	.0697459	0.31	0.756	-.1150431	.1583559
occ4	-.0927119	.0689199	-1.35	0.179	-.2277925	.0423687
treat	.1992808	.0575947	3.46	0.001	.0863973	.3121643
_cons	5.18272	.0782714	66.21	0.000	5.029311	5.336129
/ln_sig	-1.331404	.1065086	-12.50	0.000	-1.540157	-1.122651
/ln_the	.6189547	.3417248	1.81	0.070	-.0508136	1.288723
sigma	.2641063	.0281296			.2143475	.3254161
theta	1.856986	.634578			.9504559	3.62815

```
Likelihood-ratio test of theta=0: chibar2(01) =    99.88 Prob>=chibar2 = 0.000
```

Exponentiating the estimated treatment coefficient, we get 1.22 $[= \exp(0.1992808)]$. This means that the subject-specific effect of taking the treatment is to prolong the time it takes to reach a given survival probability by about 22%.

Because we used a parametric (log-normal) baseline hazard, `stcurve` can produce graphs for conditional hazards, given that the frailty takes the mean value of 1, or unconditional or marginal hazards, integrating out the frailty. The conditional hazards at the second exercise test occasion are obtained using

```
. stcurve, at1(occ2=1 occ3=0 occ4=0 treat=1)
> at2(occ2=1 occ3=0 occ4=0 treat=0) hazard
> xtitle(Time in seconds) ytitle(Hazard function)
> legend(order(1 "Treatment" 2 "Placebo"))
(option alpha1 assumed)
```

and the marginal hazards are obtained using the above command with the option
`unconditional`. The graphs are shown in figure 15.10.

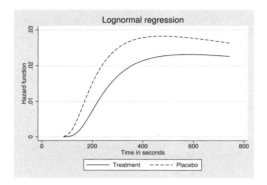 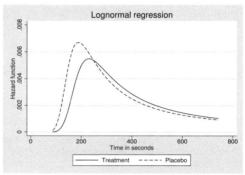

Figure 15.10: Subject-specific or conditional hazard functions (left) and population-averaged or marginal hazard functions (right) at second exercise test for treatment and placebo groups

These graphs nicely illustrate the *frailty* phenomenon. According to the fitted frailty model, the subject-specific or conditional hazard functions shown in the left panel increase until they become practically constant. In contrast, the population-averaged or marginal hazard functions in the right panel have pronounced peaks. This shape of the marginal hazard function can be explained by the fact that frail subjects tend to experience angina early and leave behind more resilient subjects.

15.10.2   Log-normal model with log-normal shared frailty

We now consider the log-normal survival model (15.10.1) with normally distributed subject-specific intercepts $\zeta_j \sim N(0, \psi)$, or in other words, log-normal frailties instead of gamma frailties. This model can be fit using `xtintreg`, the random-intercept version of the `intreg` command used earlier to fit the log-normal survival model.

Two variables are required when using `xtintreg`: the lower limit of the time to event or censoring (which we will call `dep_lo`) and the higher limit (called `dep_hi`). Because there is no left-censoring here, `dep_lo` contains the logarithm of the time to angina or censoring in seconds, whereas `dep_hi` contains the log time to angina if uncensored and Stata's missing indicator, ".", if right-censored.

```
. generate lnsecond = ln(second)

. generate dep_lo = lnsecond

. generate dep_hi = lnsecond if uncen==1
(100 missing values generated)

. replace dep_hi = . if uncen==0
(0 real changes made)
```

We can now fit the model using `xtintreg`:

```
. quietly xtset subj

. xtintreg dep_hi dep_lo occ2 occ3 occ4 treat
Random-effects interval regression          Number of obs      =         168
Group variable: subj                        Number of groups   =          21

Random effects u_i ~ Gaussian               Obs per group: min =           8
                                                           avg =         8.0
                                                           max =           8

                                            Wald chi2(4)       =       50.31
Log likelihood  =   7.7822407               Prob > chi2        =      0.0000
```

	Coef.	Std. Err.	z	P>|z|	[95% Conf.	Interval]
occ2	.0844567	.0426664	1.98	0.048	.0008321	.1680813
occ3	.0343252	.0423786	0.81	0.418	-.0487353	.1173858
occ4	-.0450026	.0422627	-1.06	0.287	-.1278359	.0378307
treat	.1889437	.0337492	5.60	0.000	.1227965	.255091
_cons	5.391495	.0918128	58.72	0.000	5.211546	5.571445
/sigma_u	.4016115	.0635735	6.32	0.000	.2770098	.5262132
/sigma_e	.1773681	.0108113	16.41	0.000	.1561783	.1985579
rho	.8367873	.0465668			.729419	.9117519

```
Observation summary:      13  left-censored observations
                         155      uncensored observations
                           0 right-censored observations
                           0      interval observations
```

The estimated treatment effect is close to the corresponding estimate for the model with a gamma shared frailty.

Important limitations of the `xtintreg` command are that random slopes are not allowed and that time-varying covariates cannot be handled. However, `gllamm` can be used to fit models with random coefficients for interval-censored data (see `gllamm` companion).

15.11   A fixed-effects approach

15.11.1   Cox regression with subject-specific baseline hazards

Instead of modeling unobserved heterogeneity via subject-specific random effects, we can use a fixed-effects approach. Cox regression models include a baseline hazard function, making them more complex than standard regression models that include only an intercept when the covariates are all zero. Whereas a subject-specific intercept is often sufficient for representing heterogeneity in standard regression models, Cox regression may require subject-specific baseline hazards. Furthermore, estimation of Cox regression models with fixed cluster-specific intercepts is problematic when censoring occurs at different times for units nested in a given cluster (Holt and Prentice 1974).

We now consider a Cox regression model with subject-specific baseline hazards $h_{0j}(t)$,

$$\ln\{h(t|\mathbf{x}_{ij})\} = \ln\{h_{0j}(t)\} + \beta_2 x_{2i} + \beta_3 x_{3i} + \beta_4 x_{4i} + \beta_5 x_{5ij} \qquad (15.14)$$

The shape of the log baseline hazards are not restricted in any way within or across subjects here, in contrast to section 15.9.1 where a common baseline hazard was multiplied by a subject-specific frailty.

The conditional or subject-specific hazards implied by the Cox model with subject-specific baseline hazards are shown in table 15.7.

Table 15.7: Conditional or subject-specific hazards implied by Cox model with subject-specific baseline hazards

Regime	Test occasion	Hazard
Placebo	1 (control)	$h_{0j}(t)$
Placebo	2	$\{h_{0j}(t)\}\exp(\beta_2)$
Placebo	3	$\{h_{0j}(t)\}\exp(\beta_3)$
Placebo	4	$\{h_{0j}(t)\}\exp(\beta_4)$
Treatment	1 (control)	$h_{0j}(t)$
Treatment	2	$\{h_{0j}(t)\}\exp(\beta_2)\exp(\beta_5)$
Treatment	3	$\{h_{0j}(t)\}\exp(\beta_3)\exp(\beta_5)$
Treatment	4	$\{h_{0j}(t)\}\exp(\beta_4)\exp(\beta_5)$

Such a model can be specified by considering each subject as a *stratum* with a separate baseline hazard function $h_{0j}(t)$ and fit using `stcox` with the `strata()` option.

```
. stcox occ2 occ3 occ4 treat, strata(subj)

        failure _d:  uncen
  analysis time _t:  second
               id:  id

Stratified Cox regr. -- Breslow method for ties

No. of subjects =          168              Number of obs    =          168
No. of failures =          155
Time at risk    =        47267
                                            LR chi2(4)       =        58.85
Log likelihood  =    -189.31811             Prob > chi2      =       0.0000
```

_t	Haz. Ratio	Std. Err.	z	P>|z|	[95% Conf. Interval]	
occ2	.5633327	.1643431	-1.97	0.049	.31801	.9979048
occ3	.7497439	.215071	-1.00	0.315	.4273049	1.315491
occ4	1.593586	.4344607	1.71	0.087	.9339245	2.719188
treat	.2594194	.0640238	-5.47	0.000	.1599296	.4208004

```
                                                            Stratified by subj
```

The estimated hazard ratio of 0.26 for the treatment represents the within-subject ratio of the hazards when treated versus not treated at a given exercise test occasion. In this sense, each subject serves as his or her own control in this analysis.

We see that estimated treatment effect is very similar to that from the model with a random intercept. This suggests that there is no endogeneity problem for the treatment variable, which is indeed true by design in this case because of randomization. This need not be the case in observational studies where different estimates would suggest endogeneity as discussed in section 3.7.4.

The approach with subject-specific baseline hazards is unproblematic when the number of occasions is determined by the design of the study as in the angina experiment. However, observational studies typically have a fixed observation period and a varying number of events observed per subject. In this case, Chamberlain (1985) pointed out that the approach is problematic because the censoring time for the last spell depends on the lengths of the preceding spells. This problem can be avoided by considering only the first $n^* > 1$ spells if all persons have experienced at least n^* spells, at the cost of a potentially large loss in efficiency. Interestingly, Allison (1996) conducted a simulation study from which he concluded that the approach based on subject-specific baseline hazards works satisfactorily for fixed observation periods and varying number of spells as long as the proportion of spells that are censored is not extreme and the number of previous events is not included as a covariate.

15.12   Different approaches to recurrent-event data

Recurrent event or multi-episode data are a special case of multivariate or multilevel survival data and can be modeled as described in the previous sections by including shared frailties in the model. There are, however, several important decisions that must be made for recurrent-event data that need not concern us in other kinds of multilevel survival data. In this section, we closely follow the review by Kelly and Lim (2000).

We will consider the following artificial dataset for three subjects A, B, and C:

```
. use http://www.stata-press.com/data/mlmus3/recurrent, clear
. list, sepby(id) noobs
```

id	number	stop	event
A	1	2	1
A	2	5	1
A	3	14	0
B	1	7	1
B	2	11	1
B	3	17	1
C	1	14	0

Subject A experienced the first event (number=1) at time 2 (stop=2, event=1), the second event (number=2) at time 5 (stop=5, event=1), and the third event (number=3) is censored at time 14 (stop=14, event=0). Subject B experiences three events at times 7, 11, and 17, and subject C experiences no event and is censored at time 14. It is important to remember to define an identifier variable that uniquely labels each "level-1" unit, here each combination of id and event number. This identifier will be used for the id() option of the stset command.

```
. egen idnum = group(id number)
```

Using id(id) instead of id(idnum) would make Stata interpret multiple rows of data per person as episodes for one duration.

Kelly and Lim (2000) describe three model components that need to be determined for recurrent-event data (besides methods for handling within-subject dependence): 1) definition of risk intervals, 2) definition of risk sets, and 3) common versus event-specific baseline hazards.

The following three subsections consider three definitions of risk intervals: total time, counting process, and gap time. For each type of risk interval, risk set definitions are discussed together with the choice between common and event-specific baseline hazard. These definitions will be discussed in the context of Cox regression, but they are equally relevant when other types of analyses are used, including discrete-time survival.

15.12.1 Total time

In this case, the *clock continues to run from the start of observation (time zero), undisturbed by event occurrences, and subjects are at risk for all events from time zero.* An example might be the risk of different kinds of infection following surgery.

In the top panel of figure 15.11, the risk intervals for events 1, 2, and 3 in the artificial dataset are shown as lines that start from zero and end when the event occurs (filled circle) or when censoring occurs (arrowhead). (The gray dummy risk intervals for subject C's second and third events will be explained below.)

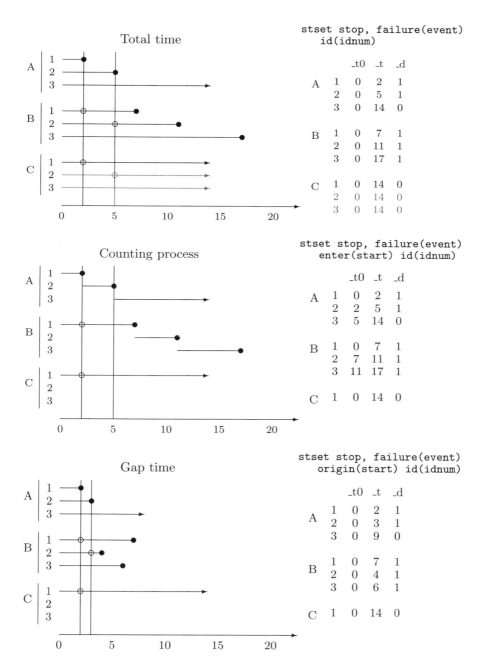

Figure 15.11: Illustration of risk intervals for total time, counting process, and gap time. Unrestricted risk sets shown as intersections between vertical and horizontal lines and restricted risk sets shown as circles (adapted from Kelly and Lim [2000]).

Total time risk intervals can be defined using the following `stset` command:

```
. stset stop, failure(event) id(idnum)
                id:  idnum
     failure event:  event != 0 & event < .
obs. time interval:  (stop[_n-1], stop]
 exit on or before:  failure
```

```
        7  total obs.
        0  exclusions
```

```
        7  obs. remaining, representing
        7  subjects
        5  failures in single failure-per-subject data
       70  total analysis time at risk, at risk from t =          0
                              earliest observed entry t =          0
                                  last observed exit t =         17
```

They can be inspected using the `list` command:

```
. list id number stop event _t0 _t _d if _st==1, sepby(id) noobs
```

id	number	stop	event	_t0	_t	_d
A	1	2	1	0	2	1
A	2	5	1	0	5	1
A	3	14	0	0	14	0
B	1	7	1	0	7	1
B	2	11	1	0	11	1
B	3	17	1	0	17	1
C	1	14	0	0	14	0

The `stset` command produces the variables `_t0`, the start times of the risk intervals; `_t`, the end times; and `_d`, the corresponding failure indicator. We only listed data that contribute information for survival analysis (for which `_st` takes the value 1).

Total time can be combined with *unrestricted risk sets*, meaning that all subjects' risk intervals contribute to the risk set for a given event, regardless of the number of events experienced by the subjects. This can be implemented by specifying a *common baseline hazard* across events. Lee, Wei, and Amato (2010) specified such a model for time to blindness in each eye due to diabetes. In the top panel of figure 15.11, unrestricted risk sets for the first two events of subject A are shown as all the intervals that intersect with the vertical lines at the corresponding event times (regardless of whether there are hollow circles). Kelly and Lim (2000) point out that this specification does not make sense because it allows subjects to make several contributions to a given risk set.

Another possibility is to use *semirestricted risk sets*, where subjects must have experienced fewer than k events to contribute to the risk set for a kth event (Wei, Lin, and Weissfeld 1989). In this case, the baseline hazard is specified as event specific by using the `strata(number)` option in the `stcox` command. Stratifying by event number means that only risk intervals for the same event number contribute to a given risk set. The risk intervals contributing to the risk sets for the first and second events of subject A are shown as hollow circles in the figure.

We see that it is now necessary to define *dummy risk intervals* for subject C (shown in gray) to make this subject contribute to the risk sets of events 2 and 3. We can create these dummy risk intervals by using the `fillin` command:

```
. fillin id number
```

This command fills in any missing combinations of the values of `id` and `number` and creates the dummy variable `_fillin` for newly created observations. We set the `stop` time for these observations equal to the previous stop time and set the failure indicator, `event`, to zero:

```
. by id (number), sort: replace stop = stop[_n-1] if _fillin==1
(2 real changes made)
. replace event = 0 if _fillin==1
(2 real changes made)
```

We also need to replace the missing values of `idnum` with unique values (because the variable appears in the `stset` command), and the easiest way to accomplish this is to drop `idnum` and define it from scratch:

```
. drop idnum
. egen idnum = group(id number)
```

The complete set of risk intervals, including the two shown in gray in figure 15.11, are then defined using

```
. stset stop, failure(event) id(idnum)
                    id:  idnum
         failure event:  event != 0 & event < .
    obs. time interval:  (stop[_n-1], stop]
     exit on or before:  failure
   ────────────────────────────────────────────────────────
        9  total obs.
        0  exclusions
   ────────────────────────────────────────────────────────
        9  obs. remaining, representing
        9  subjects
        5  failures in single failure-per-subject data
       98  total analysis time at risk, at risk from t =          0
                               earliest observed entry t =        0
                                  last observed exit t =         17
```

and listed using

```
. list id number stop event _t0 _t _d if _st==1, sepby(id) noobs
```

id	number	stop	event	_t0	_t	_d
A	1	2	1	0	2	1
A	2	5	1	0	5	1
A	3	14	0	0	14	0
B	1	7	1	0	7	1
B	2	11	1	0	11	1
B	3	17	1	0	17	1
C	1	14	0	0	14	0
C	2	14	0	0	14	0
C	3	14	0	0	14	0

15.12.2   Counting process

Here the *clock continues to run from time zero, but subjects are not at risk of the kth event until they have experienced the k−1th event.* An example would be HIV infection (event 1) and AIDS (event 2), or onset of cancer (event 1) and recurrence of cancer after treatment (event 2). Counting process risk intervals are shown in the middle panel of figure 15.11.

Counting process risk sets are defined by specifying the enter() option in stset to define the start times of the risk intervals as being equal to the end times of the previous intervals. We first generate a variable, start, equal to 0 before the first event and equal to the previous value of stop for subsequent events:

```
. by id (number), sort: generate start = stop[_n-1] if _n>1
(3 missing values generated)
. replace start = 0 if start == .
(3 real changes made)
```

We can now specify start in the enter() option of the stset command,

```
. stset stop, failure(event) enter(start) id(idnum)

                id:  idnum
     failure event:  event != 0 & event < .
obs. time interval:  (stop[_n-1], stop]
 enter on or after:  time start
 exit on or before:  failure
─────────────────────────────────────────────────────────────
        9   total obs.
        2   ignored because idnum missing
─────────────────────────────────────────────────────────────
        7   obs. remaining, representing
        7   subjects
        5   failures in single failure-per-subject data
       45   total analysis time at risk, at risk from t =          0
                               earliest observed entry t =          0
                                   last observed exit t =         17
```

and list the risk intervals,

```
. list id number stop event _t0 _t _d if _st==1, sepby(id) noobs
```

id	number	stop	event	_t0	_t	_d
A	1	2	1	0	2	1
A	2	5	1	2	5	1
A	3	14	0	5	14	0
B	1	7	1	0	7	1
B	2	11	1	7	11	1
B	3	17	1	11	17	1
C	1	14	0	0	14	0

(Note that the dummy risk intervals now do not contribute to the survival analysis because **start** equals **stop** for these intervals.) By looking at the intersections of the vertical lines for the first and second events of subject A, we can see that it is no longer problematic to specify *unrestricted risk sets* and a *common baseline hazard*. This approach was proposed by Andersen and Gill (1982).

Alternatively, we can use the **strata(number)** option to specify *restricted risk sets*, shown as hollow circles, as was done by Prentice, Williams, and Peterson (1981) in their model (2).

15.12.3   Gap time

In this case, the *clock is reset to zero for a subject every time an event occurs*. This type of risk interval makes sense if the subject is restored to a similar state after each event. The angina dataset analyzed earlier in this chapter was an example where this was deemed to be appropriate. Another example might be multiple tumor recurrences, each happening after removing the previous tumor. Gap time risk intervals are shown in the bottom panel of figure 15.11.

Gap times can be defined by specifying start in the origin() option,

```
. stset stop, failure(event) origin(start) id(idnum)
                 id:  idnum
      failure event:  event != 0 & event < .
  obs. time interval:  (stop[_n-1], stop]
   exit on or before:  failure
      t for analysis:  (time-origin)
              origin:  time start
```

```
     9   total obs.
     2   ignored because idnum missing
```

```
     7   obs. remaining, representing
     7   subjects
     5   failures in single failure-per-subject data
    45   total analysis time at risk, at risk from t =          0
                            earliest observed entry t =          0
                              last observed exit t =           14
```

which produces the following risk intervals:

```
. list id number stop event _t0 _t _d if _st==1, sepby(id) noobs
```

id	number	stop	event	_t0	_t	_d
A	1	2	1	0	2	1
A	2	5	1	0	3	1
A	3	14	0	0	9	0
B	1	7	1	0	7	1
B	2	11	1	0	4	1
B	3	17	1	0	6	1
C	1	14	0	0	14	0

In the angina data, the time variable second was already a gap time, corresponding with stop-start, so it was not necessary to use the origin() option.

In section 15.8.1, we initially used *unrestricted risk sets* but allowed the hazard function for different events (or exercise test occasions occ) to differ by a multiplicative constant by including dummy variables for test occasion in the model for the log hazard [see model (15.11)].

Later in the same section, we used *restricted risk sets* with *event-specific baseline hazards* by specifying the strata(occ) option in the stcox command [see model (15.14)]. In their model (3), Prentice, Williams, and Peterson (1981) also used gap times with restricted risk sets and event-specific baseline hazards, which would here be implemented by using the strata(number) option. The risk intervals contributing to the restricted risk sets for the first and second events of subject A are shown as hollow circles in the bottom panel of figure 15.11. See exercises 15.4 (with solutions provided) and 15.8 for analyses of real data using the different approaches to recurrent-event data discussed here.

15.13 Summary and further reading

In this chapter, we have introduced the basic notions of survival analysis and discussed various models for single-level survival modeling, such as proportional hazards models (including piecewise exponential models and Cox regression) and accelerated failure-time models (including the log-normal survival model). To accommodate clustered or multilevel survival data, we have extended proportional hazards models and accelerated failure-time models to include frailties and other subject-specific effects. We have also given an overview of important issues in survival modeling of recurrent events.

We have discussed *shared frailty models* for multivariate or multilevel survival data but have not considered *"unshared" frailty models* for single-level data where the frailty takes on a unique value for each observation. The latter models are identified only if, conditional on the frailties, a proportional hazards model is assumed, or if the hazard function is assumed to have a specific parametric form. The marginal hazard function (integrating out the frailty) then departs from the structure assumed for the conditional hazard function and thus provides information on the frailty variance. Unshared frailty models only make sense if we have a strong reason to assume proportional hazards or a parametric hazard function conditionally on the frailties.

Continuous-time survival models for competing risks have not been discussed in this chapter. In this case, there are multiple types of absorbing events, a classical example being different causes of death in mortality studies. Fortunately, it is straightforward to handle competing risks in continuous-time survival analysis if the times to the different events are conditionally independent given the covariates: simply analyze one event at a time as if it were the only event and treat the other events as censoring together with the truly censored observations.

Good introductory books on continuous-time survival analysis include Collett (2003b), Klein and Moeschberger (2003), and Hosmer, Lemeshow, and May (2008) for medicine and Allison (1984, 1995) and Box-Steffensmeier and Jones (2004) for social science. We also recommend Blossfeld, Golsch, and Rohwer (2007) and Cleves et al. (2010) for Stata-specific treatments.

Continuous-time survival modeling of clustered data is discussed in the paper by Pickles and Crouchley (1995) and the book by Skrondal and Rabe-Hesketh (2004, chap. 12). Advanced treatments of multivariate survival analysis include Duchateau and Janssen (2008) and Hougaard (2000). Models with discrete random effects, and many other issues in survival analysis, are discussed in Vermunt (1997).

The first two exercises of this chapter do not involve any clustering. The first is on reimprisonment following prison release (exercise 15.1) and the second revisits the divorce data discussed in this chapter (exercise 15.2). Exercises on recurrent events include multiple divorces (exercise 15.3), blindness in two eyes of patients suffering from diabetic retinopathy (exercise 15.6), recurrence of bladder cancer (exercise 15.4), and multiple infections (exercises 15.7 and 15.8). Exercise 15.5 is on single events for units nested in clusters, specifically, child mortality among children nested in mothers.

Exercises 15.4 and 15.8 illustrate the different approaches for modeling recurrent survival data discussed in section 15.12.

15.14   Exercises

15.1   Reimprisonment data

In this exercise, we consider reimprisonment of former prisoners after release, often called recidivism, in the state of North Carolina in the U.S.A. The main research question is whether participation in the North Carolina prisoner work release program reduces the risk of recidivism after controlling for the nature and severity of the original crime and other covariates.

Here we consider data on 1,445 prisoners provided by Wooldridge (2010). The data are part of a dataset collected by Chung, Schmidt, and Witte (1991) that includes information on prisoners released from the North Carolina prison system between July 1, 1977, and June 30, 1978. Follow-up information on these prisoners was collected through a search of North Carolina Department of Correction records in April 1984. Because of the 1-year window for selection of releases, the follow-up period ranged from 70 to 81 months. The variables were all recorded either at the beginning of the original sentence or at the time of release from prison, and no information is available on changes in the variables after release.

We will use the following variables from the dataset `recid.dta`:

- `durat`: minimum time in months from release until either return to prison or censoring due to end of follow-up period
- `ldurat`: ln(`durat`)
- `cens`: dummy variable for censoring (1: right censoring; 0: event)
- `workprg`: dummy variable for participation in North Carolina prisoner work release program during sentence (1: participation; 0: no participation)
- `priors`: number of previous incarcerations (not including the current incarceration)
- `tserved`: time served in prison (rounded to months)
- `felon`: dummy variable for conviction for felony (1: conviction for felony; 0: conviction for misdemeanor); felony means that the prison sentence is for more than 1 year, whereas misdemeanor means that the sentence is for 1 year or less
- `alcohol`: dummy variable for record indicating serious problem with alcohol (1: alcohol problem; 0: otherwise)
- `drugs`: dummy variable for record indicating use of hard drugs (1: hard drugs; 0: otherwise)
- `black`: dummy variable for being black (1: black; 0: otherwise)
- `married`: dummy variable for being married at time of release (1: married; 0: otherwise)

- `educ`: number of years of schooling completed
- `age`: age in months at time of release

1. Following Wooldridge, fit a Weibull model to the data with `workprg`, `priors`, `tserved`, `felon`, `alcohol`, `drugs`, `nonwhite`, `married`, `educ`, and `age` as covariates. Use the proportional hazards parameterization.
2. Interpret the estimated exponentiated coefficients of `workprg` and `drugs`.
3. Now fit the same model using the accelerated failure-time parameterization.
4. Interpret the estimated exponentiated coefficients of `workprg` and `drugs`.
5. Fit a log-normal accelerated failure-time model, and compare the estimated coefficients of `workprg` and `drugs` with the estimate from step 4.
6. Fit a piecewise exponential model assuming constant hazards in year 1, year 2, year 3, and the remaining time after year 3.
7. Compare the estimated exponentiated coefficients of `workprg` and `drugs` with those from step 1.

15.2 Divorce data

We mentioned in section 15.2 that the time from wedding to divorce could not be determined exactly and that the variables `lower` and `upper` represented lower- and upper-bounds for the durations. To take this *interval-censoring* into account, fit an interval-censored log-normal survival model using `intreg`.

1. Generate the lower and upper bound variables, keeping in mind that we want to fit a model to the log durations.
2. Fit the model with `heblack`, `mixrace`, `hedropout`, `hecollege`, `heolder`, and `sheolder` as covariates. (See section 15.2 on how to create these variables.)
3. Compare the estimates with those based on assuming that the divorce time is known in section 15.5.1.

15.3 Multiple divorce data

Up to now we have restricted our analysis of the divorce data provided by Lillard and Panis (2003) to the first marriage. Here we consider a larger dataset also made available by Lillard and Panis (2003) (and converted to Stata by Germán Rodríguez) that contains times from wedding to divorce for multiple marriages per respondent.

The data `divorce3.dta` contains a row of data for each marriage and most of the variables are as described in section 15.2. Additional variables we will use here are

- `id`: identifier for respondent
- `marnum`: marriage number (1, 2, etc.)
- `censor`: dummy variable for censoring due to loss to follow-up or death of spouse or respondent

1. Generate dummy variables, `second` and `thirdplus`, for the second marriage and for the third and subsequent marriages, respectively. Also create the explanatory variables `mixrace`, `hedropout`, `hecollege`, `heolder`, and `sheolder` as in section 15.2.

2. Create a variable, `dur`, by drawing a random number from the uniform distribution on the interval from `lower` to `upper`. (This can be done by drawing a number from a uniform on the interval from 0 to 1 and applying a linear transformation so that 0 becomes `lower` and 1 becomes `upper`.) Also create an appropriate identifier for the marriages (see section 15.12). Now define the data as survival data by using `stset`.

3. Fit a piecewise exponential model, cutting time at 1, 5, 9, 15, and 25 years, with the variables created in step 1 and `heblack` as covariates. Use robust standard errors for marriages clustered in respondents.

4. Now fit the same model but include a shared gamma frailty instead of using robust standard errors.

5. Compare the estimates of the marginal model in step 3 and the frailty model in step 4. In particular, comment on the change in the estimates of the coefficients of `second` and `thirdplus`.

6. What kind of risk interval did you use in the above analyses? Were the risk sets restricted or unrestricted? (See section 15.12.)

15.4 Bladder cancer data Solutions

Here we consider data from a bladder cancer study conducted by the Veterans Administration Cooperative Urological Research Group (Byar 1980). The data were previously analyzed by Wei, Lin, and Weissfeld (1989) and have been made available by Therneau and Grambsch (2000). Patients who had just had a superficial bladder tumor removed were randomly assigned to one of three treatments to prevent tumor recurrence: placebo, thiotepa, and pyridoxine. Here we consider only the placebo and thiotepa groups. The patients were followed up and information about recurrent tumors was collected, including the recurrence times from the beginning of initial treatment. Only the first four recurrence times will be analyzed here.

The data in `bladder.dta` contain the following variables:

- `id`: patient identifier
- `enum`: recurrence number
- `treat`: dummy variable for receiving thiotepa (1: thiotepa; 0: placebo)
- `number`: number of initial tumors (except 8 denotes 8 or more)
- `size`: size of largest initial tumor in centimeters
- `start`: timing of previous event in months
- `stop`: timing of event in months
- `event`: failure indicator (1: tumor recurrence; 0: censoring)

Each patient has data for each recurrence although censoring for a given recurrence implies censoring for all subsequent recurrences.

```
. list id treat number size enum start stop event if id>6&id<10, sepby(id)
> noobs
```

id	treat	number	size	enum	start	stop	event
7	0	1	1	1	0	18	0
7	0	1	1	2	18	18	0
7	0	1	1	3	18	18	0
7	0	1	1	4	18	18	0
8	0	1	3	1	0	5	1
8	0	1	3	2	5	18	0
8	0	1	3	3	18	18	0
8	0	1	3	4	18	18	0
9	0	1	1	1	0	12	1
9	0	1	1	2	12	16	1
9	0	1	1	3	16	18	0
9	0	1	1	4	18	18	0

For subject 7, the risk intervals for the second to fourth recurrences are therefore dummy risk intervals as shown in the top panel of figure 15.11. Refer to section 15.12 for this exercise. Use the `efron` option for handling ties in all Cox regression analyses below.

1. Wei, Lin, and Weissfeld (1989) specify a marginal Cox regression model based on total time and semirestricted risk sets, where the risk set for a kth event includes risk intervals for all previous events $(< k)$. They specify event-specific baseline hazards and allow the effects of `treat`, `number`, and `size` to differ between events. Fit this model.

2. Use `testparm` to test whether the coefficients of `treat` differ significantly between events (at the 5% level) and similarly for `number` and `size`.

3. Fit the model by Wei, Lin, and Weissfeld (1989) but constraining all coefficients to be the same across events.

4. In their model (2), Prentice, Williams, and Peterson (1981) use counting process risk intervals with restricted risk sets and event-specific baseline hazards. Fit this model, assuming that `treat`, `number`, and `size` have the same coefficients across events.

5. Andersen and Gill (1982) also use counting process risk intervals, but they use unrestricted risk sets and assume that all events have a common baseline hazard function. Fit this model, again assuming that `treat`, `number`, and `size` have the same coefficients across events.

6. In their model (3), Prentice, Williams, and Peterson (1981) use gap time with restricted risk sets and event-specific baseline hazards. Fit this model, assuming that `treat`, `number`, and `size` have the same coefficients across events.

7. Compare and interpret the treatment effect estimates from steps 3 to 6.

15.5 Child mortality data

Here we analyze the Guatemalan child mortality data from Pebley and Stupp (1987) and Guo and Rodríguez (1992), which were described in detail in section 14.3.

The data `mortality.dta` contain the survival times in months (variable `time`) of several children per mother (variable `momid`) together with a dummy variable, `death`, indicating whether the survival time ended in death or censoring. For instance, for mother 26, these variables are

```
. list kidid time death if momid==26, noobs
```

```
  kidid    time    death

   2601      60        0
   2602     .25        1
   2603      48        1
   2604     .25        1
   2605      60        0
```

We see that children 2601 and 2605 (children 1 and 5 of mother 26) were censored at 60 months, whereas children 2602 and 2604 died at 0.25 months and child 2603 died at 48 months.

1. Fit a piecewise exponential model assuming that the incidence rate is constant for the age bands 1–5 months, 6–11 months, 12–23 months, and 24 or more months. Use `stset` and `stsplit` to prepare the data for this analysis. Define dummy variables for the age bands, called `ageband1`, `ageband2`, etc.

2. Create the covariates used in section 14.6:

```
. generate mage2 = mage^2
. generate comp12 = f0011*(ageband4==1)
. generate comp24e = f0011*(ageband5==1)
. generate comp24l = f1223*(ageband5==1)
```

3. Use `streg` to fit a marginal piecewise exponential model with covariates `mage`, `mage2`, `border`, `p0014`, `p1523`, `p2435`, `p36up`, `pdead`, `comp12`, `comp24e`, and `comp24l`.

4. Fit the same piecewise exponential model as in step 3 but with a gamma frailty that is shared by children having the same mother.

5. Use `xtpoisson` to fit the same model as in step 4.

6. Compare the estimates with those in table 14.3 for the random-intercept complementary log-log model that treats the data as interval-censored.

15.6 Blindness data

Here will use the piecewise exponential survival model to analyze the data from exercise 14.6. The study was a randomized trial of a treatment to delay the onset of blindness in patients suffering from diabetic retinopathy. One eye of each patient was randomly selected for treatment, and patients were assessed at 4-month intervals. The endpoint used to assess the treatment effect was the occurrence of visual acuity less than 5/200 at two consecutive assessments.

The variables in the dataset `blindness.dta` are described in exercise 14.6.

1. Expand the data for analysis using the piecewise exponential model with time intervals (6,10], (10,14], . . . , (50,54], (54,58], (58,66], (66,83].

2. Use Poisson regression to fit a piecewise exponential model with dummy variables for the intervals, `treat`, `late`, and the `treat` by `late` interaction as covariates.

3. Fit the same model as in step 2 but with a normally distributed random intercept.

    a. Is there evidence for residual within-subject dependence?
    b. Interpret the estimated incidence-rate ratios for `treat`, `late`, and their interaction.

4. For the model in step 3, test the proportional hazards assumption for `treat` by including an interaction between `treat` and a variable containing the start times of the intervals. (Hint: You can use `_t0` or the variable created using the `stsplit` command in step 1.)

15.7 Randomized trial of infection prevention I

Chronic gramulomatous disease (CGD), also known as Bridges-Good syndrome or Quie syndrome, is a diverse group of immunodeficiencies. The disease usually begins in early childhood and is characterized by recurrent bouts of infection (such as pneumonia, skin abscesses, or suppurative arthritis). In the United States, about 20 new cases are diagnosed each year. In 1986, Genentech, Inc., conducted a randomized, double-blind, placebo-controlled trial with 128 CGD patients. Sixty-three patients received Genentech's humanized interferon gamma (rIFN-g) three times daily for a year, and 65 patients received placebo three times daily for a year. The primary endpoint was the time to the first serious infection, but here we analyze the times to up to seven serious infections that occurred during the study. The data were provided by Therneau and Grambsch (2000) and have previously been discussed by Fleming and Harrington (1991) among others.

The variables in `infections.dta` we will use here are

- `id`: patient identifier
- `number`: infection number
- `stop`: time of infection in days since randomization
- `start`: time of previous infection
- `infection`: indicator for infection (1: infection; 0: censoring)
- `treat`: dummy for interferon gamma versus placebo (1: rIFN-g; 0: placebo)

1. Fit a marginal Cox regression using counting process risk intervals and unrestricted risk sets with a common baseline hazard (see section 15.12). Use `treat` as the only covariate with a common coefficient across events.

2. Expand the data to clicks by using `stsplit` and fit an equivalent model using Poisson regression.

3. Fit a Poisson model with a smooth baseline hazard function, using a fifth-degree orthogonal polynomial. Is the estimate of the hazard ratio for `treat` similar to that from Cox regression?

4. Create a variable in the data representing the predicted baseline hazard function for the model from step 3. Run Cox regression again (on the expanded data) and use `stcurve` to plot the baseline hazard function. Overlay the predicted baseline hazard function for the model from step 3 by using the `addplot()` option (see section 4.8.5 for an example of using `addplot()` in the `serrbar` command).

5. Now include a frailty for subjects in the model from step 3 and fit the model first assuming a gamma frailty distribution and then a log-normal frailty distribution. Does the distributional assumption affect the estimate of the treatment effect?

See also exercise 16.9, where a three-level model is fit to these data.

15.8 Randomized trial of infection prevention II

Here we continue using the dataset `infections.dta` that was used in exercise 15.7, now focusing on different approaches to recurrent events. We will use the following additional variables:

- `age`: age of patient in years at the time of randomization
- `inherit`: pattern of inheritance (1: X-linked; 2: autosomal recessive)
- `steroids`: use of corticosteroids (1: used corticosteroids; 2: did not use corticosteroids)

Use Efron's method for handling ties throughout this exercise.

1. Wei, Lin, and Weissfeld (1989) specify a marginal Cox regression model based on total time and semirestricted risk sets, where the risk set for a kth event includes risk intervals for all previous events ($< k$). Fit such a model with `treat`, `age`, and dummy variables for X-linked inheritance and for using corticosteroids as covariates, with common regression coefficients across events.

2. In their model (2), Prentice, Williams, and Peterson (1981) use counting process risk intervals with restricted risk sets and event-specific baseline hazards. Fit this model, assuming that the covariates have the same coefficients across events.

3. Andersen and Gill (1982) also use counting process risk intervals, but they use unrestricted risk sets and assume that all events have a common baseline hazard function. Fit this model, again assuming that the covariates have the same coefficients across events.

4. In their model (3), Prentice, Williams, and Peterson (1981) use gap time with restricted risk sets and event-specific baseline hazards. Fit this model, assuming that the covariates have the same coefficients across events.

5. Compare and interpret the treatment effect estimates from steps 1 to 4.

Part VIII

Models with nested and crossed random effects

16 Models with nested and crossed random effects

16.1 Introduction

We have until now in this volume considered two-level data where units are nested in groups or clusters.

In this chapter, we first discuss higher-level multilevel or hierarchical models where units are classified by some factor (for instance, community) into top-level clusters. The units in each top-level cluster are then (sub)classified by a further factor (for instance, mother) into clusters at a lower level and so on (for instance, children nested in mothers). The factors defining the classifications are nested if a lower-level cluster can only belong to one higher-level cluster (for instance, a mother and her children can only belong to one community).

We also discuss nonhierarchical models where units are cross-classified by two or more factors. In this case, each unit can potentially belong to any combination of values of the different factors, for instance, children may be cross-classified by the elementary school and high school they attended. If the main effects of a such a cross-classification are represented by random effects, the models have crossed random effects.

For *continuous* responses, we previously discussed higher-level models with nested random effects in chapter 8 and models with crossed random effects in chapter 9. Before reading this chapter on *noncontinuous* responses, we strongly recommend that you read chapters 8 and 9.

16.2 Did the Guatemalan immunization campaign work?

Pebley, Goldman, and Rodríguez (1996) and Rodríguez and Goldman (2001) analyzed data on Guatemalan families' decisions about whether to immunize their children. Data are available from the National Survey of Maternal and Child Health (ENSMI) conducted in Guatemala in 1987. A nationally representative sample of 5,160 women aged between 15 and 44 was interviewed. The questionnaire included questions regarding factors that could affect the immunization status of children who were born in the previous 5 years and alive at the time of the interview.

Beginning in 1986, the Guatemalan government undertook a series of campaigns to immunize the population against major childhood diseases. The immunization campaign visited most of the country and often located children in their own households. The full set of recommended immunizations included three doses of DPT vaccine (against diphtheria, whooping cough, and tetanus), three doses of polio vaccine, one dose of BCG (antituberculosis), and one dose of measles vaccine.

The data considered here comprise 2,159 children aged 1–4 years for whom we have community data on health services and who received at least one immunization. The response variable of interest is whether the children received the full set of immunizations. An important explanatory variable is whether the child was at least 2 years old at the time of the interview, in which case the child was eligible to receive all immunizations during the campaign. If this variable is associated with immunization status, there is some indication that the campaign worked.

The dataset `guatemala.dta` comprises children i nested in mothers j nested in communities k. The multilevel structure of the data is displayed in figure 16.1.

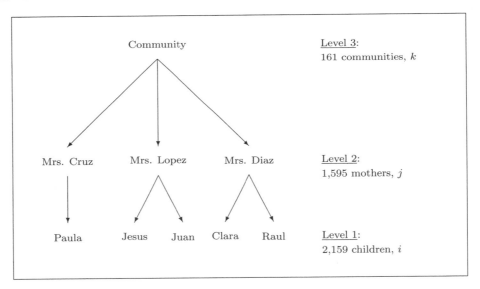

Figure 16.1: Three-level structure of Guatemalan immunization data

The dataset contains the following subset of variables from Pebley, Goldman, and Rodríguez (1996):

- Level 1 (child):

  - `immun`: indicator variable for child receiving full set of immunizations (y_{ijk})
  - `kid2p`: dummy variable for child being at least 2 years old at time of the interview and hence eligible for full set of immunizations (x_{2ijk})

- Level 2 (mother):

  - `mom`: identifier for mothers (j)
  - Ethnicity (dummy variables with Latino as reference category)

    * `indNoSpa`: mother is indigenous, not Spanish speaking (x_{3jk})
    * `indSpa`: mother is indigenous, Spanish speaking (x_{4jk})
  - Mother's education (dummy variables with no education as reference category)

    * `momEdPri`: mother has primary education (x_{5jk})
    * `momEdSec`: mother has secondary education (x_{6jk})
  - Husband's education (dummy variables with no education as reference category)

    * `husEdPri`: husband has primary education (x_{7jk})
    * `husEdSec`: husband has secondary education (x_{8jk})
    * `husEdDK`: husband's education is not known (x_{9jk})

- Level 3 (community):

  - `cluster`: identifier for communities (k)
  - `rural`: dummy variable for community being rural $(x_{10,k})$
  - `pcInd81`: percentage of population that was indigenous in 1981 $(x_{11,k})$

We read in the Guatemalan immunization data by typing

```
. use http://www.stata-press.com/data/mlmus3/guatemala
```

16.3   A three-level random-intercept logistic regression model

In section 10.6, we introduced two-level random-intercept logistic regression models where random intercepts varied between clusters. We now extend such models to include random intercepts varying at both the cluster and the supercluster levels.

16.3.1 Model specification

We specify a three-level random-intercept logistic regression model for childhood immunization with children i nested in mothers j who are nested in communities k:

$$
\begin{aligned}
\text{logit}\{\Pr(y_{ijk}=1|\mathbf{x}_{ijk},\zeta_{jk}^{(2)},\zeta_k^{(3)})\} &= \beta_1 + \beta_2 x_{2ijk} + \cdots + \beta_{11}x_{11,k} + \zeta_{jk}^{(2)} + \zeta_k^{(3)} \\
&= (\beta_1 + \zeta_{jk}^{(2)} + \zeta_k^{(3)}) + \beta_2 x_{2ijk} + \cdots + \beta_{11}x_{11,k} \quad (16.1)
\end{aligned}
$$

Here $\mathbf{x}_{ijk} = (x_{2ijk}, \ldots, x_{11,k})'$ is a vector containing all covariates, $\zeta_{jk}^{(2)} \sim N(0, \psi^{(2)})$ is a random intercept varying over mothers (level 2), and $\zeta_k^{(3)} \sim N(0, \psi^{(3)})$ is a random intercept varying over communities (level 3). The random intercepts $\zeta_{jk}^{(2)}$ and $\zeta_k^{(3)}$ are assumed to be independent of each other and independent across communities, and $\zeta_{jk}^{(2)}$ is assumed to be independent across mothers as well. Both random intercepts are assumed to be independent of the covariates $\mathbf{x}_{ijk}$. The responses y_{ijk} are independently Bernoulli distributed given the random effects and covariates.

The model can alternatively be written as a latent-response model (see section 10.2.2),

$$
y_{ijk}^* = \beta_1 + \beta_2 x_{2ijk} + \cdots + \beta_{11}x_{11,k} + \zeta_{jk}^{(2)} + \zeta_k^{(3)} + \epsilon_{ijk}
$$

where ϵ_{ijk} has a standard logistic distribution with variance $\pi^2/3$. The ϵ_{ijk} are assumed to be mutually independent and to be independent of $\zeta_{jk}^{(2)}$, $\zeta_k^{(3)}$, and $\mathbf{x}_{ijk}$. The observed binary responses are then presumed to be generated by the threshold model

$$
y_{ijk} = \begin{cases} 1 & \text{if } y_{ijk}^* > 0 \\ 0 & \text{otherwise} \end{cases}
$$

16.3.2 Measures of dependence and heterogeneity

Types of residual intraclass correlations of the latent responses

As discussed for two-level models in section 10.9.1, we can consider residual intraclass correlations for the latent responses of two children i and i'. The difference in three-level models is that we can now consider two kinds of residual intraclass correlations (as we saw for linear models in section 8.5).

For the same community k but different mothers j and j', we obtain

$$
\rho(\text{comm.}) \equiv \text{Cor}(y_{ijk}^*, y_{i'j'k}^* \mid \mathbf{x}_{ijk}, \mathbf{x}_{i'j'k}) = \frac{\psi^{(3)}}{\psi^{(2)} + \psi^{(3)} + \pi^2/3}
$$

whereas for the same mother j (and then obviously the same community k), we get

$$
\rho(\text{mother, comm.}) \equiv \text{Cor}(y_{ijk}^*, y_{i'jk}^* \mid \mathbf{x}_{ijk}, \mathbf{x}_{i'jk}) = \frac{\psi^{(2)} + \psi^{(3)}}{\psi^{(2)} + \psi^{(3)} + \pi^2/3}
$$

In a three-level model, $\psi^{(2)} > 0$ and $\psi^{(3)} > 0$, and it follows that $\rho(\text{mother, comm.}) > \rho(\text{comm.})$. This makes sense because responses for children of the same mother are more similar than responses for children from the same community having different mothers.

Types of median odds ratios

As previously discussed for two-level models in section 10.9.2, we can quantify the unobserved heterogeneity by considering the median odds ratio for pairs of randomly sampled units having the same covariate values, where the unit that has the larger random intercept is compared with the unit that has the smaller random intercept. Again, we obtain two versions of the median odds ratio for three-level models.

Comparing children of different mothers in the same community gives the median odds ratio

$$\text{OR(comm.)}_{\text{median}} = \exp\left\{\sqrt{2\psi^{(2)}}\Phi^{-1}(3/4)\right\}$$

and comparing children of different mothers from different communities gives

$$\text{OR}_{\text{median}} = \exp\left\{\sqrt{2(\psi^{(2)} + \psi^{(3)})}\Phi^{-1}(3/4)\right\}$$

There is no unexplained heterogeneity between children of the same mother (median OR=1), some unexplained heterogeneity if the mothers are different, and even more heterogeneity if the communities are also different.

16.3.3   Three-stage formulation

Retaining the distributional assumptions for the random intercepts, we can specify the same model as in (16.1) using the three-stage formulation of Raudenbush and Bryk (2002).

Following Raudenbush and Bryk, we use different letters to denote covariates at different levels: a_{ijk} for level 1, X_{jk} for level 2, and W_k for level 3. The only level-1 covariate, `kid2p`, is therefore denoted a_{1ijk}. The child-level (level-1) model can then be written as

$$\text{logit}\{\text{Pr}(y_{ijk} = 1|\pi_{0jk}, a_{1ijk})\} = \pi_{0jk} + \pi_1 a_{1ijk}$$

where the intercept π_{0jk} varies between mothers j and communities k. Denoting the seven covariates at the mother level as X_{1jk} to X_{7jk}, the mother-level (level-2) model for the intercept becomes

$$\pi_{0jk} = \beta_{00k} + \beta_{01}X_{1jk} + \cdots + \beta_{07}X_{7jk} + r_{0jk}$$

Here the only coefficient having a k subscript is the intercept β_{00k}, which therefore requires a community-level (level-3) model,

$$\beta_{00k} = \gamma_{000} + \gamma_{001}W_{1k} + \gamma_{002}W_{2k} + u_{0k}$$

where W_{1k} is `rural` and W_{2k} is `pcInd81`, the covariates at level 3.

Substituting the model for β_{00k} into the level-2 model and subsequently for π_{0jk} into the level-1 model, we obtain the reduced-form model (16.1)

$$\text{logit}\{\Pr(y_{ijk} = 1|\mathbf{x}_{ijk}, \zeta_{jk}^{(2)}, \zeta_k^{(3)})\} = \underbrace{\gamma_{000}}_{\beta_1} + \underbrace{\pi_1}_{\beta_2} \underbrace{a_{1ijk}}_{x_{2ijk}} + \underbrace{\beta_{01}}_{\beta_3} \underbrace{X_{1jk}}_{x_{3jk}} + \cdots + \underbrace{\beta_{07}}_{\beta_9} \underbrace{X_{7jk}}_{x_{9jk}}$$
$$+ \underbrace{\gamma_{001}}_{\beta_{10}} \underbrace{W_{1k}}_{x_{10,k}} + \underbrace{\gamma_{002}}_{\beta_{11}} \underbrace{W_{2k}}_{x_{11,k}} + \underbrace{r_{0jk}}_{\zeta_{jk}^{(2)}} + \underbrace{u_{0k}}_{\zeta_k^{(3)}}$$

16.4   Estimation of three-level random-intercept logistic regression models

Two Stata commands can be used to fit three-level (and higher-level) logistic regression models: xtmelogit and gllamm (xtlogit can only be used for two-level models). For the applications considered here, gllamm is faster, so we start with gllamm.

16.4.1   Using gllamm

In gllamm, the clustering variables for the different levels in the model are given in the i() option, starting from level 2 and then going up the levels. This ordering is the reverse of that used in xtmelogit (and xtmixed).

When all the random effects are random intercepts, as in the present example, gllamm does not require the eqs() option. The gllamm command for the three-level random-intercept logistic regression model (16.1) is therefore simply

```
. gllamm immun kid2p indNoSpa indSpa momEdPri momEdSec husEdPri husEdSec
> husEdDK rural pcInd81, family(binomial) link(logit) i(mom cluster) nip(5)

number of level 1 units = 2159
number of level 2 units = 1595
number of level 3 units = 161

Condition Number = 10.125574

gllamm model

log likelihood = -1328.0727
```

immun	Coef.	Std. Err.	z	P>|z|	[95% Conf. Interval]	
kid2p	1.712282	.2139083	8.00	0.000	1.293029	2.131535
indNoSpa	-.2992919	.4837166	-0.62	0.536	-1.247359	.6487753
indSpa	-.2178983	.361165	-0.60	0.546	-.9257688	.4899721
momEdPri	.3789441	.215497	1.76	0.079	-.0434222	.8013104
momEdSec	.3836723	.4605474	0.83	0.405	-.518984	1.286329
husEdPri	.4934885	.2244022	2.20	0.028	.0536683	.9333087
husEdSec	.4466857	.4008267	1.11	0.265	-.3389201	1.232291
husEdDK	-.0079424	.3485074	-0.02	0.982	-.6910043	.6751195
rural	-.8642705	.300585	-2.88	0.004	-1.453406	-.2751347
pcInd81	-1.17417	.4953427	-2.37	0.018	-2.145023	-.2033157
_cons	-1.054729	.4085558	-2.58	0.010	-1.855484	-.2539746

```
Variances and covariances of random effects
------------------------------------------------------------------------------

***level 2 (mom)

   var(1): 5.4272669 (1.3185042)

***level 3 (cluster)

   var(1): 1.1338841 (.37262628)
------------------------------------------------------------------------------
```

Here we have used only five quadrature points (by using the nip(5) option) because estimation would otherwise be slow for this sample. With as few as five points, the version of adaptive quadrature implemented in gllamm is sometimes unstable, so we have used ordinary quadrature by omitting the adapt option.

We now increase the number of quadrature points to the default of eight per dimension and use adaptive quadrature to get more accurate results. We first place the current estimates in the row matrix a and then specify the from(a) option in gllamm to use these estimates as starting values.

```
. matrix a = e(b)

. gllamm immun kid2p indNoSpa indSpa  momEdPri momEdSec husEdPri husEdSec
> husEdDK rural pcInd81, family(binomial) link(logit) i(mom cluster)
> from(a) adapt

number of level 1 units = 2159
number of level 2 units = 1595
number of level 3 units = 161

Condition Number = 9.6723285

gllamm model

log likelihood = -1328.496
```

immun	Coef.	Std. Err.	z	P>|z|	[95% Conf. Interval]	
kid2p	1.713088	.2154478	7.95	0.000	1.290818	2.135357
indNoSpa	-.299316	.4778503	-0.63	0.531	-1.235885	.6372535
indSpa	-.1576852	.3568031	-0.44	0.659	-.8570065	.5416361
momEdPri	.3840802	.2170398	1.77	0.077	-.0413101	.8094705
momEdSec	.3616156	.4738778	0.76	0.445	-.5671679	1.290399
husEdPri	.4988694	.2274981	2.19	0.028	.0529813	.9447575
husEdSec	.4377393	.4042692	1.08	0.279	-.3546138	1.230092
husEdDK	-.0091557	.3519029	-0.03	0.979	-.6988728	.6805614
rural	-.8928454	.2990444	-2.99	0.003	-1.478962	-.3067291
pcInd81	-1.154086	.4935178	-2.34	0.019	-2.121363	-.1868085
_cons	-1.027388	.4061945	-2.53	0.011	-1.823514	-.2312612

```
Variances and covariances of random effects
-------------------------------------------------------------------------

***level 2 (mom)

   var(1): 5.1872364 (1.1927247)

***level 3 (cluster)

   var(1): 1.0274006 (.31690328)
-------------------------------------------------------------------------
```

The maximum likelihood estimates have changed somewhat but not drastically. Fitting the model with 12 points per dimension (not shown) gives nearly the same results as for 8 points, indicating that the latter estimates, reported under "Est" in table 16.1, are reliable. We store these estimates for later use under the name `glri8`:

```
. estimates store glri8
```

To obtain estimated odds ratios with 95% confidence intervals, we can simply replay the results with the **eform** option:

```
. gllamm, eform

number of level 1 units = 2159
number of level 2 units = 1595
number of level 3 units = 161

Condition Number = 9.6723285

gllamm model

log likelihood = -1328.496
```

immun	exp(b)	Std. Err.	z	P>|z|	[95% Conf. Interval]	
kid2p	5.546059	1.194886	7.95	0.000	3.635758	8.46007
indNoSpa	.7413251	.3542425	-0.63	0.531	.2905774	1.891279
indSpa	.8541186	.3047522	-0.44	0.659	.4244307	1.718817
momEdPri	1.468263	.3186716	1.77	0.077	.9595316	2.246718
momEdSec	1.435647	.6803213	0.76	0.445	.5671293	3.634237
husEdPri	1.646858	.3746571	2.19	0.028	1.05441	2.57219
husEdSec	1.549201	.6262943	1.08	0.279	.7014443	3.421546
husEdDK	.9908861	.3486957	-0.03	0.979	.4971454	1.974986
rural	.4094889	.1224554	-2.99	0.003	.2278742	.7358499
pcInd81	.3153457	.1556287	-2.34	0.019	.1198682	.8296026
_cons	.3579408	.1453936	-2.53	0.011	.1614573	.7935322

```
Variances and covariances of random effects
------------------------------------------------------------------------------

***level 2 (mom)

    var(1): 5.1872364 (1.1927247)

***level 3 (cluster)

    var(1): 1.0274006 (.31690328)
------------------------------------------------------------------------------
```

The estimated conditional odds ratios with confidence intervals are also reported under "OR" in table 16.1. The explanatory variable of main interest **kid2p** has a large estimated odds ratio of 5.55, adjusted for the other observed covariates and given the random intercepts at the mother and community levels. There is thus some evidence of an effect of the government campaign on child immunization, although it should be emphasized that the study is observational and thus vulnerable to confounding.

Table 16.1: Maximum likelihood estimates for three-level random-intercept logistic model (using eight-point adaptive quadrature in `gllamm`)

	Est	(SE)	OR	(95% CI)
Fixed part				
β_1 [_cons]	−1.03	(0.41)		
β_2 [kid2p]	1.71	(0.22)	5.55	(3.64, 8.46)
β_3 [indNoSpa]	−0.30	(0.48)	0.74	(0.29, 1.89)
β_4 [indSpa]	−0.16	(0.36)	0.85	(0.42, 1.72)
β_5 [momEdPri]	0.38	(0.22)	1.47	(0.96, 2.25)
β_6 [momEdSec]	0.36	(0.47)	1.44	(0.57, 3.63)
β_7 [husEdPri]	0.50	(0.23)	1.65	(1.05, 2.57)
β_8 [husEdSec]	0.44	(0.40)	1.55	(0.70, 3.42)
β_9 [husEdDK]	−0.01	(0.35)	0.99	(0.50, 1.97)
β_{10} [rural]	−0.90	(0.30)	0.41	(0.23, 0.74)
β_{11} [pcInd81]	−1.15	(0.49)	0.32	(0.12, 0.83)
Random part				
$\psi^{(2)}$	5.19			
$\psi^{(3)}$	1.03			
Log likelihood		−1328.50		

To interpret the random part of the model, it is instructive to consider the estimated residual intraclass correlations of the latent responses. For children of the same mother, $\widehat{\rho}(\text{mother, comm.})$ is obtained as

```
. display (1.0274+5.1872)/(1.0274+5.1872+_pi^2/3)
.65386089
```

and for children of different mothers in the same community, $\widehat{\rho}(\text{comm.})$ is obtained as

```
. display 1.0274/(1.0274+5.1872+_pi^2/3)
.10809653
```

Comparing children of different mothers from different communities, we obtain the estimated median odds ratio

```
. display exp(sqrt(2*(1.0274+5.1872))*invnorm(3/4))
10.782434
```

and comparing children of different mothers from the same community, we obtain

```
. display exp(sqrt(2*(5.1872))*invnorm(3/4))
8.7800782
```

These odds ratios are large compared with the estimated odds ratio for `kid2p`.

16.4.2 Using xtmelogit

In xtmelogit, the clustering variables for the different levels in the model are given starting from the top level and then going down the levels, just as in xtmixed. In the current three-level application, we first specify random intercepts for communities using || cluster: and then random intercepts for mothers using || mom:.

Because estimation of this model is time consuming, we start by using the computationally efficient Laplace method, which is obtained in xtmelogit by specifying the intpoints(1) or laplace option (even this command will take a long time):

```
. xtmelogit immun kid2p indNoSpa indSpa momEdPri momEdSec husEdPri husEdSec
> husEdDK rural pcInd81 || cluster: || mom:, intpoints(1)
Mixed-effects logistic regression              Number of obs     =     2159
```

Group Variable	No. of Groups	Observations per Group			Integration Points
		Minimum	Average	Maximum	
cluster	161	1	13.4	55	1
mom	1595	1	1.4	3	1

```
                                           Wald chi2(10)     =      93.95
Log likelihood = -1359.709                 Prob > chi2       =     0.0000
```

immun	Coef.	Std. Err.	z	P>|z|	[95% Conf. Interval]	
kid2p	1.28234	.158357	8.10	0.000	.9719658	1.592714
indNoSpa	-.2017537	.3312773	-0.61	0.543	-.8510452	.4475378
indSpa	-.0893714	.2481248	-0.36	0.719	-.5756871	.3969443
momEdPri	.262748	.1495222	1.76	0.079	-.0303101	.5558061
momEdSec	.2554452	.3288252	0.78	0.437	-.3890403	.8999307
husEdPri	.3687133	.1566719	2.35	0.019	.0616421	.6757846
husEdSec	.3216275	.2795014	1.15	0.250	-.2261852	.8694402
husEdDK	.0096085	.2430354	0.04	0.968	-.4667321	.4859491
rural	-.6601202	.2075705	-3.18	0.001	-1.066951	-.2532896
pcInd81	-.8580105	.3431571	-2.50	0.012	-1.530586	-.1854348
_cons	-.7624847	.2861215	-2.66	0.008	-1.323272	-.2016969

Random-effects Parameters	Estimate	Std. Err.	[95% Conf. Interval]	
cluster: Identity				
sd(_cons)	.7207785	.1005431	.5483599	.9474101
mom: Identity				
sd(_cons)	1.111227	.1579589	.841019	1.468248

```
LR test vs. logistic regression:     chi2(2) =     88.72   Prob > chi2 = 0.0000
Note: LR test is conservative and provided only for reference.
Note: log-likelihood calculations are based on the Laplacian approximation.
```

The estimates from the Laplace method are placed in the row matrix a for subsequent use as starting values:

```
. matrix a = e(b)
```

We now perform maximum likelihood estimation using adaptive quadrature with five integration points in xtmelogit by specifying the intpoints(5) option. We can save time by using the Laplace estimates just placed in the matrix a as starting values instead of letting xtmelogit compute its own starting values. This is accomplished by using the from() option to pass the starting values to xtmelogit and the refineopts(iterate(0)) option to specify 0 iterations for "refining" the starting values:

```
. xtmelogit immun kid2p indNoSpa indSpa momEdPri momEdSec husEdPri husEdSec
> husEdDK rural pcInd81 || cluster: || mom:, intpoints(5) from(a)
> refineopts(iterate(0))
Mixed-effects logistic regression              Number of obs     =       2159
```

Group Variable	No. of Groups	Observations per Group Minimum	Average	Maximum	Integration Points
cluster	161	1	13.4	55	5
mom	1595	1	1.4	3	5

```
                                          Wald chi2(10)     =      91.18
Log likelihood = -1329.993                Prob > chi2       =     0.0000
```

immun	Coef.	Std. Err.	z	P>|z|	[95% Conf. Interval]	
kid2p	1.640996	.2003109	8.19	0.000	1.248394	2.033598
indNoSpa	-.293059	.4533644	-0.65	0.518	-1.181637	.595519
indSpa	-.1521803	.3383808	-0.45	0.653	-.8153945	.511034
momEdPri	.3615499	.2047625	1.77	0.077	-.0397772	.762877
momEdSec	.3411994	.4484829	0.76	0.447	-.5378109	1.22021
husEdPri	.4748191	.2145763	2.21	0.027	.0542574	.8953809
husEdSec	.4164453	.3824737	1.09	0.276	-.3331894	1.16608
husEdDK	-.0073808	.3328712	-0.02	0.982	-.6597963	.6450347
rural	-.8526129	.2837777	-3.00	0.003	-1.408807	-.2964189
pcInd81	-1.098836	.4679617	-2.35	0.019	-2.016024	-.1816477
_cons	-.9821477	.3854029	-2.55	0.011	-1.737524	-.2267719

Random-effects Parameters	Estimate	Std. Err.	[95% Conf. Interval]	
cluster: Identity				
sd(_cons)	.9768667	.1458175	.7290813	1.308864
mom: Identity				
sd(_cons)	2.093691	.2102444	1.719636	2.54911

```
LR test vs. logistic regression:     chi2(2) =    148.15    Prob > chi2 = 0.0000
Note: LR test is conservative and provided only for reference.
```

The maximum likelihood estimates based on five-point adaptive quadrature are placed in the row matrix a for use as starting values:

```
. matrix a = e(b)
```

The estimated standard deviation $\sqrt{\widehat{\psi^{(2)}}}$ of the mother-level random intercept and the estimated effects of the covariates are substantially larger than the estimates obtained from the Laplace method. It is evident that the rather crude Laplace method should be used with considerable caution.

However, the five-point estimates using xtmelogit differ from the eight-point solution using gllamm, so it appears that more integration points are required. We therefore use eight-point adaptive quadrature with the five-point estimates in the vector a as starting values (this takes some time to run):

```
. xtmelogit immun kid2p indNoSpa indSpa momEdPri momEdSec husEdPri husEdSec
> husEdDK rural pcInd81 || cluster: || mom:, intpoints(8) from(a)
> refineopts(iterate(0))
```

Mixed-effects logistic regression   Number of obs   =   2159

Group Variable	No. of Groups	Observations per Group Minimum	Average	Maximum	Integration Points
cluster	161	1	13.4	55	8
mom	1595	1	1.4	3	8

```
                                              Wald chi2(10)    =     82.76
Log likelihood = -1328.437                    Prob > chi2      =    0.0000
```

immun	Coef.	Std. Err.	z	P>|z|	[95% Conf. Interval]	
kid2p	1.715209	.2156215	7.95	0.000	1.292599	2.137819
indNoSpa	-.300494	.4779275	-0.63	0.530	-1.237215	.6362266
indSpa	-.158379	.3573174	-0.44	0.658	-.8587083	.5419502
momEdPri	.3851237	.2173339	1.77	0.076	-.0408429	.8110903
momEdSec	.3623701	.4742332	0.76	0.445	-.5671098	1.29185
husEdPri	.4999467	.22774	2.20	0.028	.0535845	.9463089
husEdSec	.4393949	.4048559	1.09	0.278	-.354108	1.232898
husEdDK	-.0091831	.3521292	-0.03	0.979	-.6993437	.6809775
rural	-.8959982	.3001188	-2.99	0.003	-1.48422	-.3077761
pcInd81	-1.15788	.494736	-2.34	0.019	-2.127545	-.1882149
_cons	-1.027295	.4065234	-2.53	0.012	-1.824066	-.2305236

Random-effects Parameters	Estimate	Std. Err.	[95% Conf. Interval]	
cluster: Identity				
sd(_cons)	1.015412	.1566953	.7503911	1.374033
mom: Identity				
sd(_cons)	2.281914	.2604568	1.824498	2.854007

```
LR test vs. logistic regression:    chi2(2) =    151.27   Prob > chi2 = 0.0000
Note: LR test is conservative and provided only for reference.
```

These estimates are close to the estimates from gllamm.

The likelihood-ratio test for the null hypothesis that both random intercept variances are zero that is reported at the bottom of the output is conservative, that is, the p-value is too large. Because the p-value is tiny although it is conservative, we can reject the null hypothesis. As discussed in display 8.1, the asymptotic null distribution is $\frac{1}{4}\chi^2(0) + \frac{1}{2}\chi^2(1) + \frac{1}{4}\chi^2(2)$, so we could also obtain the asymptotic p-value as

```
display chi2tail(1,151.27)/2 + chi2tail(2,151.27)/4
```

Note that $\frac{1}{4}\chi^2(0)$ is a mass at 0 and does not contribute to the tail probability.

16.5   A three-level random-coefficient logistic regression model

We have already seen that the communities vary in their overall levels of immunization, due to both fixed effects and random effects. It would be interesting to investigate whether the effect of the campaign also varies between communities. This can be achieved by including a random coefficient $\zeta_{2k}^{(3)}$ of kid2p (x_{2ijk}) at level 3. To keep the model simple, we retain only kid2p and the community-level covariates:

$$
\begin{aligned}
\text{logit}\{\Pr(y_{ijk}=1|\mathbf{x}_{ijk}, \zeta_{jk}^{(2)}, \zeta_{1k}^{(3)}, \zeta_{2k}^{(3)})\} &= \beta_1 + \beta_2 x_{2ijk} + \beta_{10} x_{10,ijk} + \beta_{11} x_{11,ijk} \\
&\quad + \zeta_{jk}^{(2)} + \zeta_{1k}^{(3)} + \zeta_{2k}^{(3)} x_{2ijk} \\
&= (\beta_1 + \zeta_{jk}^{(2)} + \zeta_{1k}^{(3)}) + (\beta_2 + \zeta_{2k}^{(3)}) x_{2ijk} \\
&\quad + \beta_{10} x_{10,ijk} + \beta_{11} x_{11,ijk} \qquad (16.2)
\end{aligned}
$$

The random intercept $\zeta_{1k}^{(3)}$ and the random coefficient $\zeta_{2k}^{(3)}$ at the community level have a bivariate normal distribution with zero means and covariance matrix

$$
\mathbf{\Psi}^{(3)} = \begin{bmatrix} \psi_{11}^{(3)} & \psi_{12}^{(3)} \\ \psi_{21}^{(3)} & \psi_{22}^{(3)} \end{bmatrix}, \qquad \psi_{21}^{(3)} = \psi_{12}^{(3)} \qquad (16.3)
$$

The random effects at the community level $\zeta_{1k}^{(3)}$ and $\zeta_{2k}^{(3)}$ are assumed to be independent of the random intercept $\zeta_{jk}^{(2)}$ at the mother level. All random effects are assumed to be independent of the covariates $\mathbf{x}_{ijk}$.

It is unusual to include a random coefficient for a given variable at a higher level and not at the lower level(s). Here we have done so because the "treatment" was applied at the community level, and we believe that its effect will vary more between communities than between mothers within communities. Furthermore, individual mothers do not provide much information on the mother-specific effect of kid2p because they do not have many children of each type.

16.6   Estimation of three-level random-coefficient logistic regression models

16.6.1   Using gllamm

We first fit a three-level random-intercept logistic regression model that only includes the treatment variable x_{2ijk} and the community-level variables $x_{10,k}$ and $x_{11,k}$ as covariates,

$$\text{logit}\{\Pr(y_{ijk}=1|\mathbf{x}_{ijk},\zeta_{jk}^{(2)},\zeta_{k}^{(3)})\} = \beta_1 + \beta_2 x_{2ijk} + \beta_{10} x_{10,k} + \beta_{11} x_{11,k} + \zeta_{jk}^{(2)} + \zeta_{k}^{(3)}$$

using the previous estimates for the three-level random-intercept model with the full set of covariates as starting values. These estimates are retrieved using

```
. estimates restore glri8
```

We then copy the starting values into the matrix a,

```
. matrix a = e(b)
```

and pass that to gllamm using the from() option. We must also specify the skip option, because a contains extra parameters (the regression coefficients of the covariates we have dropped):

```
. gllamm immun kid2p rural pcInd81, family(binomial) link(logit)
> i(mom cluster) from(a) skip adapt eform

number of level 1 units = 2159
number of level 2 units = 1595
number of level 3 units = 161

Condition Number = 5.4620661

gllamm model

log likelihood = -1335.0426
```

immun	exp(b)	Std. Err.	z	P>|z|	[95% Conf. Interval]	
kid2p	5.369864	1.151166	7.84	0.000	3.527661	8.174097
rural	.3450857	.0979057	-3.75	0.000	.1978923	.6017623
pcInd81	.1905651	.0678821	-4.65	0.000	.0948053	.3830485
_cons	.8051822	.2467615	-0.71	0.480	.4416002	1.468112

```
Variances and covariances of random effects
------------------------------------------------------------------------------

***level 2 (mom)

    var(1): 5.2136851 (1.2063284)

***level 3 (cluster)

    var(1): 1.0333637 (.31347077)
------------------------------------------------------------------------------
```

We store the estimates for later use under the name `glri0`:

```
. estimates store glri0
```

The maximum likelihood estimates are shown under "Random intercept" in table 16.2. The estimate of the conditional odds ratio for `kid2p` has not changed considerably compared with the estimate for the full model in table 16.1, suggesting that discarding the level-2 covariates does not seriously distort the estimate.

We then turn to estimation of model (16.2), which has a random coefficient $\zeta_{2k}^{(3)}$ for `kid2p` at the community level, as well as random intercepts $\zeta_{jk}^{(2)}$ and $\zeta_{1k}^{(3)}$ at the mother and community levels, respectively. We first specify equations for the intercept(s) and slope:

```
. generate cons = 1
. eq inter: cons
. eq slope: kid2p
```

The new model has two extra parameters: the variance $\psi_{22}^{(3)}$ of the random coefficient and the covariance $\psi_{21}^{(3)}$ between the random intercept and the random coefficient at the community level. We can therefore use the matrix containing the previous estimates as starting values if we add two more values:

```
. matrix a = e(b)
. matrix a = (a,.2,0)
```

As in two-level random-coefficient models, we must use the `nrf()` option to specify the number of random effects. However, in the three-level case, we must specify two numbers of random effects, `nrf(1 2)`, for levels 2 and 3. In the `eqs()` option, we must then specify one equation for level 2 and two equations for level 3. To speed up estimation (although it will still be slow), we reduce the number of quadrature points per dimension for the two random effects at level 3 from the default of eight to four by using the `nip()` option. (More points are often required at the lowest level than at the higher levels.)

```
. gllamm immun kid2p rural pcInd81, family(binomial) link(logit) i(mom cluster)
> nrf(1 2) eqs(inter inter slope) nip(8 4 4) from(a) copy adapt eform

number of level 1 units = 2159
number of level 2 units = 1595
number of level 3 units = 161

Condition Number = 7.1155374

gllamm model

log likelihood = -1330.8285
```

immun	exp(b)	Std. Err.	z	P>|z|	[95% Conf. Interval]	
kid2p	6.729718	1.975344	6.50	0.000	3.785711	11.96317
rural	.3296138	.0989118	-3.70	0.000	.1830516	.5935225
pcInd81	.1769139	.0667759	-4.59	0.000	.0844261	.3707207
_cons	.6861712	.2465649	-1.05	0.295	.3392864	1.387709

```
Variances and covariances of random effects
-----------------------------------------------------------------------

***level 2 (mom)

    var(1): 5.8320415 (1.4186368)

***level 3 (cluster)

    var(1): 2.4200309 (1.0925729)
    cov(2,1): -1.5234084 (.94663254) cor(2,1): -.72984351

    var(2): 1.8003307 (.98477224)
-----------------------------------------------------------------------
```

We store the estimates under the name `glrc`:

```
. estimates store glrc
```

The estimates of the conditional odds ratios and the residual between-mother variance (at level 2) have not changed considerably compared with the three-level random-intercept model. At the community level (level 3), the estimated elements of the covariance matrix of the random intercept and random slope are given under ***level 3 (cluster). The random-intercept variance, now interpretable as the residual between-community variance of the latent responses for children who were too young to be immunized at the time of the campaign (kid2p=0), has increased considerably to $\widehat{\psi}_{11}^{(3)} = 2.42$. The variance of the slope of kid2p can be interpreted as the residual variability in the effectiveness of the campaign across communities and is estimated as $\widehat{\psi}_{22}^{(3)} = 1.80$. The estimated correlation between the random intercepts and slopes is

$$\frac{\widehat{\psi}_{21}^{(3)}}{\sqrt{\widehat{\psi}_{11}^{(3)} \widehat{\psi}_{22}^{(3)}}} = -0.73$$

which suggests that for given covariate values, the immunization campaign was less effective in communities where immunization rates are already high for children who were too young to be immunized during the campaign (kid2p=0). The estimates are also shown in table 16.2 under "Random coefficient". Practically the same estimates (not shown) are obtained with eight quadrature points per dimension.

The three-level random-intercept logistic model, which is nested in the random-coefficient logistic regression model, is rejected at the 5% significance level (using a conservative likelihood-ratio test because the null hypothesis is on the border of parameter space).

```
. lrtest glrc glri0
Likelihood-ratio test                        LR chi2(2)  =      8.43
(Assumption: glri0 nested in glrc)           Prob > chi2 =    0.0148
```

As a next step, we could include cross-level interactions between kid2p and the community-level covariates to try to explain the variability in the effect of kid2p between communities, but we will not pursue this here.

Table 16.2: Maximum likelihood estimates from gllamm for three-level random-intercept and random-coefficient logistic regression models

	Random intercept		Random coefficient	
	Est	(95% CI)	Est	(95% CI)
Fixed part				
Conditional odds ratios				
$\exp(\beta_2)$ [kid2p]	5.37	(3.53, 8.17)	6.73	(3.79, 11.96)
$\exp(\beta_{10})$ [rural]	0.35	(0.20, 0.60)	0.33	(0.18, 0.59)
$\exp(\beta_{11})$ [pcInd81]	0.19	(0.09, 0.38)	0.18	(0.08, 0.37)
Random part				
$\psi^{(2)}$	5.21		5.83	
$\psi_{11}^{(3)}$	1.03		2.42	
$\psi_{22}^{(3)}$			1.80	
$\psi_{21}^{(3)}$			-1.52	
Log likelihood	$-1,335.04$		$-1,330.83$	

16.6.2 Using xtmelogit

We can fit the three-level random-coefficient model (16.2) by using the following xtmelogit command (this takes a long time to run):

```
. xtmelogit immun kid2p rural pcInd81 || cluster: kid2p, covariance(unstructured)
> || mom:, intpoints(4 8) or
Mixed-effects logistic regression              Number of obs      =      2159
```

| Group Variable | No. of Groups | Observations per Group | | | Integration Points |
		Minimum	Average	Maximum	
cluster	161	1	13.4	55	4
mom	1595	1	1.4	3	8

```
                                         Wald chi2(3)       =      60.78
Log likelihood = -1330.7479              Prob > chi2        =     0.0000
```

immun	Odds Ratio	Std. Err.	z	P>|z|	[95% Conf. Interval]	
kid2p	6.758302	1.986326	6.50	0.000	3.798934	12.02302
rural	.3276506	.0988741	-3.70	0.000	.1813622	.5919368
pcInd81	.1751774	.0666157	-4.58	0.000	.0831355	.3691216
_cons	.6885445	.2472419	-1.04	0.299	.3406304	1.391812

Random-effects Parameters	Estimate	Std. Err.	[95% Conf. Interval]	
cluster: Unstructured				
sd(kid2p)	1.342271	.3678758	.7844257	2.29683
sd(_cons)	1.555728	.350135	1.00083	2.41828
corr(kid2p,_cons)	-.7284215	.137793	-.9052754	-.3363832
mom: Identity				
sd(_cons)	2.424341	.2929176	1.913147	3.072125

```
LR test vs. logistic regression:      chi2(4) =     163.69   Prob > chi2 = 0.0000
Note: LR test is conservative and provided only for reference.
```

We store the estimates under the name xtrc:

```
. estimates store xtrc
```

In the random part for the community level, specified after the first double pipe, ||, the syntax cluster: kid2p is used to include a random slope for kid2p, and a random intercept is included by default. The covariance(unstructured) option is used to estimate the covariance matrix in (16.3) without any restrictions. Without this option, the covariance is set to zero, which is rarely recommended in practice (see section 4.4.2). The random part for the mother level is specified after the second double pipe, ||, where mom: means that there should be only a random intercept. We specified the intpoints(4 8) option to use four integration points for each random effect at level 3 and eight integration points at level 2.

16.7   Prediction of random effects

16.7.1   Empirical Bayes prediction

At the time of writing this book, gllapred (after estimation with gllamm) is the only command that can produce empirical Bayes predictions of the random effects for multilevel logistic models (predict for xtmelogit produces empirical Bayes modal predictions). After retrieving the gllamm estimates for the three-level random-coefficient logistic regression model,

```
. estimates restore glrc
```

we use gllapred with the u option,

```
. gllapred zeta, u
(means and standard deviations will be stored in zetam1 zetas1 zetam2 zetas2
> zetam3 zetas3)
```

and list the predicted random effects for the first nine children in the dataset:

```
. sort cluster mom
. list cluster mom zetam1 zetam2 zetam3 in 1/9, sepby(mom) noobs
```

cluster	mom	zetam1	zetam2	zetam3
1	2	1.0097228	.15487673	.04783502
36	185	-2.0246433	-.42102637	-.01032988
36	186	-2.0246433	-.42102637	-.01032988
36	187	-2.0246433	-.42102637	-.01032988
36	188	-.69007018	-.42102637	-.01032988
36	188	-.69007018	-.42102637	-.01032988
36	189	1.1806865	-.42102637	-.01032988
36	190	2.2574648	-.42102637	-.01032988
36	190	2.2574648	-.42102637	-.01032988

Here zetam1 is the predicted random intercept $\widetilde{\zeta}_{jk}^{(2)}$ for mothers, zetam2 is the predicted random intercept $\widetilde{\zeta}_{1k}^{(3)}$ for communities, and zetam3 is the predicted random slope $\widetilde{\zeta}_{2k}^{(3)}$ for communities. The order of these random effects is the same as in the eqs() option and always from the lowest to the highest level.

The random intercept for mothers can vary between mothers but is constant across multiple children of the same mother (mothers 188 and 190). Mothers 185, 186, and 187 all have the same predictions because they have the same covariate values and responses. The predicted random intercepts and slopes at the community level are constant across children and mothers from the same community (clusters 1 and 36). We can plot the predicted random intercepts $\widetilde{\zeta}_{1k}^{(3)}$ and slopes $\widetilde{\zeta}_{2k}^{(3)}$ for the communities by using

```
. egen pick_com = tag(cluster)
. twoway scatter zetam3 zetam2 if pick_com==1, xtitle(Intercept) ytitle(Slope)
```

The resulting graph is shown in figure 16.2.

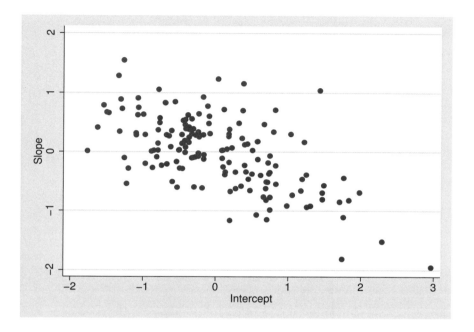

Figure 16.2: Empirical Bayes predictions of community-level random slopes versus community-level random intercepts; based on three-level random-coefficient logistic regression model

Approximate comparative standard errors for the predicted random effects are in the variables zetas1, zetas2, and zetas3.

16.7.2 Empirical Bayes modal prediction

The predict command for xtmelogit provides the *modes* of the posterior distributions instead of the means. We first restore the estimates for the three-level random-coefficient regression model from xtmelogit,

```
. estimates restore xtrc
```

and then use the predict command with the reffects option:

```
. predict comms commi mother, reffects
```

Here we have chosen variable names `comms` and `commi` for the slopes and intercepts at the community level, respectively, and `mother` for the mother-level random intercepts, to remind us which random effects have been stored in which variable (the `predict` command also produces informative variable labels). `xtmelogit` starts from the highest level, and the random intercept always comes last, exactly as in the output.

We list the empirical Bayes modal predictions by using

```
. sort cluster mom
. list cluster mom mother commi comms in 1/9, sepby(mom) noobs
```

cluster	mom	mother	commi	comms
1	2	.6343669	.0970523	.0302855
36	185	-1.786311	-.4229505	-.0808258
36	186	-1.786311	-.4229505	-.0808258
36	187	-1.786311	-.4229505	-.0808258
36	188	-.7136105	-.4229505	-.0808258
36	188	-.7136105	-.4229505	-.0808258
36	189	.8373247	-.4229505	-.0808258
36	190	1.886907	-.4229505	-.0808258
36	190	1.886907	-.4229505	-.0808258

These predictions are similar but not identical to the empirical Bayes counterparts listed (in the same order) earlier. Approximate comparative standard errors can be obtained using the `reses` option.

16.8   Different kinds of predicted probabilities

16.8.1   Predicted population-averaged or marginal probabilities: New clusters

At the time of writing this book, marginal or population-averaged probabilities can be predicted for multilevel logistic regression models only by using `gllapred` after estimation using `gllamm`. Here we have to integrate out all three random effects (the random intercept at level 2 and the random intercept and random coefficient at level 3).

The command for obtaining predicted marginal probabilities of immunization is the same as for two-level models discussed in section 10.13.1:

```
. estimates restore glrc
(results glrc are active now)
. gllapred margp, mu marginal
(mu will be stored in margp)
```

These probabilities can be interpreted as the probabilities of immunization for randomly chosen children of randomly chosen mothers from randomly chosen communities, with given covariate values.

16.8.2   Predicted median or conditional probabilities

Median probabilities are just conditional probabilities given that the random effects are zero. We can obtain median probabilities of immunization by setting the random effects equal to their median values of zero,

```
. generate z1 = 0
. generate z2 = 0
. generate z3 = 0
```

and then using `gllapred` with the `us()` option to base the predictions on these particular values for the random effects:

```
. gllapred condp, mu us(z)
(mu will be stored in condp)
```

Plotting both these conditional probabilities and the marginal probabilities from the previous section by using

```
. label define r 0 "Urban" 1 "Rural"
. label values rural r
. twoway (line condp pcInd81 if kid2p==0, lpatt(solid) sort)
> (line condp pcInd81 if kid2p==1, lpatt(solid) sort)
> (line margp pcInd81 if kid2p==0, lpatt(dash) sort)
> (line margp pcInd81 if kid2p==1, lpatt(dash) sort),
> by(rural) legend(order(1 "Conditional" 3 "Marginal"))
> xtitle(Percentage Indigenous) ytitle(Probability)
```

we obtain the graph in figure 16.3. It is clear that the marginal effects of both percentage indigenous (slopes of dashed curves) and the child being eligible for vaccination during the campaign (vertical distances between dashed curves) are attenuated compared with the conditional effects (solid curves) as usual.

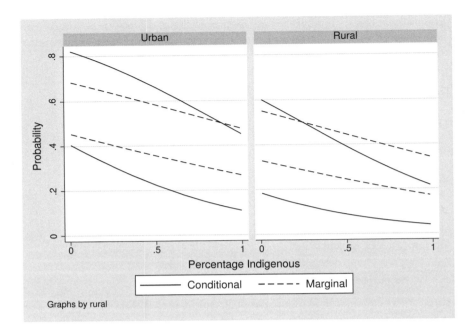

Figure 16.3: Predicted median or conditional probabilities of immunization with random effects set to zero (solid curves) and marginal probabilities of immunization (dashed curves). Curves higher up in each graph correspond to `kid2p = 1` and curves lower down to `kid2p = 0`. Based on three-level random-coefficient logistic regression model.

16.8.3 Predicted posterior mean probabilities: Existing clusters

We now consider predicting the probability of immunization for a child of a particular mother in a particular community. This prediction will benefit from the data we already have for the mother's other children and for the other children in the community. The information these data provide about the community and mother-level random effects is expressed by the posterior distribution of the random effects, given the immunization status and covariates of all children in the community. We therefore integrate the conditional probability of immunization for the child over the posterior distribution of the random effects, as shown for two-level models in (13.11).

The `gllapred` command is also the same as for two-level models:

```
. gllapred postp, mu
(mu will be stored in postp)
```

We now list the predicted posterior mean probabilities together with the empirical Bayes predictions of the random effects:

```
. sort cluster mom
. list cluster mom kid2p zetam1 zetam2 zetam3 postp in 1/9, sepby(mom) noobs
```

cluster	mom	kid2p	zetam1	zetam2	zetam3	postp
1	2	1	1.0097228	.15487673	.04783502	.82746602
36	185	1	-2.0246433	-.42102637	-.01032988	.34772187
36	186	1	-2.0246433	-.42102637	-.01032988	.34772187
36	187	1	-2.0246433	-.42102637	-.01032988	.34772187
36	188	1	-.69007018	-.42102637	-.01032988	.55892132
36	188	1	-.69007018	-.42102637	-.01032988	.55892132
36	189	1	1.1806865	-.42102637	-.01032988	.79839256
36	190	0	2.2574648	-.42102637	-.01032988	.70573537
36	190	1	2.2574648	-.42102637	-.01032988	.90750043

For community 36, all covariates are constant except that `kid2p` changes from 0 to 1 for mother 190 (the other covariates are community specific). Mothers 185, 186, and 187 have a low predicted random intercept of -2.02 and correspondingly a low predicted posterior mean probability of immunization of 0.35, whereas mother 190 has a large predicted random intercept of 2.26 and a correspondingly large predicted probability of 0.91 for the child with `kid2p = 1`. For the other child with `kid2p = 0`, the predicted probability of 0.71 is lower as expected.

16.9 Do salamanders from different populations mate successfully?

We now consider the famous salamander mating data from three experiments conducted by S. J. Arnold and P. A. Verrell; see Verrell and Arnold (1989) for the background of the experiments. The data were first introduced into the statistical literature by McCullagh and Nelder (1989) and then analyzed by Karim and Zeger (1992), Breslow and Clayton (1993), and many others. The purpose of the experiments considered here was to investigate the extent to which mountain dusky salamanders from two geographically isolated populations in North Carolina, U.S.A., would cross-breed. The two populations were named after the location where the salamanders were collected, the Rough Butt Bald in the Great Balsam Mountains ("roughbutt") and the Whiteside Mountain in the Highlands Plateau ("whiteside").

The first experiment in the summer of 1986 used two groups of 20 salamanders. Each group comprised five roughbutt males (RBM), five whiteside males (WSM), five roughbutt females (RBF), and five whiteside females (WSF). Within each group, 60 male–female pairs were formed so that each salamander had three potential partners from the same population and three partners from the other population. This design is shown under

"Experiment 1" in table 16.3, where the pairs are indicated by a 1 (successful mating) or 0 (unsuccessful mating). Two further similar experiments were conducted in the fall of 1987 as shown under "Experiment 2" and "Experiment 3" in table 16.3. Experiment 2 actually used the same salamanders as experiment 1, with salamander 21 corresponding to salamander 1, salamander 31 to salamander 11, etc. Experiment 3 used a new set of salamanders.

Table 16.3: Salamander mating data (layout adapted from Vaida and Meng [2005])

Experiment 1

Group 1

		RBM					WSM				
		01	02	03	04	05	06	07	08	09	10
RBF	01	1			1	1	0			1	1
	02	1		1		1		1	1		1
	03	1	1	1			0		1	1	
	04		1		1	1		1		1	0
	05		1	1	1		1	1	1		
WSF	06	0	0		0			0	1		0
	07	1	0			1	1			1	1
	08			0	0	0	1	1			1
	09	1		1	0		1			1	1
	10		0	1		0		1	1	0	

Group 2

		RBM					WSM				
		11	12	13	14	15	16	17	18	19	20
RBF	11		0	1	1		1			1	1
	12	0		0	1			0	0	0	
	13	1			1	0	0	0			1
	14	1	0			1		1	1	0	
	15		0	0		0	1	0	1		
WSF	16		1		0	0			1	1	1
	17	0	0		0		0	1			0
	18	0	0	0			1	0		0	
	19			1	0	0			1	1	1
	20	0			0	0	0	0	1		

Experiment 2

Group 3

		RBM					WSM				
		21	22	23	24	25	26	27	28	29	30
RBF	21	1			1	1	0			0	0
	22	0		1		1		1	0		1
	23	1	0	1			1		1	1	
	24		0		1	0		0		0	0
	25		0	1	1		0	0	0		
WSF	26	1	0		1			1	1		1
	27	0	0			0	0			1	1
	28			0	0	0	0	0			0
	29	1		1	1		1		1	1	
	30		0	1		0	1	1	1		

Group 4

		RBM					WSM				
		31	32	33	34	35	36	37	38	39	40
RBF	31		0	0	1		0			1	1
	32	1		1	1			0	1	0	
	33	0			0	0	1	0			1
	34	1	1			1		1	0	1	
	35		0	1		0	1	0	1		
WSF	36		0		0	0		1	1	0	
	37	0	0		0		1	0			1
	38	0	0	0			1	1		1	
	39			1	0	0			1	1	0
	40	0			0	0	0	0	0		

Experiment 3

Group 5

		RBM					WSM				
		41	42	43	44	45	46	47	48	49	50
RBF	41	1			1	1	0			1	1
	42	0		0		0		0	1		0
	43	1	1	1			1		1	0	
	44		1		1	0		0		0	1
	45		1	0	1		0	0	1		
WSF	46	0	0		0			0	1		0
	47	0	0			0	0			1	1
	48			0	0	0	0	1			1
	49	0			0	0	0		1	1	
	50		0	0		0		0	1	1	

Group 6

		RBM					WSM				
		51	52	53	54	55	56	57	58	59	60
RBF	51		1	1	1		1			0	1
	52	0		1	0			0	0	0	
	53	1			0	1	1	1			1
	54	0	0			1		1	1	0	
	55	1	1		1		1	0	1		
WSF	56		0		0	0		1	0	1	
	57	0	0		0		1	1			1
	58	1	0	1			1	0		0	
	59			1	0	1			0	0	1
	60	0			0	1	1	1	1		

The dataset `salamander.dta` has the following variables:

- y: indicator for successful mating (1: successful; 0: unsuccessful)
- male: male identifier as in table 16.3 (1–60)
- female: female identifier as in table 16.3 (1–60)
- experiment: experiment number (1–3)
- group: identifiers of six groups of 20 salamanders as in table 16.3 with groups {1,2}, {3,4}, and {5,6} used in experiments 1, 2, and 3, respectively
- rbm, wsm, rbf, wsf: dummy variables for salamander being of the given type: roughbutt males, whiteside males, roughbutt females, and whiteside females, respectively

We read in the salamander data by typing

```
. use http://www.stata-press.com/data/mlmus3/salamander, clear
```

Because the design is complicated, we should first make sure that we understand the data correctly by, for instance, producing the cross-tabulation for the first groups for experiment 1.

```
. table female male if group==1, contents(mean y)
```

female	male 1	2	3	4	5	6	7	8	9	10
1	1			1	1	0			1	1
2	1		1		1		1	1		1
3	1	1	1			0		1	1	
4		1		1	1		1		1	0
5		1	1	1		1	1	1		
6	0	0		0			0	1		0
7	1	0			1	1			1	1
8			0	0	0	1	1			1
9	1		1	0		1		1	1	
10		0	1		0			1	1	0

16.10   Crossed random-effects logistic regression

We now consider Model A from Karim and Zeger (1992) that treated the salamanders from experiments 1 and 2 as independent. A logistic regression model with crossed random effects was specified for successful mating. The model included the covariates wsm (x_{2i}), wsf (x_{3j}), and their interaction, as well as random intercepts ζ_{1i} for males i ($i = 1, \ldots, I$) and ζ_{2j} for females j ($j = 1, \ldots, J$)

$$\text{logit}\{\Pr(y_{ij} = 1 | x_{2i}, x_{3j}, \zeta_{1i}, \zeta_{2j})\} = \beta_1 + \beta_2 x_{2i} + \beta_3 x_{3j} + \beta_4 x_{2i} x_{3j} + \zeta_{1i} + \zeta_{2j}$$

The random intercepts are assumed to be distributed as $\zeta_{1i} \sim N(0, \psi_1)$ and $\zeta_{2j} \sim N(0, \psi_2)$, to be independent of each other, and to be independent of the covariates x_{2i}

and x_{3j}. The random intercepts are crossed because each male i has the same value of his random intercept ζ_{1i} across all females and each female j has the same value of her random intercept ζ_{2j} across all males.

In section 9.3.3, we described in detail how to fit models with crossed random effects by defining the entire dataset to be a cluster at level 3, specifying level-3 random coefficients of dummy variables for the levels of one of the factors, say, males, and defining females to be clusters at level 2 with level-2 random intercepts. The random part of the model is expressed as

$$\zeta_{1i} + \zeta_{2j} \equiv \sum_{r=1}^{I} \zeta_{ra}^{(3)} d_{ri} + \zeta_{j}^{(2)}$$

where a denotes the entire dataset ("all") and $\zeta_{ra}^{(3)}$ is the random coefficient of the dummy variable d_{ri} for male i (equal to 1 if $r = i$ and 0 otherwise) so that

$$\sum_{r=1}^{I} \zeta_{ra}^{(3)} d_{ri} = \zeta_{ia}^{(3)} \equiv \zeta_{1i}$$

The syntax `_all: R.male` makes it easy to specify random coefficients for male dummy variables varying at level 3 (the entire dataset). The complete syntax for the random part of the model is `|| _all: R.male || female:`.

Unfortunately, this model has 60 random coefficients and 1 random intercept, and computation time for generalized linear mixed models increases exponentially with the number of random effects. We will therefore exploit the fact that salamanders are nested within six groups with no matings occurring across groups. This allows us to use the trick described in section 9.8 with `group` as the highest level in the model instead of `_all`. We then only need 10 random coefficients for males (10 males per group); the first random coefficient multiplies a dummy variable for the salamander being the first male in each group (male salamanders 1, 11, 21, 31, 41, and 51); the second is for the second male in each group (male salamanders 2, 12, 22, 32, 42, and 52); and so forth. These sets of six salamanders do not share the same value of the random coefficient because they are in different groups and the random coefficient is nested in groups.

A model with 10 random coefficients and 1 random intercept is still computationally demanding to fit. We will therefore initially use one integration point, which corresponds to the Laplace approximation. We generally do not recommend relying on this approach alone without conducting some simulations to make sure that the estimates are reasonable.

We first relabel the males so that the sets of six salamanders (for example, the set of first salamanders in the six groups) described above have the same label (we do the same for females although this is not necessary):

```
. generate m = male - (group-1)*10
. generate f = female - (group-1)*10
```

To list the new labels, we use the **egen** function **tag()** to define a dummy variable, **pickmale**, for picking out one observation per male:

```
. egen pickmale = tag(male)
```

We can then sort and display the data for the first two sets of six males:

```
. sort m group
. list group male m if pickmale==1&m<3, sepby(m) noobs
```

group	male	m
1	1	1
2	11	1
3	21	1
4	31	1
5	41	1
6	51	1
1	2	2
2	12	2
3	22	2
4	32	2
5	42	2
6	52	2

Consider the data for the first set of six males with **m** equal to 1. A dummy variable for this set of six males will multiply the random coefficient $\zeta_{1k}^{(3)}$, where k denotes **group** (and replaces the subscript a we used previously before implementing the trick). Importantly, the random coefficient $\zeta_{1k}^{(3)}$ takes on a different value for each male in the set, depending on the group it belongs to (namely, $\zeta_{11}^{(3)}$, $\zeta_{12}^{(3)}$, $\zeta_{13}^{(3)}$, $\zeta_{14}^{(3)}$, $\zeta_{15}^{(3)}$, or $\zeta_{16}^{(3)}$).

We start by generating the interaction term,

```
. generate ww = wsf*wsm
```

and fit the model using `xtmelogit` with the `laplace` option:

```
. xtmelogit y wsm wsf ww || group: R.m || f:, laplace
Mixed-effects logistic regression               Number of obs      =        360
```

Group Variable	No. of Groups	Observations per Group			Integration Points
		Minimum	Average	Maximum	
group	6	60	60.0	60	1
f	60	6	6.0	6	1

```
                                                Wald chi2(3)       =      37.57
Log likelihood = -209.27659                     Prob > chi2        =     0.0000
```

y	Coef.	Std. Err.	z	P>|z|	[95% Conf. Interval]	
wsm	-.7020417	.4614756	-1.52	0.128	-1.606517	.2024339
wsf	-2.90421	.5608211	-5.18	0.000	-4.003399	-1.805021
ww	3.588444	.6390801	5.62	0.000	2.33587	4.841018
_cons	1.008221	.3937705	2.56	0.010	.2364454	1.779997

Random-effects Parameters	Estimate	Std. Err.	[95% Conf. Interval]	
group: Identity				
sd(R.m)	1.020296	.2483604	.6331794	1.64409
f: Identity				
sd(_cons)	1.083683	.2530921	.6856555	1.712769

```
LR test vs. logistic regression:     chi2(2) =     27.02    Prob > chi2 = 0.0000
Note: LR test is conservative and provided only for reference.
Note: log-likelihood calculations are based on the Laplacian approximation.
```

Having reduced the dimensionality of integration to 10 at level 3 and 1 at level 2 (compared with 60 at level 3 and 1 at level 2 without the trick), it becomes feasible to use adaptive quadrature in `xtmelogit` with two integration points. To speed things up, we use the estimates from the Laplace approximation as starting values. The `refineopts(iterate(0))` option is used to prevent `xtmelogit` from computing its own starting values:

```
. matrix a = e(b)
. xtmelogit y wsm wsf ww || group: R.m || f:, intpoints(2) from(a)
> refineopts(iterate(0))
```

Mixed-effects logistic regression   Number of obs   =   360

Group Variable	No. of Groups	Observations per Group			Integration Points
		Minimum	Average	Maximum	
group	6	60	60.0	60	2
f	60	6	6.0	6	2

Log likelihood = -208.39862

Wald chi2(3)   =   37.40
Prob > chi2   =   0.0000

y	Coef.	Std. Err.	z	P>|z|	[95% Conf. Interval]	
wsm	-.6946282	.4668329	-1.49	0.137	-1.609604	.2203474
wsf	-2.910889	.5661132	-5.14	0.000	-4.020451	-1.801328
ww	3.587045	.6378697	5.62	0.000	2.336843	4.837246
_cons	1.003062	.3996581	2.51	0.012	.2197465	1.786377

Random-effects Parameters	Estimate	Std. Err.	[95% Conf. Interval]	
group: Identity				
sd(R.m)	1.058889	.2481503	.6689157	1.676213
f: Identity				
sd(_cons)	1.114577	.2522591	.7152533	1.736841

LR test vs. logistic regression: chi2(2) =   28.78   Prob > chi2 = 0.0000
Note: LR test is conservative and provided only for reference.

This did not take long, so we increase the number of integration points to three (this command took a very long time to run!):

```
. matrix a=e(b)
. xtmelogit y wsm wsf ww || group: R.m || f:, intpoints(3) from(a)
> refineopts(iterate(0))
Mixed-effects logistic regression               Number of obs      =         360
```

Group Variable	No. of Groups	Observations per Group Minimum	Average	Maximum	Integration Points
group	6	60	60.0	60	3
f	60	6	6.0	6	3

```
                                                Wald chi2(3)       =       37.07
Log likelihood = -207.71151                     Prob > chi2        =      0.0000
```

y	Coef.	Std. Err.	z	P>|z|	[95% Conf. Interval]	
wsm	-.6962912	.4750503	-1.47	0.143	-1.627373	.2347903
wsf	-2.946427	.5785634	-5.09	0.000	-4.080391	-1.812464
ww	3.621843	.6442771	5.62	0.000	2.359083	4.884603
_cons	1.013944	.4097109	2.47	0.013	.2109251	1.816962

Random-effects Parameters	Estimate	Std. Err.	[95% Conf. Interval]	
group: Identity				
sd(R.m)	1.100838	.2558109	.6981081	1.735898
f: Identity				
sd(_cons)	1.161527	.2610426	.7477042	1.804385

```
LR test vs. logistic regression:     chi2(2) =     30.15   Prob > chi2 = 0.0000
Note: LR test is conservative and provided only for reference.
```

The null hypothesis that the male and female random-effects variances are both zero is rejected even using the conservative test. As discussed in display 8.1, the asymptotic null distribution is $\frac{1}{4}\chi^2(0) + \frac{1}{2}\chi^2(1) + \frac{1}{4}\chi^2(2)$, so we could also obtain the p-value as

```
. display chi2tail(1,30.15)/2 + chi2tail(2,30.15)/4
9.094e-08
```

The estimates using the Laplace approximation and adaptive quadrature with three integration points (AQ-3pt) are given in table 16.4 together with estimates using Monte Carlo EM (MCEM) from Vaida and Meng (2005), Markov chain Monte Carlo (MCMC) with noninformative priors from Karim and Zeger (1992), and penalized quasilikelihood (PQL) from Breslow and Clayton (1993). Treating the estimates from MCEM as the gold standard (because they are maximum likelihood estimates employing Monte Carlo integration), the Laplace estimates are considerably better than the PQL estimates, and the adaptive quadrature estimates with just three quadrature points are nearly identical to the MCEM estimates.

Table 16.4: Different estimates for the salamander mating data

	Laplace		AQ-3pt		MCEM	MCMC		PQL	
	Est	(SE)	Est	(SE)	Est	Est	(SE)*	Est	(SE)
Fixed part									
β_1 [_cons]	1.00	(0.39)	1.01	(0.41)	1.02	1.03	(0.43)	0.79	(0.32)
β_2 [wsm]	−0.70	(0.46)	−0.70	(0.48)	−0.70	−0.69	(0.50)	−0.54	(0.39)
β_3 [wsf]	−2.90	(0.56)	−2.95	(0.58)	−2.96	−3.01	(0.60)	−2.29	(0.43)
β_4 [wsm×wsf]	3.59	(0.64)	3.62	(0.64)	3.63	3.74	(0.68)	2.82	(0.50)
Random part									
$\sqrt{\psi_1}$ [Males]	1.02		1.10		1.11	1.16		0.79	
$\sqrt{\psi_2}$ [Females]	1.08		1.16		1.18	1.22		0.84	

*Range of 90% CI divided by 3.3

When the random intercepts are zero, $\zeta_{1i} = \zeta_{2j} = 0$, the odds of successful mating for salamander pairs where both male and female are roughbutts (RBM−RBF) are estimated as

```
. lincom _cons, or
 ( 1)  [eq1]_cons = 0
```

y	Odds Ratio	Std. Err.	z	P>|z|	[95% Conf. Interval]	
(1)	2.75645	1.129348	2.47	0.013	1.23482	6.153139

For RBM−WSF, we get

```
. lincom _cons + wsf, or
 ( 1)  [eq1]wsf + [eq1]_cons = 0
```

y	Odds Ratio	Std. Err.	z	P>|z|	[95% Conf. Interval]	
(1)	.1447882	.0675077	-4.14	0.000	.0580576	.3610831

For WSM−RBF, we get

```
. lincom _cons + wsm, or
 ( 1)  [eq1]wsm + [eq1]_cons = 0
```

y	Odds Ratio	Std. Err.	z	P>|z|	[95% Conf. Interval]	
(1)	1.373899	.5370912	0.81	0.416	.6385558	2.956042

And for WSM–WSF, we get

```
. lincom _cons + wsm + wsf + ww, or
( 1)  [eq1]wsm + [eq1]wsf + [eq1]ww + [eq1]_cons = 0
```

y	Odds Ratio	Std. Err.	z	P>|z|	[95% Conf. Interval]	
(1)	2.699504	1.099259	2.44	0.015	1.215256	5.996529

Within the same population, the odds of successful mating are high for both white-sides and roughbutts. Across populations, the odds are considerably lower, particularly if the male is a roughbutt and the female a whiteside. However, for a given type of pairing, the estimated median odds ratio comparing the more fertile of two randomly chosen pairs with the less fertile pair is large,

```
. display exp(sqrt(2*(1.10^2+1.16^2))*invnormal(3/4))
4.5946103
```

indicating that there is considerable unexplained heterogeneity.

16.11   Summary and further reading

We have applied models with nested random effects to a three-level dataset on immunization of children in Guatemala where children are nested in mothers who are nested in communities. Different kinds of median odds ratios and residual intraclass correlations between latent responses were described for dichotomous responses. Intraclass correlations between latent responses can also be used in cumulative logit and probit models for ordinal responses and the idea of median odds ratios, median incidence-rate ratios, etc., is applicable for all response types.

We also considered a model with crossed random effects for male and female salamanders, where each male salamander mated with six female salamanders and each female salamander mated with six male salamanders. Although many fundamental issues regarding nested and crossed random effects remain the same as in the continuous case, we have emphasized some of the special challenges in modeling noncontinuous responses.

Raudenbush and Bryk (2002) and Goldstein (2011) discuss generalized linear mixed models with nested and crossed random effects. Guo and Zhao (2000) is a review paper on two- and three-level models for binary responses. We also recommend an excellent encyclopedia entry by Rasbash (2005) on crossed random-effects models and multiple-membership models.

The exercises cover all the response types discussed in this volume. Exercises 16.1, 16.2, 16.3, 16.10, and 16.11 are on binary responses; exercise 16.4 is on an ordinal response; exercise 16.8 is on a nominal response; exercises 16.6 and 16.7 are on counts; exercise 16.5 is on discrete-time survival; and exercise 16.9 is on continuous-time survival. Exercise 16.10 is about multilevel and random-item item-response models and involves crossed and nested random effects. In exercise 16.11, a biometrical genetic model is fit to binary twin data.

16.12 Exercises

16.1 Dairy-cow data

Consider again the data on dairy cows from Dohoo et al. (2001) and Dohoo, Martin, and Stryhn (2010) that was used in exercise 8.8 (see also exercise 10.5).

1. Fit a two-level random-intercept logistic regression model for the response variable `fscr`, an indicator for conception at the first insemination attempt (first service). Include a random intercept for cow and the covariates `lncfs`, `ai`, and `heifer`.

2. Now extend the model by including a random intercept for herds as well. Use `xtmelogit` or `gllamm` with five integration points to speed up estimation. Is there any evidence for unobserved heterogeneity in fertility between herds?

16.2 Tower-of-London data Solutions

Rabe-Hesketh, Toulopoulou, and Murray (2001a) analyzed data on patients with schizophrenia, their relatives, and controls. Cognitive performance was assessed by the Tower of London (a computerized task), which was repeated at three levels of difficulty, starting with the easiest and ending with the hardest. The data have a three-level structure with measurements at occasion i for person j in family k. We will consider the dichotomous response y_{ijk}, equal to 1 if the tower was completed in the minimum number of moves and 0 otherwise.

The dataset `towerl.dta` contains the following variables:

- `famnum`: family identifier
- `id`: person identifier
- `dtlm`: indicator for completing the task in the minimum number of moves
- `level`: level of difficulty of the Tower of London (x_{ijk})
- `group`: group (1: controls; 2: relatives; 3: schizophrenics)

1. Fit the two-level random-intercept model (random intercept for persons):

$$\text{logit}\{\Pr(y_{ijk}=1 \mid \mathbf{x}_{ijk}, \zeta_{jk}^{(2)})\} = \beta_0 + \beta_1 x_{ijk} + \beta_2 g_{2ijk} + \beta_3 g_{3ijk} + \zeta_{jk}^{(2)}$$

where g_{2ijk} and g_{3ijk} are dummy variables for groups 2 and 3, respectively, and $\zeta_{jk}^{(2)} \sim N(0, \psi^{(2)})$ is independent of the covariates $\mathbf{x}_{ijk}$. Here and throughout the exercise, `level` is treated as continuous.

2. Fit the three-level random-intercept model (random intercepts for persons and families):

$$\text{logit}\{\Pr(y_{ijk}=1 \mid \mathbf{x}_{ijk}, \zeta_{jk}^{(2)}, \zeta_k^{(3)})\} = \beta_0 + \beta_1 x_{ijk} + \beta_2 g_{2ijk} + \beta_3 g_{3ijk} + \zeta_{jk}^{(2)} + \zeta_k^{(3)}$$

where $\zeta_{jk}^{(2)} \sim N(0, \psi^{(2)})$ is independent of $\zeta_k^{(3)} \sim N(0, \psi^{(3)})$ and both random effects are assumed independent of $\mathbf{x}_{ijk}$.

3. Compare the models in steps 1 and 2 using a likelihood-ratio test, but retain the three-level model even if the null hypothesis is not rejected at the 5% level.

4. Include a group (controls, relatives, schizophrenics) by level of difficulty interaction in the three-level model. Test the interaction using both a Wald test and a likelihood-ratio test.

5. For the model in step 4, obtain predicted marginal or population-averaged probabilities using `gllapred`. (This requires fitting the model in `gllamm`.) Plot the probabilities against the levels of difficulty with different curves for the three groups.

16.3 Antibiotics data

Acute respiratory tract infection (ARI) is a common disease among children, pneumonia being a leading cause of death in young children in developing countries. In China, the standard medication for ARI is antibiotics, which has led to concerns about antibiotics misuse and resultant drug resistance. As a response, the World Health Organization (WHO) introduced a program of case management for ARI in children under 5 years old in China in the 1990s.

Here we consider data on physicians' prescribing behavior of antibiotics in two Chinese counties, only one of which was in the WHO program. These data have previously been analyzed by Yang (2001); see also Skrondal and Rabe-Hesketh (2003b) and Skrondal and Rabe-Hesketh (forthcoming).

Medical records were examined for medicine prescribed and a correct diagnosis determined from symptoms and clinical signs. The antibiotic prescription was defined as abuse if there were no clinical indications.

The dataset `antibiotics.dta` has the following variables:

- Level 1 (child):

  - `abuse`: classification of prescription (1: correct use; 2: abuse of one antibiotic; 3: abuse of several antibiotics)
  - `age`: age in years (0–4)
  - `temp`: body temperature in Centigrade, centered at 36 degrees
  - `Paymed`: dummy variable for patient paying for his or her own medication
  - `Selfmed`: dummy variable for self-medication before seeing doctor
  - `Wrdiag`: dummy variable for diagnosis classified as wrong

- Level 2 (doctor):

  - `doc`: doctor identifier
  - `DRed`: doctor's education (ordinal with six categories from self-taught to medical school)

- Level 3 (hospital):
  - `hosp`: hospital identifier
  - `WHO`: dummy variable for hospital being in the WHO program

1. Recode `abuse` into a dichotomous variable equal to 1 if there was abuse of one or more antibiotics and 0 otherwise.

2. Fit a three-level logistic regression model for the dichotomized variable `abuse` with random intercepts for doctors and hospitals and no covariates. Use adaptive quadrature with five quadrature points.

3. Add the dummy variable for hospital being in the WHO program to the model. Obtain and interpret the estimated odds ratio for the WHO program, as well as its 95% confidence interval. Does this analysis suggest that the program has been effective?

4. Include all the other covariates in the model and reassess the apparent effect of the WHO program. Comment on the change in the estimated random-intercept variances between the model without covariates (step 2) and the model with all the covariates (step 4).

16.4 Smoking-intervention data

In this exercise, we use the dataset `tvsfpors.dta` from Flay et al. (1988) that was described in exercise 11.3.

1. Investigate how tobacco and health knowledge is influenced by the interventions by fitting a three-level proportional odds model with students nested in classes nested in schools. Include `cc`, `tv`, their interaction, and `prethk` as covariates. Use five-point adaptive quadrature.

2. Does this model fit better than the two-level models with a random intercept either at the class level or at the school level?

16.5 Cigarette data

Use the dataset `cigarette.dta` from Flay et al. (1988) that was described in exercise 14.7.

1. Expand the data to person–period data.

2. Fit the discrete-time survival model that assumes the continuous-time hazards to be proportional. Use dummy variables for periods; include `cc`, `tv`, their interaction, and `male` as explanatory variables; and specify random intercepts for classes and schools. (Five quadrature points at each level should suffice.)

3. Interpret the exponentials of the estimated regression coefficients.

4. Perform likelihood-ratio tests of the null hypotheses that each of the variance components is zero using a 5% level of significance.

5. ❖ Perform a likelihood-ratio test of the joint null hypothesis that both variance components are zero using a 5% level of significance (see display 8.1).

16.6 Health-care reform data

This exercise is based on the health-care reform dataset drvisits.dta from Winkelmann (2004) that was discussed in section 13.4.

1. Fit the random-intercept Poisson model specified on page 696.
2. Write down and fit the three-level model that also includes an occasion and person-specific random intercept in addition to a person-specific random intercept. Use five-point adaptive quadrature.
3. Does the three-level model appear to fit better than the two-level model?
4. ❖ Consider the total random part in the three-level model, and compare the variances for 1996 and 1998 and the covariance with the corresponding estimates for the random-coefficient model fit in section 13.8 (see also section 13.8.3). Comment on the difference between the random-coefficient and three-level models. Is one more general than the other?

16.7 Skin-cancer data

Here we consider the dataset skincancer.dta from Langford, Bentham, and McDonald (1998) that was described in exercise 13.7.

1. Use gllamm to fit a Poisson model for the number of male deaths using the log expected number of male deaths as an offset (see page 724) and uv and squared uv as covariates. Include random intercepts for county, region, and nation. This is a four-level model requiring the option i(county region nation). To speed up estimation in gllamm, use eight quadrature points at level 2 and five points at levels 3 and 4 (the nip(8 5 5) option).
2. Write down the model, and interpret the estimates.
3. Obtain empirical Bayes predictions for the nations.
4. Instead of using a random intercept for nation, include dummy variables for all the nations and exclude the constant (use xtmelogit). How do the estimated regression coefficients of the nation dummy variables compare with the empirical Bayes predictions from step 3? How do the estimated standard errors for these regression coefficients compare with the standard errors for the empirical Bayes predictions?

16.8 British election data

In exercise 12.4, several models are fit to British national election data from 1987 and 1992. In addition to the variables described in exercise 12.4, the dataset elections.dta contains the variable constit, an identifier for the constituency (the district for which a member of parliament is elected) where the respondent lives.

1. Create a variable chosen equal to 1 for the party voted for (rank equal to 1) and zero for the other parties, and standardize lrdist and inflation to have mean 0 and variance 1.

2. Consider a discrete-choice model for party voted for with yr87 and yr92 as the only covariates with party-specific coefficients. Use Conservatives as base outcome and fit the model using clogit and gllamm.

3. Extend the model to include a random slope for lrdist at the voter and constituency levels and fit the extended model in gllamm. Use five quadrature points to speed up estimation.

4. Test whether the constituency-level random slope is needed.

16.9 Randomized trial of infection prevention

1. Solve exercise 15.7.

2. Extend the model with a lognormal frailty at the subject level (fit in step 5 of exercise 15.7) by adding another lognormal frailty at the center level (variable center). Fit the model using xtmepoisson.

3. Is there evidence for between-center variability in the baseline hazards?

16.10 Item-response data

Here we consider a dataset collected by Doolaard (1999) and previously analyzed by Fox and Glas (2001) and Vermunt (2008).

The data are from an 18-item math test taken by 2,156 students belonging to 97 schools in the Netherlands. We will fit a standard one-parameter logistic item response (IRT) model, also known as a *Rasch model* (see also exercises 10.4 and 11.2) as well as more advanced IRT models.

The variables in the dataset cito.dta that we will use here are

- y1, y2, ..., y18: binary item responses for items 1 through 18 (1: correct; 0: incorrect)
- school: school identifier

1. Create a person identifier and reshape the data to long form.

2. Fit a Rasch model, that is, a logistic regression model with a random intercept for person, fixed coefficients for dummy variables for the 18 items, and no fixed intercept (this is fastest using xtlogit). The coefficients of the item dummies are minus the item difficulties, and the random intercept is the person ability.

3. Use the predict command with the xb option to store estimates of minus the difficulties in the data.

4. Now consider a model with random effects of items and persons. The random-item model for the log odds that person j responds correctly to item i can be written as

$$\text{logit}\{\Pr(y_{ij} = 1|\zeta_{1i}, \zeta_{2j})\} \;=\; \beta + \zeta_{1i} + \zeta_{2j}$$

where ζ_{1i} and ζ_{2j} are uncorrelated normally distributed random item effects and random person effects, respectively, both with zero mean and with variances ψ_1 and ψ_2. The random item effects and random person effects are assumed to be independent of each other. Fit the model using `xtmelogit` with the `laplace` option.

5. Predict the random effects and add the estimated fixed intercept to the predicted item effects.

6. Produce a scatterplot to compare the estimated fixed item effects from step 2 with the predicted random item effects from step 5. Include a $y = x$ line in the graph.

7. Now add a random intercept for school and fit the model using `xtmelogit` with the `laplace` option.

8. For a given item, what is the estimated median odds ratio comparing randomly chosen students from different schools?

16.11 ❖ Twin hayfever data

In this exercise, we consider a biometrical genetic model for twin data on hayfever. Twin models for binary responses are typically probit models and the model for the latent response, referred to as "liability" in this context, is analogous to the model for twin data with observed continuous responses (see exercises 2.3 and 8.10). The latent response y_{ij}^* underlying hayfever status for twin i in twin-pair j is modeled as

$$y_{ij}^* = \mu + A_{ij} + D_{ij} + C_{ij} + \epsilon_{ij}$$

where A_{ij} are additive genetic effects, D_{ij} are dominance genetic effects, C_{ij} is a common environment effect, and ϵ_{ij} is a unique environment effect. The terms are independent of each other and have variances σ_A^2, σ_D^2, σ_C^2, and 1, respectively. The genetic effects A_{ij} and D_{ij} are each perfectly correlated between members of the same twin pair for MZ (monozygotic or identical) twins and have correlations $1/2$ and $1/4$, respectively, for DZ (dizygotic or fraternal) twins. The shared environment effect C_{ij} is perfectly correlated for both types of twins, whereas ϵ_{ij} is uncorrelated for both types of twins. The mean and variance of y_{ij}^* is constant. Because the full model is not identifiable using twin data, the ACE or ADE models are usually estimated (omitting either D_{ij} or C_{ij} from the model).

Rabe-Hesketh, Skrondal, and Gjessing (2008) show that the covariance structure implied by the biometrical model can be induced using the following three-level model:

$$y_{ikj}^* = \beta_1 + \zeta_{kj}^{(2)} + \zeta_j^{(3)} + \epsilon_{ikj} \tag{16.4}$$

where k is an artificially created identifier that equals the twin pair identifier j for MZ twins and the person identifier i for DZ twins. The variance ψ_3 of $\zeta_j^{(3)}$ is shared by both members of the twin pair for both MZ and DZ twins. In contrast, the variance ψ_2 of $\zeta_{kj}^{(2)}$ is shared by members of the twin pair only for MZ twins, producing the additional covariance for MZ twins while keeping the total variance constant across twin types. The covariances are therefore

$$\mathrm{Cov}(y_{ij}, y_{i'j}) \;=\; \begin{cases} \psi_2 + \psi_3 = \sigma_A^2 + \sigma_D^2 + \sigma_C^2 & \text{for MZ twins} \\ \psi_3 = \sigma_A^2/2 + \sigma_D^2/4 + \sigma_C^2 & \text{for DZ twins} \end{cases}$$

For the ACE model ($\sigma_D^2 = 0$), we have

$$\sigma_A^2 = 2\psi_2 \qquad \sigma_C^2 = \psi_3 - \psi_2 \qquad \sigma_e^2 = 1$$

and for the ADE model ($\sigma_C^2 = 0$), we have

$$\sigma_A^2 = 3\psi_3 - \psi_2 \qquad \sigma_D^2 = 2(\psi_2 - \psi_3) \qquad \sigma_e^2 = 1$$

The data in `twin_hay` are from Hopper et al. (1990) and contain the following variables:

- `pair`: twin-pair identifier (j)
- `member`: twin identifier (1,2 for each twin-pair) (i)
- `M`: dummy variable for monozygotic (identical) twin pair (1: monozygotic; 2: dizygotic)
- `h`: hayfever status (1: present; 0: absent)
- `male`: dummy variable for being male
- `freq`: number of twin-pairs having the same pattern of values in `M`, `h`, and `male`

The data are in collapsed or aggregated form with `freq` representing the number of twin-pairs having a given set of values of `M`, `h`, and `male`. For MZ twins, there are six patterns of `h` and `male` (`h` equal to 00, 11, 01 for `male` equal to 00 and 11) and for DZ twins there are the same six patterns for same-sex twins, plus four patterns for mixed-sex twins (with {`h`, `male`} equal to {00,01}, {11,01}, {01,01}, {10,01}), giving 6+6+4=16 patterns and 32 rows of data (representing 7,614 individuals). Fitting the model with the data in such a collapsed form is extremely efficient.

1. Create the artificial identifier k described above and rename `freq` to `num3` so that this variable can be specified as level-3 (twin-pair level) frequency weights in `gllamm`. (`gllamm` requires that frequency weight variables end in a number that denotes the level that the weights pertain to.)
2. Fit the `probit` model in (16.4) using `gllamm`. Use the option `weight(num)` to specify that the level-3 weights are in `num3`. Note that the last character of the variable name is omitted in the `weight()` option, and that `gllamm` looks for `num1`, `num2`, and `num3`, assuming that the weights at a given level are 1 if the corresponding variable is not found.
3. Calculate the estimated variance components for the liability according to the ADE model.
4. Calculate the estimated heritability according to the ADE model, where the heritability h^2 is defined as

$$h^2 \;=\; \frac{\sigma_A^2 + \sigma_D^2}{\sigma_A^2 + \sigma_D^2 + 1}$$

A Syntax for gllamm, eq, and gllapred: The bare essentials

Here we describe the features of `gllamm` and `gllapred` that are most important for multilevel and longitudinal modeling. We also describe the `eq` command that is required for fitting random-coefficient models in `gllamm`. The full set of options for `gllamm`, `gllapred`, and `gllasim` is given in appendices B, C, and D, respectively. See Stata's *Longitudinal-Data/Panel-Data Reference Manual* for [XT] **xtmixed**.

Title

> **gllamm** — Generalized linear latent and mixed models

Syntax

> gllamm *depvar* [*indepvars*] [*if*] [*in*], i(*varlist*) [*options*]

options	Description
Model	
*i(*varlist*)	cluster identifiers from level 2 and up
<u>nr</u>f(#,...,#)	number of random effects at each level
<u>eq</u>s(*eqnames*)	equation for each random effect
<u>f</u>amily(*familyname*)	distribution of *depvar*, given the random effects; default is family(gaussian)
<u>l</u>ink(*linkname*)	link function; default is canonical link for family() specified
Model 2	
<u>noc</u>onstant	suppress constant or intercept
<u>o</u>ffset(*varname*)	include *varname* in model with coefficient constrained to 1
Integration method	
<u>adap</u>t	use adaptive quadrature; default is ordinary quadrature
<u>ni</u>p(#)	number of integration points
Reporting	
eform	report exponentiated coefficients

*i(*varlist*) is required.

915

familyname	Description
gaussian	Gaussian (normal)
poisson	Poisson
binomial	Bernoulli

linkname	Description
identity	identity
log	log
logit	logit
probit	probit
ologit	ordinal logit, proportional odds
oprobit	ordinal probit

Description

gllamm fits a wide range of models. Here we consider multilevel generalized linear models.

In the two-level case, units i (for example, students) are nested in clusters j (for example, schools). In this case, the i() option specifies a single variable, the identifier for the clusters. The conditional expectation or mean μ_{ij} of the response y_{ij} (*depvar*) given the covariates and random effects is linked to the linear predictor ν_{ij} via a *link function* $g(\cdot)$, specified using the link() option,

$$g(E[y_{ij}|\nu_{ij}]) \equiv g(\mu_{ij}) = \nu_{ij}$$

The distribution of y_{ij} given its mean is a member of the *exponential family*, specified using the family() option. In a simple random-coefficient model, the *linear predictor* has the form

$$\nu_{ij} = \beta_1 + \beta_2 x_{ij} + \zeta_{1j}^{(2)} + \zeta_{2j}^{(2)} x_{ij}$$

where x_{ij} is a covariate, β_1 is a *constant* or intercept, and β_2 is the slope or regression coefficient of x_{ij}. This *fixed part* of the model is specified using *indepvars* (the constant is not explicitly specified but included by default). $\zeta_{1j}^{(2)}$ is a *random intercept*, whereas $\zeta_{2j}^{(2)}$ is a *random slope* or *random coefficient*. These two *random effects* at level 2 represent deviations for cluster j from the mean intercept β_1 and slope β_2, respectively. The fact that there are two random effects is declared using the nrf() option. The eqs() option is used to specify two corresponding equations previously defined using the eq command. Each equation consists of a single variable, containing the values multiplying the random effect, 1 and x_{ij}, respectively.

In a three-level model for units i nested in clusters j nested in superclusters k (for example, students in classes in schools), an example of a linear predictor would be

$$\nu_{ijk} = \beta_1 + \beta_2 x_{ijk} + \zeta_{1jk}^{(2)} + \zeta_{2jk}^{(2)} x_{ijk} + \zeta_k^{(3)}$$

where $\zeta_k^{(3)}$ is the deviation for supercluster k from the mean intercept. The three-level structure can be communicated to gllamm by specifying the identifiers for levels 2 and 3 in the i() option. The nrf() option now requires two numbers of random effects, one for level 2 and one for level 3, here nrf(2 1). The eqs() option expects three equation names, first two equations for level 2 and then another equation for level 3. Here the first and last equation names could be the same.

Options

_____| Model |_____

i(*varlist*) gives the variables that define the hierarchical, nested clusters from the lowest level (finest clusters) to the highest level, for example, i(student class school).

nrf(#,...,#) specifies the number of random effects at each level, that is, for each variable in i(*varlist*). The default is nrf(1,...,1).

eqs(*eqnames*) specifies the equation names (defined before running gllamm using eq; see page 918). In multilevel models, these equations simply define the variables multiplying the random effects. The number of equations per level is specified in the nrf() option.

family(*family*) specifies the family to be used for the distribution of the response given the mean. The default is family(gaussian).

link(*link*) specifies the link function to be used, linking the conditional expectation of the response to the linear predictor.

_____| Model 2 |_____

noconstant, offset(*varname*); see [R] **estimation options**.

_____| Integration method |_____

adapt specifies adaptive quadrature. The default is ordinary quadrature.

nip(#) specifies the number of integration points.

_____| Reporting |_____

eform displays the exponentiated coefficients and corresponding standard errors and confidence intervals. For family(binomial) link(logit) (that is, logistic regression), exponentiated coefficients are odds ratios; for family(poisson) link(log) (that is, Poisson regression), exponentiated coefficients are rate ratios.

Title

> **eq** — Equations

Syntax

> eq [define] *eqname*: *varlist*

Description

eq [define] can be used to define an equation having the name *eqname* for use with estimation commands such as gllamm. The equation is just a linear combination of the variables in *varlist*. For instance,

    eq slope: x

implies the equation λx, whereas

    eq longer: x1 x2 x3 x4

implies the equation $\lambda_1 x_1 + \lambda_2 x_2 + \lambda_3 x_3 + \lambda_4 x_4$. If used to specify a component of a model in an estimation command, the λ parameters will typically be estimated. In gllamm, the eqs() option sets the coefficient of the first variable to 1. The thresh() and peqs() options add an intercept or constant to the linear combination, giving $\lambda_0 + \lambda_1 x_1 + \lambda_2 x_2 + \lambda_3 x_3 + \lambda_4 x_4$ in the previous example.

Title

gllapred — predict command for gllamm

Syntax

gllapred *varname* [*if*] [*in*] [, *statistic option*]

statistic	Description
u	empirical Bayes predictions of random effects
ustd	standardized empirical Bayes predictions
xb	fixed part of linear predictor
linpred	linear predictor with empirical Bayes predictions of random effects plugged in
mu	mean of response; by default, posterior mean
pearson	Pearson residual; by default, posterior mean

option	Description
marginal	combined with mu, gives marginal or population-averaged mean

Description

gllapred is the prediction command for gllamm.

For numerical integration, gllapred uses the same number of integration points as the preceding gllamm command and the same method (adaptive or nonadaptive quadrature).

For predictions of quantities that are linear in the random effects (u, ustd, and linpred), posterior means or empirical Bayes predictions are substituted for the random effects. These are means of the posterior distributions of the random effects given the observed responses, with parameter estimates plugged in. With the u option, standard deviations of the posterior distributions are also returned. The ustd option returns the empirical Bayes predictions divided by their approximate sampling standard deviations. The sampling standard deviation is approximated by the square root of the difference between the estimated prior variance ψ and the posterior variance. For linear models, this approximation is exact (except that the parameter estimates are plugged in).

For predictions of quantities that are nonlinear functions of the random effects, the expectation of the nonlinear function is returned, by default, with respect to the posterior distribution of the random effects. The marginal option specifies that the expectation should be taken with respect to the prior distribution of the random effects (not conditioning on the observed responses).

Only one of the statistics may be requested at a time. With the u and ustd options, *varname* is the prefix used for the variables that will contain the predictions.

Options

u returns posterior means (empirical Bayes predictions) and posterior standard deviations of the random effects in *varname*m1, *varname*m2, etc., and *varname*s1, *varname*s2, etc., respectively, where the order of the random effects is the same as in the call to gllamm.

ustd returns standardized empirical Bayes predictions of the random effects in *varname*m1, *varname*m2, etc. Each empirical Bayes prediction is divided by the square root of the difference between the estimated prior and posterior variances, which equals the sampling standard deviation in linear models but only approximates it in other models.

xb returns the fixed-effects part of the linear predictor in *varname*.

linpred returns the linear predictor, including the fixed- and random-effects part, where empirical Bayes predictions are substituted for the random effects.

mu returns the expectation of the response, for example, the predicted probability in the case of dichotomous responses. By default, the expectation is with respect to the posterior distribution of the random effects; also see the marginal option.

pearson returns Pearson residuals. By default, the posterior expectation with respect to the random effects is returned.

marginal together with the mu option gives the expectation of the response with respect to the prior distribution of the random effects. This is useful for looking at the marginal or population-averaged effects of covariates.

B Syntax for gllamm

Here we give the full syntax for gllamm with all available options.

Title

> **gllamm** — Generalized linear latent and mixed models

Syntax

gllamm *depvar* [*indepvars*] [*if*] [*in*], i(*varlist*) [*options*]

options	Description
Linear predictor	
*i(*varlist*)	cluster identifiers from level 2 and up
nrf(#,...,#)	number of random effects at each level
eqs(*eqnames*)	equation for each random effect
† frload(#,...,#)	free first factor loading
Linear predictor 2	
noconstant	suppress constant or intercept
offset(*varname*)	include *varname* in model with coefficient constrained to 1
Response model	
family(*familynames*)	distributions of *depvar*, given the random effects; default is family(gaussian)
† fv(*varname*)	variable assigning distributions to units or observations
link(*linknames*)	link functions; default is canonical link for family() specified (if only one *familyname* is given)
† lv(*varname*)	variable assigning links to units or observations
† denom(*varname*)	variable containing binomial denominator if link(binomial) is used; default denominator is 1
s(*eqname*)	equation for log of residual standard deviation
thresh(*eqnames*)	equations for threshold models
† ethresh(*eqnames*)	equations for alternative threshold models
† expanded(*varname*...)	combined with link(mlogit) specifies that data are in expanded form
† basecategory(#)	reference category for multinomial logit models
† composite(*varname*...)	composite link

*i(*varlist*) is required.

options (cont'd)	Description
Random-effects distribution	
† <u>nocorrel</u>	uncorrelated random effects
ip(*string*)	continuous or discrete random effects
Structural model	
<u>geqs</u>(*eqnames*)	equations for regressions of random effects on covariates
† <u>bmatrix</u>(*matrix*)	matrix for regressions among random effects
† <u>peqs</u>(*eqname*)	equation for multinomial logit model for probabilities of discrete distribution
Weights	
<u>weight</u>(*varname*)	frequency weights at different levels
<u>pweight</u>(*varname*)	inverse probability sampling weights at different levels
Constraints	
† <u>constraints</u>(*clist*)	linear parameter constraints
Integration method	
<u>adapt</u>	use adaptive quadrature; default is ordinary quadrature
ip(m)	use spherical integration rules; default is cartesian product
<u>nip</u>(#)	number of integration points
SE/Robust	
<u>robust</u>	standard errors based on sandwich estimator
<u>cl</u>uster(*varname*)	adjust standard errors for intragroup correlation
Reporting	
eform	report exponentiated coefficients
† <u>l</u>evel(#)	confidence level for confidence intervals; default is 95%
† <u>trace</u>	report model details and more detailed iteration log
† <u>nolog</u>	suppress iteration log
† <u>nodis</u>play	do not display estimates
† allc	display all parameters as they are estimated (sometimes transformation of parameters usually reported)
lf0(# #)	specify number of parameters and log likelihood of nested model for likelihood-ratio test
† <u>eval</u>	evaluate log likelihood for parameters given in from(*matrix*)
† <u>init</u>	fit the model where random part is set to 0
† <u>noest</u>	do not fit the model
† <u>dots</u>	display dots each time the log likelihood is evaluated

options (cont'd)	Description
Starting values	
<u>fr</u>om(*matrix*)	row matrix of starting values
copy	ignore equation and column names of matrix specified in from(*matrix*), relying on values being in correct order
skip	matrix of starting values has extra parameters
† long	matrix of starting values is for model without constraints (with constraints() option)
† <u>se</u>arch(#)	number of values to try in search for starting values for random part of model
Max options	
† <u>ite</u>rate(#)	maximum number of Newton–Raphson iterations
† adoonly	use ado-version of gllamm
Gâteaux derivative	
gateaux(# # #)	range and number of steps for Gâteaux derivative search

† Options not used elsewhere in this book.

familyname	Description
<u>gaussian</u>	Gaussian (normal)
<u>poisson</u>	Poisson
<u>binom</u>ial	Bernoulli or binomial (with denom(*varname*))
† <u>gamma</u>	gamma

† Options not used elsewhere in this book.

linkname	Description
<u>iden</u>tity	identity
<u>log</u>	log
† <u>reciprocal</u>	reciprocal (power −1)
<u>logit</u>	logit
<u>probit</u>	probit
† <u>s</u>probit	scaled probit
cll	complementary log-log
<u>ologit</u>	ordinal logit, proportional odds
<u>oprobit</u>	ordinal probit
<u>soprobit</u>	scaled ordinal probit
† <u>ocll</u>	ordinal complementary log-log
† <u>mlogit</u>	multinomial logit

† Options not used elsewhere in this book.

Description

gllamm fits generalized linear latent and mixed models (GLLAMMs). The models include random effects or latent variables varying at different levels. Here we use *random effect* instead of *latent variable* because we regard the terms more or less as synonyms and because this book is essentially about random-effects models. GLLAMMs consist of a linear predictor, response model, random-effects distribution, and structural model.

Linear predictor

For an L-level model, the most general form of the linear predictor is

$$
\begin{aligned}
\nu \;=\; & \beta_1 x_1 + \cdots + \beta_p x_p \\
& + \eta_1^{(2)}(\lambda_{11}^{(2)} z_{11}^{(2)} + \cdots + \lambda_{1q_{12}}^{(2)} z_{1q_{12}}^{(2)}) + \cdots \\
& + \eta_{M_2}^{(2)}(\lambda_{M_2,1}^{(2)} z_{M_2,1}^{(2)} + \cdots + \lambda_{M_2,1}^{(2)} z_{M_2,q_{M_2,2}}^{(2)}) + \cdots \\
& + \eta_1^{(L)}(\lambda_{11}^{(L)} z_{11}^{(L)} + \cdots + \lambda_{1q_{1L}}^{(L)} z_{1q_{1L}}^{(L)}) + \cdots \\
& + \eta_{M_L}^{(L)}(\lambda_{M_L,1}^{(L)} z_{M_L,1}^{(L)} + \cdots + \lambda_{M_L,1}^{(L)} z_{M_L,q_{M_L,L}}^{(L)}), \\
& \text{where } \lambda_{m1}^{(l)} = 1, \text{ for } l = 2, \ldots, L, \ m = 1, \ldots, M_l
\end{aligned}
$$

where we have omitted subscripts for the units at the different levels $(ijk\ldots)$ to simplify notation. Here the fixed part in the first line has the usual form. In the random part, $\eta_m^{(l)}$ is a random effect at level l, where l goes from 2 to L and, at each level, m goes from 1 to M_l. Each random effect is multiplied by a linear combination of covariates $z_{mk}^{(l)}$ with coefficients $\lambda_{mk}^{(l)}$, $k = 1, \ldots, q_{ml}$. These coefficients sometimes function as factor loadings in measurement models. In ordinary random-coefficient models, each term in parentheses collapses to a single variable $z_m^{(l)}$.

Using vector notation and summation signs, we can write the linear predictor more compactly as

$$
\nu \;=\; \mathbf{x}'\boldsymbol{\beta} + \sum_{l=2}^{L} \sum_{m=1}^{M_l} \eta_m^{(l)} \mathbf{z}_m^{(l)'} \boldsymbol{\lambda}_m^{(l)}, \quad \lambda_{m1}^{(l)} = 1
$$

In gllamm, the i() option is used to define the levels 2 to L of the model, and the nrf() option is used to specify the numbers of random effects M_l at each level. The eqs() option is used to specify equations defining the linear combinations $\mathbf{z}_m^{(l)'} \boldsymbol{\lambda}_m^{(l)}$ multiplying the random effects.

Response model

The response model is a generalized linear model with unit-specific link function and distribution from the exponential family, specified using the link(), lv(), family(), and fv() options. In addition, the variance of the response, given the linear predictor, can depend on covariates as can the thresholds in ordinal response models (via the `s()` and `thresh()` options, respectively).

Random-effects distribution

The random effects can be either continuous or discrete. In the continuous case, the following applies to the disturbances if there is a structural model (see below). The random effects are multivariate normal with zero means. Random effects at the same level may be correlated, but there are no correlations across levels.

In the discrete case, the M_l random effects at level l take on discrete values that can be thought of as locations in M_l dimensions. The locations together with the associated probabilities are parameters of the model. Random effects at different levels are again assumed to be mutually independent. Discrete random effects can be used for latent-class or finite-mixture models or for *nonparametric maximum likelihood* estimation. The `ip()` option is used to specify whether the random effects are continuous or discrete.

Structural model

Continuous case

In the structural model, each random effect can be regressed on covariates w,

$$\eta_m^{(l)} = \gamma_{m1}^{(l)} w_{m1}^{(l)} + \cdots + \gamma_{m,r_{ml}}^{(l)} w_{m,r_{ml}}^{(l)} + \zeta_m^{(l)}$$

where $\zeta_m^{(l)}$ is a disturbance or residual. These linear combinations of covariates are defined as equations and passed to `gllamm` using the `geqs()` option.

The structural model also allows random effects to be regressed on other random effects. These relations, together with the regressions on observed covariates given above, are easiest to express using a column vector η for all random effects from levels 2 to L and ζ for the corresponding disturbances

$$\eta = \mathbf{B}\eta + \mathbf{\Gamma}\mathbf{w} + \zeta$$

where $\mathbf{B}$ (specified using the `bmatrix()` option) is an upper triangular matrix of regression coefficients allowing random effects to be regressed on other random effects. The term $\mathbf{\Gamma}\mathbf{w}$ represents the regressions of random effects on covariates written more explicitly above.

Discrete case

In the discrete case, the `peqs()` option can be used to specify multinomial logit models for the probabilities of the different locations for the random effects.

Options

i(*varlist*) gives the variables that define the hierarchical, nested clusters from the lowest level (finest clusters) to the highest level, for example, i(student class school).

nrf(#,...,#) specifies the number of random effects M_l at each level l, that is, for each variable in i(*varlist*). The default is nrf(1,...,1).

eqs(*eqnames*) specifies the equation names (defined before running gllamm) for the linear combinations $\mathbf{z}_m^{(l)\prime}\boldsymbol{\lambda}_m^{(l)}$ multiplying the random effects. The equations for the level-2 random effects are listed first, followed by those for the level-3 random effects, etc., the number of equations per level being specified in the nrf() option. If required, constants should be explicitly included in the equation definitions using variables equal to 1. If the option is not used, the random effects are assumed to be random intercepts, and only one random effect is allowed per level. The first coefficient $\lambda_{m1}^{(l)}$ is set to one, unless the frload() option is specified. The other coefficients are estimated together with the (co)variances of the random effects.

frload(#,...,#) lists the random effects for which the first coefficient $\lambda_{m1}^{(l)}$ should be freely estimated instead of set to 1. It is up to the user to define appropriate constraints to identify the model. Here the random effects are referred to as 1, 2, 3, etc., in the order in which they are defined by the eqs() option.

noconstant, offset(*varname*); see [R] **estimation options**.

family(*familynames*) specifies the family (or families) to be used for the response probabilities (or densities) given the random effects. The default is family(gaussian). Several families may be given, in which case, the variable allocating families to units or observations must be given using fv(*varname*).

fv(*varname*) is required if several families are specified in the family() option. The variable indicates which family applies to which unit or observation. A value of one refers to the first family specified in family(), etc.

denom(*varname*) gives the variable containing the binomial denominator for the responses whose family was specified as binomial. The default denominator is 1.

s(*eqname*) specifies that the log of the standard deviation (or of the coefficient of variation) at level 1 for normally (or gamma) distributed responses (or the scale for the sprobit or soprobit links) is given by the linear combination of covariates defined by *eqname*. This allows for heteroskedasticity at level 1. For example, if dummy variables for groups are used in the definition of *eqname*, different residual variances are estimated for different groups.

link(*linknames*) specifies the link functions linking the conditional means to the linear predictors. If a single family is specified, the default link is the canonical link. Several links may be given, in which case, the variable assigning links to units or observations must be given using lv(*varname*).

lv(*varname*) is the variable whose values indicate which link applies to which unit or observation. See fv(*varname*) for details.

thresh(*eqnames*) specifies equations for the thresholds for ordinal responses. One equation is specified for each ordinal response, and constants are automatically added. This option allows the effects of some covariates to differ between the categories of the ordinal outcome rather than assuming a constant effect—the parallel-regression assumption, or with the ologit link, the proportional odds assumption. Variables used in the model for the thresholds generally should not appear in the fixed part of the linear predictor.

ethresh(*eqnames*) is the same as thresh(*eqnames*), except that a different parameterization is used for the threshold model. To ensure that $\kappa_{s-1} \leq \kappa_s$, the model is $\kappa_s = \kappa_{s-1} + \exp(\mathbf{x}'\boldsymbol{\alpha})$, for $s = 2, \ldots, S-1$, where S is the number of response categories.

expanded(*varname varname string*) is used with the mlogit link and specifies that the data have been expanded as illustrated below:

A		B		
choice		response	altern	selected
1		1	1	1
2		1	2	0
		1	3	0
		2	1	0
		2	2	1
		2	3	0

The variable choice is the multinomial response (possible values 1, 2, 3), response labels the original lines of data, altern gives the possible responses or alternatives, and selected indicates the response that was given. The syntax would be expanded(response selected m), and altern would be used as the dependent variable. This expanded form allows the user to have alternative-specific covariates, apply different random effects to different alternatives, and have different alternative sets for different individuals. The third argument is o if one set of coefficients should be estimated for the explanatory variables and m if one set of coefficients is to be estimated for each category of the response except the reference category.

basecategory(*#*) specifies the value of the response to be used as the reference category when the mlogit link is used. This option is ignored if the expanded() option is used with the third argument equal to m.

composite(*varname varname* ...) specifies that a composite link be used. The first
   variable is a cluster identifier (cluster below) so that linear predictors within the
   cluster can be combined into a single composite link. The second variable (ind
   below) indicates to which response the composite links defined by the subsequent
   weight variables belong. Observations with ind=0 have a missing link. The remaining variables (c1 and c2 below) specify weights for the composite links. The
   composite link based on the first weight variable will go to where ind=1, etc.

Example:

Data setup with form of inverse link   Interpretation of
h_i determined by link() and lv():   composite(cluster ind c1 c2)

cluster	ind	c1	c2	inverse link			cluster	composite link
1	1	1	0	h_1			1	h_1 - h_2
1	2	-1	1	h_2			1	h_2 + h_3
1	0	0	1	h_3	==>		1	missing
2	1	1	0	h_4			2	h_4 + h_5
2	2	1	1	h_5			2	h_5 + 2*h_6
2	0	0	2	h_6			2	missing

─────┌ Random-effects distribution └──

nocorrel may be used to constrain all correlations to zero if there are several random
   effects at any of the levels and if these are modeled as multivariate normal.

ip(*string*) requests that the random effects be multivariate normal (Gaussian) if *string*
   is g and be discrete with freely estimated mass points if *string* is f. The default
   is Gaussian quadrature. With the ip(f) option, only nip-1 mass-point locations
   are estimated, the last being determined by setting the mean of the mass-point
   distribution to 0. The ip(fn) option can be specified to estimate all nip masses
   freely—the user must then make sure that the mean is not modeled in the linear
   predictor, for example, by specifying the noconstant option. See integration method
   for a description of ip(m).

─────┌ Structural model └──

geqs(*eqnames*) specifies equations for regressions of random effects on explanatory variables. The second character of the equation name indicates which random effect is
   regressed on the variables used in the equation definition; for example, f1:a b means
   that the first random effect is regressed on a and b (without a constant).

bmatrix(*matrix*) specifies a square matrix **B** of regression coefficients for the dependence of the random effects on other random effects. The matrix must be upper
   diagonal and have number of rows and columns equal to the total number of random effects. Elements of *matrix* are 1 where coefficients in **B** should be estimated
   and 0 where they should be set to 0.

peqs(*eqname*) can be used with the ip(f) or ip(fn) option to allow the (prior) probabilities of the discrete distribution to depend on covariates via a multinomial logit model. A constant is automatically included in addition to the covariates specified in the eq command.

◯ ⌐ Weights ⌐ ──

weight(*wt*) specifies that variables *wt*1, *wt*2, etc., contain frequency weights. The suffixes (1,2,...) in the variable names should not be included in the weight() option. They determine at what level each weight applies. If only some of the weight variables exist, for example, only level-2 weights, the other weights are assumed to be equal to 1. For example, if the level-1 units are occasions (or panel waves) in longitudinal data and the level-2 units are individuals, and the only variable used in the analysis is a binary variable, result, then we can collapse dataset A into dataset B by defining level-1 weights as follows:

A				B			
ind	occ	result		ind	occpat	result	wt1
1	1	0		1	1	0	2
1	2	0		2	2	0	1
2	3	0		2	3	1	1
2	4	1		3	4	0	1
3	5	0		3	5	1	1
3	6	1					

The two occasions for individual 1 in dataset A have the same result. The first row in B therefore represents two occasions (occasions 1 and 2), as indicated by wt1. The variable occpat labels the unique patterns of responses at level 1.

The two individuals 2 and 3 in dataset B have the same pattern of results over the measurement occasions (both have two occasions with values 0 and 1). We can therefore collapse the data into dataset C by using level-2 weights:

B				C				
ind	occpat	result	wt1	indpat	occpat	result	wt1	wt2
1	1	0	2	1	1	0	2	1
2	2	0	1	2	2	0	1	2
2	3	1	1	2	3	1	1	2
3	4	0	1					
3	5	1	1					

The variable indpat labels the unique patterns of responses at level 2, and wt2 indicates that indpat 1 in dataset C represents one individual and indpat 2 represents two individuals; that is, all the data for individual 2 are replicated once. Collapsing the data in this way can make gllamm run faster.

pweight(*varname*) specifies that variables *varname*1, *varname*2, etc., contain inverse probability sampling weights for levels 1, 2, etc. As far as the estimates and log likelihood are concerned, the effect of specifying these weights is the same as for frequency weights, but the standard errors will be different. Robust standard errors will automatically be provided. This pseudolikelihood approach should be used with caution if the sampling weights apply to units at a lower level than the highest level in the multilevel model. The weights are not rescaled; scaling is the responsibility of the user.

⌐ Constraints ⌐

constraint(*clist*) specifies the constraint numbers of the linear constraints to be applied. Constraints are defined using the constraint() command; see [R] **constraint**. To find out the equation names needed to specify the constraints, run gllamm with the noest and trace options.

⌐ Integration method ⌐

adapt specifies adaptive quadrature; the default is ordinary quadrature.

ip(m) specifies spherical quadrature that can be more efficient than the default cartesian product quadrature when there are several random effects.

nip(#,...,#) specifies the number of integration points or masses to be used for each integral or summation. When quadrature is used, a value may be given for each random effect. When freely estimated masses are used, a value may be given for each level of the model. If only one argument is given, the same number of integration points will be used for each summation. The default value is 8. When used with ip(m), nip() specifies the degree d of the approximation (corresponds in accuracy approximately to $(d+1)/2$ points per dimension for cartesian product quadrature). Only certain values are available for spherical quadrature: for two random effects, 5, 7, 9, 11, and 15; for more than two random effects, 5 and 7.

⌐ SE/Robust ⌐

robust specifies that the Huber/White/sandwich estimator of the covariance matrix of the parameter estimates is to be used. If a model has been fit without the robust option, the robust standard errors can be obtained by simply typing gllamm, robust.

cluster(*varname*) specifies that the highest-level units of the GLLAMM model are nested in even higher-level clusters, where *varname* contains the cluster identifier. Robust standard errors will be provided that take this clustering into account. If a model has been fit without this option, the robust standard errors for clustered data can be obtained using the command gllamm, cluster(*varname*).

eform displays the exponentiated coefficients and corresponding standard errors and confidence intervals. For family(binomial) link(logit) (that is, logistic regression), exponentiated coefficients are odds ratios; for family(poisson) link(log) (that is, Poisson regression), exponentiated coefficients are rate ratios.

level(#) specifies the confidence level as a percentage for confidence intervals of the fixed coefficients. The default is 95.

trace displays details of the model being fit, as well as details of the maximum likelihood iterations.

nolog suppresses output for maximum likelihood iterations.

nodisplay suppresses output of the estimates but still shows the iteration log, unless nolog is used.

allc causes all estimated parameters to be displayed in a regression table, which may be transformations of the parameters usually reported, in addition to the usual output.

lf0(##) gives the number of parameters and the log likelihood for a likelihood-ratio test to compare the model to be fit with a simpler model. A likelihood-ratio chi-squared test is only performed if the lf0() option is used.

eval causes the program to evaluate the log likelihood for values passed to gllamm using the from(matrix) option.

init sets random part of model to zero so that only the fixed part is fit. This is also how initial values or starting values for the fixed part are computed in gllamm.

noest is used to prevent the program from carrying out the estimation. This may be used with the trace option to check that the model is correct and get the information needed to set up a matrix of initial values. Global macros are available that are normally deleted. Particularly useful may be M_initf and M_initr, matrices for the parameters (fixed part and random part, respectively).

dots causes a dot to be printed (if used together with trace) each time the likelihood-evaluation program is called by ml. This helps the user to assess how long gllamm is likely to take to run and reassures the user that it is making some progress when it is very slow.

from(matrix) specifies the matrix (one row) to be used as starting values. The column names and equation names must be correct (see help matrix), unless the copy option is used. The parameter values given may be previous estimates, obtained using e(b). This is useful if new covariates are added or if the number of integration points (or locations in a discrete distribution) is increased. The skip option must be used if the model to be fit contains fewer parameters than are included in matrix, for instance, if covariates are dropped.

copy; see from(*matrix*).

skip; see from(*matrix*).

long can be used with the from(*matrix*) option when parameter constraints are used to indicate that the matrix of initial values corresponds to the unconstrained model; that is, it has more elements than will be fit.

search(#) causes the program to search for initial values for the random-effects variances at level 2 (in range 0 to 3). The argument specifies the number of random searches. This option may only be used with ip(g) and when from(*matrix*) is not used.

____| Max options |_____

iterate(#) specifies the maximum number of iterations. With the adapt option, using the iterate(#) option will cause gllamm to skip the Newton–Raphson iterations usually performed at the end without updating the quadrature locations. iterate(0) is like eval, except that standard errors are computed.

adoonly causes gllamm to use only ado-code instead of internalized code. gllamm will be faster if it uses internalized versions of some of the functions available from Stata 7 (if updated on or after 26 October 2001).

____| gateaux derivative |_____

gateaux(# # #) can be used with the ip(f) or ip(fn) options to increase the number of mass points by one from a previous solution with parameter estimates specified using from(*matrix*). The number of parameters and log likelihood of the previous solution must be specified using the lf0(# #) option. The program searches for the location of the new mass point by placing a small mass at the location given by the first argument and moving it to the second argument in the number of steps specified by the third argument. (If there are several random effects, this search is done in each dimension, resulting in a regular grid of search points.) If the maximum increase in likelihood is greater than 0, the location corresponding to this maximum is used as the initial value of the new location; otherwise, the program stops. If the program stops, this suggests that the nonparametric maximum likelihood estimator has been obtained.

C   Syntax for gllapred

After you fit a model using gllamm, you can use gllapred to obtain predictions of various quantities.

Title

gllapred — predict command for gllamm

Syntax

gllapred *varname* [*if*] [*in*] [, *statistic options*]

statistic	Description
u	empirical Bayes predictions of random effects; disturbances if there is a structural model
corr	posterior correlations between random effects or disturbances
fac	empirical Bayes predictions of random effects
ustd	standardized empirical Bayes predictions
xb	fixed part of linear predictor
linpred	linear predictor with empirical Bayes predictions of random effects plugged in
mu	mean of response; by default, posterior mean
pearson	Pearson residual; by default, posterior mean
deviance	deviance residual; by default, posterior mean
anscombe	Anscombe residual; by default, posterior mean
cooksd	Cook's distance for top-level clusters
p	posterior probabilities if random effects are discrete
s	standard deviations if s() option was used
ll	log-likelihood contributions from top-level clusters

options	Description
<u>marg</u>inal	combined with mu, gives marginal or population-averaged mean
us(*varname*)	substitute specific values for random effects in *varname*1, etc.
<u>ab</u>ove(#,...,#)	for ordinal responses with mu option, return probability that *y* exceeds # (several values if more than one ordinal link)
<u>out</u>come(#)	for mlogit link with mu option, probability that *y*=#
nooffset	suppress offset
fsample	predict for full sample, not just estimation sample
<u>from</u>(*matrix*)	use parameters in *matrix* instead of estimated parameters
<u>ad</u>apt	use adaptive quadrature even if gllamm did not
adoonly	use ado-version of gllapred

Description

See description for gllapred on page 919.

Options

u returns posterior means (empirical Bayes predictions) and posterior standard deviations of the random effects in *varname*m1, *varname*m2, etc., and *varname*s1, *varname*s2, etc., respectively, where the order of the random effects is the same as in the call to gllamm. In the case of continuous random effects, the integration method (ordinary versus adaptive quadrature) and the number of quadrature points used is the same as in the previous call to gllamm. If the gllamm model includes equations for the random effects (geqs or bmatrix), the posterior means and standard deviations of the disturbances ζ are returned.

corr returns posterior correlations of the random effects in *varname*c21, etc. This option only works together with the u option. If there is a structural model, posterior correlations of the disturbances are returned.

fac returns posterior means (empirical Bayes predictions) and posterior standard deviations of the random effects in *varname*m1, *varname*m2, etc., and *varname*s1, *varname*s2, etc. If there is a structural model, predictions of the random effects on the left-hand side of the structural equations are returned.

xb returns the fixed-effects part of the linear predictor in *varname* including the offset (if there is one), unless the nooffset option is used.

ustd returns standardized posterior means (empirical Bayes predictions) of the disturbances in *varname*m1, *varname*m2, etc. Each posterior mean is divided by the square root of the difference between the prior and posterior variances, which approximates the sampling standard deviation.

cooksd returns Cook's distances for the top-level units in *varname*.

linpred returns the linear predictor including the fixed- and random-effects part where posterior means (empirical Bayes predictions) are substituted for the random effects.

mu returns the expectation of the response, for example, the predicted probability in the case of dichotomous responses. By default, the expectation is with respect to the posterior distribution of the random effects; also see the `marginal` and `us()` options. The offset is included (if there is one in the `gllamm` model), unless the `nooffset` option is specified.

pearson returns Pearson residuals. By default, the posterior expectation with respect to the random effects is returned. The `us()` option can be used to obtain the conditional residual when specific values are substituted for the random effects.

deviance returns deviance residuals. By default, the posterior expectation with respect to the random effects is returned. The `us()` option can be used to obtain the conditional residual when specific values are substituted for the random effects.

anscombe returns Anscombe residuals. By default, the posterior expectation with respect to the random effects is returned. The `us()` option can be used to obtain the conditional residual when specific values are substituted for the random effects.

p can only be used for two-level models fit using the `ip(f)` or `ip(fn)` option. `gllapred` returns the posterior probabilities in *varname*1, *varname*2, etc., giving the probabilities of classes 1, 2, etc. `gllapred` also displays the (prior) probability and location matrices to help interpret the posterior probabilities.

s returns the scale. This is useful if the `s()` option was used in `gllamm` to specify level-1 heteroskedasticity.

ll returns the log-likelihood contributions of the highest-level (level-L) units.

marginal together with the `mu` option gives the expectation of the response with respect to the prior distribution of the random effects. This is useful for looking at the marginal or population-averaged effects of covariates.

us(*varname*) specifies values for the random effects to calculate conditional quantities, such as the conditional mean of the responses (`mu` option), given the values of the random effects. Here *varname* specifies the prefix for the variables, and `gllapred` will look for *varname*1, *varname*2, etc.

above(#,...,#) returns probabilities that *depvar* exceeds # (ordinal responses). If there are several ordinal responses, a different value can be specified for each ordinal response or a single value given for all ordinal responses.

outcome(#) specifies the outcome for which the predicted probability should be returned (`mu` option) if there is a nominal response. This option is not necessary if the `expanded()` option was used in `gllamm` because in this case predicted probabilities are returned for all outcomes.

nooffset excludes the offset from the predictions (with the `xb`, `linpred`, or `mu` options). It will only make a difference if the `offset()` option was used in `gllamm`.

fsample causes gllapred to return predictions for the full sample (except units or observations excluded in the if and in qualifiers), not just the estimation sample. The returned log likelihood may be missing because gllapred will not exclude observations with missing values on any of the variables used in the likelihood calculation. It is up to the user to exclude these observations using if or in.

from(*matrix*) specifies a row matrix of parameter values for which the predictions should be made. The column and equation names will be ignored. Without this option, the parameter estimates from the last gllamm model will be used.

adapt specifies that numerical integration should be performed using adaptive quadrature instead of ordinary quadrature. This option is not necessary if estimation in gllamm used adaptive quadrature.

adoonly causes gllamm to use only ado-code. This option is not necessary if gllamm was run with the adoonly option.

D   Syntax for gllasim

After you fit a model using `gllamm`, you can use `gllasim` to simulate responses from the model.

Title

gllasim — simulate command for gllamm

Syntax

gllasim *varname* $\left[\,if\,\right]$ $\left[\,in\,\right]$ $\left[\,,\ statistic\ options\,\right]$

statistic	Description
y	response
u	random effects (or disturbances if there is a structural model)
fac	random effects
<u>lin</u>pred	linear predictor
mu	mean of response (substituting simulated values for random effects)

options	Description
us(*varname*)	substitute specific values for random effects in *varname*1, etc.
<u>ab</u>ove(#,...,#)	for ordinal responses with `mu` option, return probability that y exceeds # (several values if more than one ordinal link)
<u>out</u>come(#)	for `mlogit` link with `mu` option, probability that $y=\#$
<u>noof</u>fset	suppress offset
fsample	simulate for full sample, not just estimation sample
<u>fr</u>om(*matrix*)	use parameters in *matrix* instead of estimated parameters
adoonly	use ado-version of `gllasim`

Description

gllasim is the simulation command for gllamm. The command is somewhat similar to gllapred, except that random effects are simulated from the prior random-effects distribution instead of predicted as the posterior means.

By default, the response is simulated. If other statistics are requested, the response can be simulated as well if the y option is used. With the u, fac, linpred, and mu options, *varname* is just the prefix of the variable names in which results are stored.

Options

y returns simulated responses in *varname*. This option is only necessary if u, fac, linpred, or mu is also specified.

u returns simulated random effects in *varname*p1, *varname*p2, etc., where the order of the random effects is the same as in the call to gllamm (in the order of the equations in the eqs() option). If the gllamm model includes equations for the random effects (geqs or bmatrix), the simulated disturbances are returned.

fac returns the simulated random effects in *varname*p1, *varname*p2, etc., instead of the disturbances if the gllamm model includes equations for the random effects (geqs() or bmatrix() options in gllamm), that is, the random effects on the left-hand side of the structural model.

linpred returns the linear predictor, including the fixed and simulated random parts in *varname*p. The offset is included (if there is one in the gllamm model), unless the nooffset option is specified.

mu returns the expected value of the response conditional on the simulated values for the random effects, for example, a probability if the responses are dichotomous.

us(*varname*) specifies that, instead of simulating the random effects, gllasim should use the variables in *varname*1, *varname*2, etc.

above(#,...,#) returns probabilities that *depvar* exceeds # (with the mu option) if there are ordinal responses. A single number can be given for all ordinal responses.

outcome(#) specifies the outcome for which the predicted probability should be returned (mu option) if there is a nominal response and the expanded() option has not been used in gllamm (with the expanded() option, predicted probabilities are returned for all outcomes).

nooffset can be used with the linpred and mu options to exclude the offset from the simulated value. This will only make a difference if the offset() option was used in gllamm.

fsample causes gllasim to simulate values for the full sample (except observations excluded in the if and in qualifiers), not just the estimation sample.

from(*matrix*) specifies a matrix of parameters for which the simulations should be made. The column and equation names will be ignored. Without this option, the parameter estimates from the last `gllamm` model will be used.

adoonly causes `gllasim` to use only ado-code. This option is not necessary if `gllamm` was run with the adoonly option.

References

Acitelli, L. K. 1997. Sampling couples to understand them: Mixing the theoretical with the practical. *Journal of Social and Personal Relationships* 14: 243–261.

Agresti, A. 2002. *Categorical Data Analysis*. 2nd ed. Hoboken, NJ: Wiley.

———. 2010. *Analysis of Ordinal Categorical Data*. 2nd ed. Hoboken, NJ: Wiley.

Agresti, A., J. G. Booth, J. P. Hobert, and B. Caffo. 2000. Random-effects modeling of categorical response data. *Sociological Methodology* 30: 27–80.

Agresti, A., and R. Natarajan. 2001. Modeling clustered ordered categorical data: A survey. *International Statistical Review* 69: 345–371.

Aitkin, M. 1978. The analysis of unbalanced cross-classifications. *Journal of the Royal Statistical Society, Series A* 41: 195–223.

Allison, P. D. 1982. Discrete-time methods for the analysis of event histories. In *Sociological Methodology 1982*, ed. S. Leinhardt, 61–98. San Francisco: Jossey-Bass.

———. 1984. *Event History Analysis: Regression for Longitudinal Event Data*. Newbury Park, CA: Sage.

———. 1995. *Survival Analysis Using SAS: A Practical Guide*. Cary, NC: SAS Institute.

———. 1996. Fixed-effects partial likelihood for repeated events. *Sociological Methods & Research* 25: 207–222.

Allison, P. D., and R. P. Waterman. 2002. Fixed-effects negative binomial regression models. In *Sociological Methodology 2002*, ed. R. M. Stolzenberg, 247–265. Oxford: Blackwell.

Amin, S., I. Diamond, and F. A. Steele. 1998. Contraception and religiosity in Bangladesh. In *Continuing Demographic Transition*, ed. G. W. Jones, R. M. Douglas, J. C. Caldwell, and R. M. D'Souza, 268–289. Oxford: Oxford University Press.

Andersen, P. K., and R. D. Gill. 1982. Cox's regression model for counting processes: A large sample study. *Annals of Statistics* 10: 1100–1120.

Barber, J. S., S. Murphy, W. G. Axinn, and J. Maples. 2000. Discrete-time multilevel hazard analysis. In *Sociological Methodology 2000*, ed. R. M. Stolzenberg, 201–235. Oxford: Blackwell.

Blossfeld, H.-P., K. Golsch, and G. Rohwer, ed. 2007. *Event History Analysis with Stata*. Mahwha, NJ: Erlbaum.

Box-Steffensmeier, J. M., and B. S. Jones. 2004. *Event History Modeling: A Guide for Social Scientists*. Cambridge: Cambridge University Press.

Breslow, N. E., and D. G. Clayton. 1993. Approximate inference in generalized linear mixed models. *Journal of the American Statistical Association* 88: 9–25.

Breslow, N. E., and N. E. Day. 1987. *Statistical Methods in Cancer Research: Vol. 2—The Design and Analysis of Cohort Studies*. Lyon: IARC.

Brody, R. A., and B. I. Page. 1972. Comment: The assessment of policy voting. *American Political Science Review* 66: 450–458.

Byar, D. P. 1980. The veterans administration study of chemoprophylaxis for recurrent stage I bladder tumors: Comparison of placebo, pyroxidene, and topical thiotepa. In *Bladder Tumors and Other Topics in Urological Oncology*, ed. M. Pavone-Macaluso, P. H. Smith, and F. Edsmyr, 363–370. New York: Plenum.

Cameron, A. C., and P. K. Trivedi. 1998. *Regression Analysis of Count Data*. Cambridge: Cambridge University Press.

———. 2005. *Microeconometrics: Methods and Applications*. Cambridge: Cambridge University Press.

Capaldi, D. M., L. Crosby, and M. Stoolmiller. 1996. Predicting the timing of first sexual intercourse for at-risk adolescent males. *Child Development* 67: 344–359.

Chamberlain, G. 1985. Heterogeneity, omitted variable bias, and duration dependence. In *Longitudinal Analysis of Labor Market Data*, ed. J. J. Heckman and B. Singer, 3–38. Cambridge: Cambridge University Press.

Chen, Z., and L. Kuo. 2001. A note on the estimation of the multinomial logit model with random effects. *American Statistician* 55: 89–95.

Chung, C.-F., P. Schmidt, and A. D. Witte. 1991. Survival analysis: A survey. *Journal of Quantitative Criminology* 7: 59–98.

Clayton, D. G., and J. Kaldor. 1987. Empirical Bayes estimates of age-standardized relative risks for use in disease mapping. *Biometrics* 43: 671–681.

Cleves, M., W. W. Gould, R. G. Gutierrez, and Y. Marchenko. 2010. *An Introduction to Survival Analysis Using Stata*. 3rd ed. College Station, TX: Stata Press.

Collett, D. 2003a. *Modelling Binary Data*. 2nd ed. London: Chapman & Hall/CRC.

———. 2003b. *Modelling Survival Data in Medical Research*. 2nd ed. London: Chapman & Hall/CRC.

Danahy, D. T., D. T. Burwell, W. S. Aranov, and R. Prakash. 1976. Sustained hemo-dynamic and antianginal effect of high dose oral isosorbide dinitrate. *Circulation* 55: 381–387.

Davis, C. S. 1991. Semi-parametric and non-parametric methods for the analysis of repeated measurements with applications to clinical trials. *Statistics in Medicine* 10: 1995–1980.

————. 2002. *Statistical Methods for the Analysis of Repeated Measurements*. New York: Springer.

De Backer, M., C. De Vroey, E. Lesaffre, I. Scheys, and P. De Keyser. 1998. Twelve weeks of continuous oral therapy for toenail onychomycosis caused by dermatophytes: A double-blind comparative trial of terbinafine 250 mg/day versus itraconazole 200 mg/day. *Journal of the American Academy of Dermatology* 38: 57–63.

De Boeck, P., and M. Wilson, ed. 2004. *Explanatory Item Response Models: A Generalized Linear and Nonlinear Approach*. New York: Springer.

Diggle, P. J., P. J. Heagerty, K.-Y. Liang, and S. L. Zeger. 2002. *Analysis of Longitudinal Data*. 2nd ed. Oxford: Oxford University Press.

Dohoo, I. R., W. Martin, and H. Stryhn. 2010. *Veterinary Epidemiologic Research*. 2nd ed. Charlottetown, Canada: VER Inc.

Dohoo, I. R., E. Tillard, H. Stryhn, and B. Faye. 2001. The use of multilevel models to evaluate sources of variation in reproductive performance in dairy cattle in Reunion Island. *Preventive Veterinary Medicine* 50: 127–144.

Doolaard, S. 1999. Schools in Change or School in Chain. PhD diss., University of Twente, The Netherlands.

Duchateau, L., and P. Janssen. 2008. *The Frailty Model*. New York: Springer.

Embretson, S. E., and S. P. Reise. 2000. *Item Response Theory for Psychologists*. Mahwah, NJ: Erlbaum.

Everitt, B. S., and A. Pickles. 2004. *Statistical Aspects of the Design and Analysis of Clinical Trials*. Rev. ed. London: Imperial College Press.

Fahrmeir, L., and G. Tutz. 2001. *Multivariate Statistical Modelling Based on Generalized Linear Models*. 2nd ed. New York: Springer.

Fitzmaurice, G. M. 1998. Regression models for discrete longitudinal data. In *Statistical Analysis of Medical Data: New Developments*, ed. B. S. Everitt and G. Dunn, 175–201. London: Arnold.

Fitzmaurice, G. M., N. M. Laird, and J. H. Ware. 2011. *Applied Longitudinal Analysis*. 2nd ed. Hoboken, NJ: Wiley.

Flay, B. R., B. R. Brannon, C. A. Johnson, W. B. Hansen, A. L. Ulene, D. A. Whitney-Saltiel, L. R. Gleason, S. Sussman, M. D. Gavin, K. M. Glowacz, D. F. Sobol, and D. C. Spiegel. 1988. The television, school, and family smoking cessation and prevention project: I. Theoretical basis and program development. *Preventive Medicine* 17: 585–607.

Fleming, T. R., and D. P. Harrington. 1991. *Counting Processes and Survival Analysis.* New York: Wiley.

Fox, J. 1997. *Applied Regression Analysis, Linear Models, and Related Methods.* Thousand Oaks, CA: Sage.

Fox, J.-P., and C. A. W. Glas. 2001. Bayesian estimation of a multilevel IRT model using Gibbs sampling. *Psychometrika* 66: 269–286.

Gail, M. H., S. Wieand, and S. Piantadosi. 1984. Biased estimates of treatment effect in randomized experiments with nonlinear regressions and omitted covariates. *Biometrika* 71: 431–444.

Gelman, A., and J. Hill. 2007. *Data Analysis Using Regression and Multilevel/Hierarchical Models.* Cambridge: Cambridge University Press.

Gibbons, R. D., and D. Hedeker. 1994. Application of random-effects probit regression models. *Journal of Consulting and Clinical Psychology* 62: 285–296.

Gibbons, R. D., D. Hedeker, C. Waterneaux, and J. M. Davis. 1988. Random regression models: A comprehensive approach to the analysis of longitudinal psychiatric data. *Psychopharmacology Bulletin* 24: 438–443.

Goldstein, H. 2011. *Multilevel Statistical Models.* 4th ed. Chichester, UK: Wiley.

Greene, W. H. 2012. *Econometric Analysis.* 7th ed. Upper Saddle River, NJ: Prentice Hall.

Greenland, S. 1994. Alternative models for ordinal logistic regression. *Statistics in Medicine* 13: 1665–1677.

Grilli, L. 2005. The random-effects proportional hazards model with grouped survival data: A comparison between the grouped continuous and continuation ratio versions. *Journal of the Royal Statisticial Society, Series A* 168: 83–94.

Guo, G. 1993. Event-history analysis for left-truncated data. In *Sociological Methodology 1993*, ed. P. V. Marsden, 217–243. Oxford: Blackwell.

Guo, G., and G. Rodríguez. 1992. Estimating a multivariate proportional hazards model for clustered data using the EM algorithm, with an application to child survival in Guatemala. *Journal of the American Statistical Association* 87: 969–976.

Guo, G., and H. Zhao. 2000. Multilevel modeling of binary data. *Annual Review of Sociology* 26: 441–462.

Hall, B. H., Z. Griliches, and J. A. Hausman. 1986. Patents and R and D: Is there a lag? *International Economic Review* 27: 265–283.

Hamerle, A. 1991. On the treatment of interrupted spells and initial conditions in event history analysis. *Sociological Methods & Research* 19: 388–414.

Hardin, J. W., and J. M. Hilbe. 2003. *Generalized Estimating Equations*. Boca Raton, FL: Chapman & Hall/CRC.

Hausman, J. A., B. H. Hall, and Z. Griliches. 1984. Econometric models for count data with an application to the patents–R & D relationship. *Econometrica* 52: 909–938.

Heagerty, P. J., and B. F. Kurland. 2001. Misspecified maximum likelihood estimates and generalized linear mixed models. *Biometrika* 88: 973–985.

Heath, A., R. Jowell, J. K. Curtice, J. A. Brand, and J. C. Mitchell. 1993. *British General Election Panel Study, 1987–1992 (SN: 2983)*. Colchester, UK: Economic and Social Data Service.

Hedeker, D. 1999. MIXNO: A computer program for mixed-effects logistic regression. *Journal of Statistical Software* 4: 1–92.

———. 2005. Generalized linear mixed models. In *Encyclopedia of Statistics in Behavioral Science*, ed. B. S. Everitt and D. Howell, 729–738. London: Wiley.

———. 2008. Multilevel models for ordinal and nominal variables. In *Handbook of Multilevel Analysis*, ed. J. de Leeuw and E. Meijer, 237–274. New York: Springer.

Hedeker, D., and R. D. Gibbons. 1996. MIXOR: A computer program for mixed-effects ordinal regression analysis. *Computer Methods and Programs in Biomedicine* 49: 157–176.

———. 2006. *Longitudinal Data Analysis*. Hoboken, NJ: Wiley.

Hedeker, D., R. D. Gibbons, M. du Toit, and Y. Cheng. 2008. *SuperMix: Mixed Effects Models*. Lincolnwood, IL: Scientific Software International.

Hedeker, D., O. Siddiqui, and F. B. Hu. 2000. Random-effects regression analysis of correlated grouped-time survival data. *Statistical Methods in Medical Research* 9: 161–179.

Hemingway, H., C. Langenberg, J. Damant, C. Frost, K. Pyörälä, and E. Barrett-Connor. 2008. Prevalence of angina in women versus men: A systematic review and meta-analysis of international variations across 31 countries. *Circulation* 117: 1526–1536.

Hensher, D. A., J. M. Rose, and W. H. Greene. 2005. *Applied Choice Analysis: A Primer*. Cambridge: Cambridge University Press.

Hilbe, J. M. 2011. *Negative Binomial Regression*. 2nd ed. Cambridge: Cambridge University Press.

Hole, A. R. 2007. Fitting mixed logit models by using maximum simulated likelihood. *Stata Journal* 7: 388–401.

Holt, J. D., and R. L. Prentice. 1974. Survival analyses in twin studies and matched pairs experiments. *Biometrika* 61: 17–30.

Hopper, J. L., M. C. Hannah, G. T. Macaskill, J. D. Mathews, and D. C. Rao. 1990. Twin concordance for a binary trait: III. A binary analysis of hay fever and asthma. *Genetic Epidemiology* 7: 277–289.

Hosmer, D. W., Jr., and S. Lemeshow. 2000. *Applied Logistic Regression*. 2nd ed. New York: Wiley.

Hosmer, D. W., Jr., S. Lemeshow, and S. May. 2008. *Applied Survival Analysis: Regression Modeling of Time to Event Data*. 2nd ed. New York: Wiley.

Hougaard, P. 2000. *Analysis of Multivariate Survival Data*. New York: Springer.

Hough, R. L., S. Harmon, H. Tarke, S. Yamashiro, R. Quinlivan, P. Landau-Cox, M. S. Hurlburt, P. A. Wood, R. Milone, V. Renker, A. Crowell, and E. Morris. 1997. Supported independent housing: Implementation issues and solutions in the San Diego Project. In *Mentally Ill and Homeless: Special Programs for Special Needs*, ed. W. R. Breakey and J. W. Thompson, 95–117. The Netherlands: OPA Amsterdam.

Huq, N. M., and J. Cleland. 1990. *Bangladesh Fertility Survey 1989 (Main Report)*. Dhaka: National Institute of Population Research and Training.

Hurlburt, M. S., P. A. Wood, and R. L. Hough. 1996. Providing independent housing for the homeless mentally ill: A novel approach to evaluating long-term longitudinal housing patterns. *Journal of Community Psychology* 24: 291–310.

Huster, W. J., R. Brookmeyer, and S. G. Self. 1989. Modelling paired survival data with covariates. *Biometrics* 45: 145–156.

Jain, D. C., N. J. Vilcassim, and P. K. Chintagunta. 1994. A random-coefficients logit brand-choice model applied to panel data. *Journal of Business & Economic Statistics* 12: 317–328.

Jenkins, S. P. 1995. Easy estimation methods for discrete-time duration models. *Oxford Bulletin of Economics and Statistics* 57: 129–136.

Johnson, V. E., and J. H. Albert. 1999. *Ordinal Data Modeling*. New York: Springer.

Jones, B., and M. G. Kenward. 2003. *Design and Analysis of Cross-Over Trials*. 2nd ed. Boca Raton, FL: Chapman & Hall/CRC.

Karim, M. R., and S. L. Zeger. 1992. Generalized linear models with random effects; salamander mating revisited. *Biometrics* 48: 631–644.

Kelly, P. J., and L. Lim. 2000. Survival analysis for recurrent event data: An application to childhood infectious diseases. *Statistics in Medicine* 19: 13–33.

Kenny, D. A., D. A. Kashy, and W. L. Cook. 2006. *Dyadic Data Analysis*. New York: Guilford Press.

Klein, J. P., and M. L. Moeschberger. 2003. *Survival Analysis: Techniques for Censored and Truncated Data*. 2nd ed. New York: Springer.

Koch, G. G., G. J. Carr, I. A. Amara, M. E. Stokes, and T. J. Uryniak. 1990. Categorical data analysis. In *Statistical Methodology in the Pharmaceutical Sciences*, ed. D. A. Berry, 389–473. New York: Marcel Dekker.

Lancaster, T. 1990. *The Econometric Analysis of Transition Data*. Cambridge: Cambridge University Press.

Langford, I. H., G. Bentham, and A.-L. McDonald. 1998. Multi-level modelling of geographically aggregated health data: A case study on malignant melanoma mortality and UV exposure in the European community. *Statistics in Medicine* 17: 41–57.

Langford, I. H., and T. Lewis. 1998. Outliers in multilevel data. *Journal of the Royal Statistical Society, Series A* 161: 121–160.

Larsen, K., and J. Merlo. 2005. Appropriate assessment of neighborhood effects on individual health: Integrating random and fixed effects in multilevel logistic regression. *American Journal of Epidemiology* 161: 81–88.

Larsen, K., J. H. Petersen, E. Budtz-Jørgensen, and L. Endahl. 2000. Interpreting parameters in the logistic regression model with random effects. *Biometrics* 56: 909–914.

Lawson, A. B., A. Biggeri, D. Böhning, E. Lesaffre, J.-F. Viel, and R. Bertollini, ed. 1999. *Disease Mapping and Risk Assessment for Public Health*. New York: Wiley.

Lawson, A. B., W. J. Browne, and C. L. Vidal Rodeiro. 2003. *Disease Mapping with WinBUGS and MLwiN*. New York: Wiley.

Lee, E. W., L. J. Wei, and D. A. Amato. 2010. Cox-type regression analysis for large numbers of small groups of correlated failure time observations. In *Survival Analysis: State of the Art*, ed. J. P. Klein and P. K. Goel, 237–247. Dordrecht, The Netherlands: Kluwer.

Lesaffre, E., and B. Spiessens. 2001. On the effect of the number of quadrature points in a logistic random effects model: An example. *Journal of the Royal Statistical Society, Series C* 50: 325–335.

Leyland, A. H. 2001. Spatial analysis. In *Multilevel Modelling of Health Statistics*, ed. A. H. Leyland and H. Goldstein, 143–157. Chichester, UK: Wiley.

Lillard, L. A. 1993. Simultaneous equations for hazards: Marriage duration and fertility timing. *Journal of Econometrics* 56: 189–217.

Lillard, L. A., and C. W. A. Panis. 1996. Marital status and mortality: The role of health. *Demography* 33: 313–327.

Lillard, L. A., and C. W. A. Panis, ed. 2003. *aML User's Guide and Reference Manual.* Los Angeles, CA: EconWare.

Lipsitz, S., and G. M. Fitzmaurice. 2009. Generalized estimating equations for longitudinal data analysis. In *Longitudinal Data Analysis*, ed. G. M. Fitzmaurice, M. Davidian, G. Verbeke, and G. Molenberghs, 43–78. Boca Raton, FL: Chapman & Hall/CRC.

Long, J. S. 1997. *Regression Models for Categorical and Limited Dependent Variables.* Thousand Oaks, CA: Sage.

Long, J. S., P. D. Allison, and R. McGinnis. 1993. Rank advancement in academic careers: Sex differences and the effects of productivity. *American Sociological Review* 58: 703–722.

Long, J. S., and J. Freese. 2006. *Regression Models for Categorical and Limited Dependent Variables using Stata.* 2nd ed. College Station, TX: Stata Press.

Lorr, M., and C. J. Klett. 1966. *Inpatient Multidimensional Psychiatric Scale (IMPS).* Palo Alto, CA: Consulting Psychologists Press.

Machin, D., T. M. Farley, B. Busca, M. J. Campbell, and C. d'Arcangues. 1988. Assessing changes in vaginal bleeding patterns in contracepting women. *Contraception* 38: 165–179.

Mare, R. D. 1994. Discrete-time bivariate hazards with unobserved heterogeneity: A partially observed contingency table approach. In *Sociological Methodology 1994*, ed. P. V. Marsden, 341–383. Oxford: Blackwell.

McCullagh, P., and J. A. Nelder. 1989. *Generalized Linear Models.* 2nd ed. London: Chapman & Hall/CRC.

McCulloch, C. E., S. R. Searle, and J. M. Neuhaus. 2008. *Generalized, Linear, and Mixed Models.* 2nd ed. Hoboken, NJ: Wiley.

McKnight, B., and S. K. van den Eeden. 1993. A conditional analysis for two-treatment multiple-period crossover designs with binomial or Poisson outcomes and subjects who drop out. *Statistics in Medicine* 12: 825–834.

Molenberghs, G., and G. Verbeke. 2005. *Models for Discrete Longitudinal Data.* New York: Springer.

O'Connell, A. A., J. Goldstein, H. J. Rogers, and C. Y. J. Peng. 2008. Multilevel logistic models for dichotomous and ordinal data. In *Multilevel Modeling of Educational Data*, ed. A. A. O'Connell and D. B. McCoach, 199–242. Charlotte, NC: Information Age Publishing.

O'Connell, A. A., and D. B. McCoach, ed. 2008. *Multilevel Modeling of Educational Data*. Charlotte, NC: Information Age Publishing.

OECD. 2000. *Manual for the PISA 2000 Database*. Paris: OECD. http://www.pisa.oecd.org/dataoecd/53/18/33688135.pdf.

Pebley, A. R., N. Goldman, and G. Rodríguez. 1996. Prenatal and delivery care and childhood immunization in Guatemala: Do family and community matter? *Demography* 33: 231–247.

Pebley, A. R., and P. W. Stupp. 1987. Reproductive patterns and child mortality in Guatemala. *Demography* 24: 43–60.

Pickles, A., and R. Crouchley. 1994. Generalizations and applications of frailty models for survival and event data. *Statistical Methods in Medical Research* 3: 263–278.

———. 1995. A comparison of frailty models for multivariate survival data. *Statistics in Medicine* 14: 1447–1461.

Prentice, R. L., B. J. Williams, and A. V. Peterson. 1981. On the regression analysis of multivariate failure time data. *Biometrika* 68: 373–379.

Quine, S. 1973. Achievement orientation of aboriginal and white Australian adolescents. PhD diss., Australian National University, Canberra, Australia.

Rabe-Hesketh, S., A. Pickles, and A. Skrondal. 2003. Correcting for covariate measurement error in logistic regression using nonparametric maximum likelihood estimation. *Statistical Modelling* 3: 215–232.

Rabe-Hesketh, S., and A. Skrondal. 2006. Multilevel modelling of complex survey data. *Journal of the Royal Statistical Society, Series A* 169: 805–827.

———. 2009. Generalized linear mixed-effects models. In *Longitudinal Data Analysis*, ed. G. M. Fitzmaurice, M. Davidian, G. Verbeke, and G. Molenberghs, 79–106. Boca Raton, FL: Chapman & Hall/CRC.

Rabe-Hesketh, S., A. Skrondal, and H. K. Gjessing. 2008. Biometrical modeling of twin and family data using standard mixed model software. *Biometrics* 64: 280–288.

Rabe-Hesketh, S., A. Skrondal, and A. Pickles. 2002. Reliable estimation of generalized linear mixed models using adaptive quadrature. *Stata Journal* 2: 1–21.

———. 2004. GLLAMM manual. Working Paper 160, Division of Biostatistics, University of California–Berkeley. http://www.bepress.com/ucbbiostat/paper160/.

———. 2005. Maximum likelihood estimation of limited and discrete dependent variable models with nested random effects. *Journal of Econometrics* 128: 301–323.

Rabe-Hesketh, S., T. Toulopoulou, and R. M. Murray. 2001a. Multilevel modeling of cognitive function in schizophrenic patients and their first degree relatives. *Multivariate Behavioral Research* 36: 279–298.

Rabe-Hesketh, S., S. Yang, and A. Pickles. 2001b. Multilevel models for censored and latent responses. *Statistical Methods in Medical Research* 10: 409–427.

Randall, J. H. 1989. The analysis of sensory data by generalized linear model. *Biometrical Journal* 31: 781–793.

Rasbash, J. 2005. Cross-classified and multiple membership models. In *Encyclopedia of Statistics in Behavioral Science*, ed. B. S. Everitt and D. Howell, 441–450. London: Wiley.

Rasbash, J., F. A. Steele, W. J. Browne, and H. Goldstein. 2009. *A User's Guide to MLwiN Version 2.10*. Bristol: Centre for Multilevel Modelling, University of Bristol. http://www.bristol.ac.uk/cmm/software/mlwin/download/manual-print.pdf.

Raudenbush, S. W., and C. Bhumirat. 1992. The distribution of resources for primary education and its consequences for educational achievement in Thailand. *International Journal of Educational Research* 17: 143–164.

Raudenbush, S. W., and A. S. Bryk. 2002. *Hierarchical Linear Models: Applications and Data Analysis Methods*. 2nd ed. Thousand Oaks, CA: Sage.

Raudenbush, S. W., A. S. Bryk, Y. F. Cheong, and R. Congdon. 2004. *HLM 6: Hierarchical Linear and Nonlinear Modeling*. Lincolnwood, IL: Scientific Software International.

Revelt, D., and K. E. Train. 2000. Customer-specific taste parameters and mixed logit: households' choice of electricity supplier. Working paper E00-274, University of California–Berkeley, Department of Economics. http://129.3.20.41/eps/em/papers/0012/0012001.pdf.

Rock, D. A., and J. M. Pollack. 2002. Early childhood longitudinal study-kindergarten class of 1998–99 (ECLS—K), psychometric report for kindergarten through first grade. Working paper 2002-05, National Center for Education Statistics. http://nces.ed.gov/pubs2002/200205.pdf.

Rodríguez, G., and I. Elo. 2003. Intra-class correlation in random-effects models for binary data. *Stata Journal* 3: 32–46.

Rodríguez, G., and N. Goldman. 2001. Improved estimation procedures for multilevel models with binary response: A case-study. *Journal of the Royal Statistical Society, Series A* 164: 339–355.

Ross, E. A., and D. Moore. 1999. Modeling clustered, discrete, or grouped time survival data with covariates. *Biometrics* 55: 813–819.

Senn, S. 2002. *Cross-Over Trials in Clinical Research*. 2nd ed. New York: Wiley.

Singer, J. D. 1993. Are special educators' career paths special? Results from a 13-year longitudinal study. *Exceptional Children* 59: 262–279.

Singer, J. D., and J. B. Willett. 1993. It's about time: Using discrete-time survival analysis to study duration and the timing of events. *Journal of Educational Statistics* 18: 155–195.

———. 2003. *Applied Longitudinal Data Analysis: Modeling Change and Event Occurrence*. Oxford: Oxford University Press.

Skrondal, A., and S. Rabe-Hesketh. 2003a. Generalized linear mixed models for nominal data. In *Proceedings of the Joint Statistical Meeting*, 3931–3936. Alexandria, VA: American Statistical Association.

———. 2003b. Some applications of generalized linear latent and mixed models in epidemiology: Repeated measures, measurement error and multilevel modeling. *Norwegian Journal of Epidemiology* 13: 265–278.

———. 2003c. Multilevel logistic regression for polytomous data and rankings. *Psychometrika* 68: 267–287.

———. 2004. *Generalized Latent Variable Modeling: Multilevel, Longitudinal, and Structural Equation Models*. Boca Raton, FL: Chapman & Hall/CRC.

———. 2007. Redundant overdispersion parameters in multilevel models for categorical responses. *Journal of Educational and Behavioral Statistics* 32: 419–430.

———. 2009. Prediction in multilevel generalized linear models. *Journal of the Royal Statistical Society, Series A* 172: 659–687.

Skrondal, A., and S. Rabe-Hesketh, ed. 2010. *Multilevel Modelling, Vol. III—Multilevel Generalized Linear Models*. London: Sage.

Skrondal, A., and S. Rabe-Hesketh. Forthcoming. Generalized linear latent and mixed models. In *Statistical Methods in Clinical and Epidemiological Research*, ed. M. Veierød, S. Lydersen, and P. Laake. Oslo: Gyldendal Akademisk.

Smans, M., C. S. Muir, and P. Boyle. 1993. *Atlas of Cancer Mortality in the European Economic Community*. Lyon, France: IARC Scientific Publications.

SOEP Group. 2001. The German Socio-Economic Panel (SOEP) after more than 15 years—Overview. *Proceedings of the 2000 Fourth International Conference of German Socio-Economic Panel Study Users (GSOEP2000), Vierteljahrshefte zur Wirtschaftsforschung* 70: 7–14.

Spielberger, C. D. 1988. *State-Trait Anger Expression Inventory (STAXI): Professional Manual. Research Edition*. Tampa, FL: Psychological Assessment Resources.

Stryhn, H., J. Sanchez, P. Morley, C. Booker, and I. R. Dohoo. 2006. Interpretation of variance parameters in multilevel Poisson regression models. In *Proceedings of the 11th Symposium of the International Society for Veterinary Epidemiology and Economics*, 702–704. Cairns, Australia.

Thall, P. F., and S. C. Vail. 1990. Some covariance models for longitudinal count data with overdispersion. *Biometrics* 46: 657–671.

Therneau, T. M., and P. M. Grambsch. 2000. *Modeling Survival Data: Extending the Cox Model.* New York: Springer.

Train, K. E. 2009. *Discrete Choice Methods with Simulation.* 2nd ed. Cambridge: Cambridge University Press.

Train, K. E., and G. Sonnier. 2005. Mixed logit with bounded distributions of correlated partworths. In *Applications of Simulation Methods in Environmental and Resource Economics,* ed. R. Scarpa and A. Alberini, 117–134. Dordrecht, The Netherlands: Springer.

Tutz, G., and W. Hennevogl. 1996. Random effects in ordinal regression models. *Computational Statistics & Data Analysis* 22: 537–557.

Vaida, F., and X.-L. Meng. 2005. Two slice-EM algorithms for fitting generalized linear mixed models with binary response. *Statistical Modelling* 5: 229–242.

Vansteelandt, K. 2000. Formal models for contextualized personality psychology. PhD diss., Katholieke Universiteit Leuven, Belgium.

Vella, F., and M. Verbeek. 1998. Whose wages do unions raise? A dynamic model of unionism and wage rate determination for young men. *Journal of Applied Econometrics* 13: 163–183.

Vermunt, J. K. 1997. *Log-Linear Models for Event Histories.* Thousand Oaks, CA: Sage.

———. 2008. Multilevel latent variable modeling: An application in education testing. *Austrian Journal of Statistics* 37: 285–299.

Verrell, P. A., and S. J. Arnold. 1989. Behavioral observations of sexual isolation among allopatric populations of the mountain dusky salamander, *Desmognathus Ochrophaeus. Evolution* 43: 745–755.

Vittinghoff, E., S. C. Shiboski, D. V. Glidden, and C. E. McCulloch. 2005. *Regression Methods in Biostatistics: Linear, Logistic, Survival, and Repeated Measures Models.* New York: Springer.

von Bortkiewicz, L. 1898. *Das Gesetz der Kleinen Zahlen.* Leipzig: Teubner.

Ware, J. H., D. W. Dockery, A. Spiro, III, F. E. Speizer, and B. G. Ferris, Jr. 1984. Passive smoking, gas cooking, and respiratory health of children living in six cities. *American Review of Respiratory Disease* 129: 366–374.

Wei, L. J., D. Y. Lin, and L. Weissfeld. 1989. Regression analysis of multivariate incomplete failure time data by modeling marginal distributions. *Journal of the American Statistical Association* 84: 1065–1073.

Wiggins, R. D., K. Ashworth, C. A. O'Muircheartaigh, and J. I. Galbraith. 1990. Multilevel analysis of attitudes to abortion. *Statistician* 40: 225–234.

Williams, R. 2010. Fitting heterogeneous choice models with oglm. *Stata Journal* 10: 540–567.

Winkelmann, R. 2004. Health care reform and the number of doctor visits—an econometric analysis. *Journal of Applied Econometrics* 19: 455–472.

———. 2008. *Econometric Analysis of Count Data.* 5th ed. New York: Springer.

Wooldridge, J. M. 2005. Simple solutions to the initial conditions problem in dynamic, nonlinear panel data models with unobserved heterogeneity. *Journal of Applied Econometrics* 20: 39–54.

———. 2010. *Econometric Analysis of Cross Section and Panel Data.* 2nd ed. Cambridge, MA: MIT Press.

Yang, M. 2001. Multinomial regression. In *Multilevel Modelling of Health Statistics*, ed. A. H. Leyland and H. Goldstein, 107–125. Chichester, UK: Wiley.

Zheng, X., and S. Rabe-Hesketh. 2007. Estimating parameters of dichotomous and ordinal item response models with gllamm. *Stata Journal* 7: 313–333.

Author index

Subject index